Structural design of masonry

Structural design of masonry
Second Edition

Andrew Orton

Longman
London and New York

Longman Scientific & Technical,
Longman Group UK Limited,
Longman House, Burnt Mill, Harlow,
Essex, CM20 2JE, England
and Associated Companies, throughout the world.

First published 1986
Second edition 1992
Second impression 1993

ISBN 0 582 091012

British Library Cataloguing in Publication Data
A CIP record for this book is available from the British
Library

Set by 4 in Compugraghic Plantin 10/11 pt
Produced by Longman Singapore Publishers (Pte) Ltd
Printed in Singapore

Contents

v

Contents

Introduction

On account of its cost, availability and great flexibility in use, masonry has become the most popular of the durable structural materials in the history of building. For example, at present, more than three-quarters of the world's population is housed in masonry buildings. It is also a material in which the structural principles and design methods governing its use are relatively easy to understand and to apply, even for those with a limited background in structural mechanics. This book is intended both as a general introduction to masonry, and for the designer, as a practical guide to the calculation of masonry buildings using BS5628: *Part 1 Structural use of unreinforced masonry* and BS5628: *Part 2 Structural use of reinforced and prestressed masonry*. These codes of practice are based on limit state design, in which the masonry structure is said to reach a 'limit state' of strength, deflection or cracking when it is no longer acceptable in these respects. The codes' basic method is simple calculation checks at each limit state, and their use is recommended even for small buildings, such as houses, as against the prescriptive rules generally given in building regulations which are very narrow in scope and do not set out proper criteria for the design of masonry. The BS 5628 code follows general principles of building design and incorporates the results of research and experience too. An understanding of it, and the background to it, enables the designer to make realistic checks on the safety of masonry buildings and also allows him to exploit fully the potential of masonry as a building material. There are two particular features of this book: firstly the use of design procedures and examples as a way of presenting information and, secondly, the comparatively generous amounts of space given to appendices which contain outline drawing details and design data, including information about bending moments in masonry wall panels with openings and one-way and two-way spanning masonry wall panels with only partial fixity at some edges.

Chapters 1, 2, 3 and 4 are introductory chapters about masonry elements, masonry buildings, their foundations and the structural principles involved in masonry design. Chapters 5, 6, 7, 8, 9, 10 and 11 are about the practical design of masonry buildings and elements under vertical and horizontal loads. The topics covered in these chapters include walls, columns, wall panels, freestanding walls, concentrated loads, composite beams, stability, shear walls, connections and accidental damage with an emphasis on the overall design of the building. The chapters present the material as briefly as possible and in the sequence which would normally be followed in design calculations. An effort has been made to clarify the interpretation of points in Codes of Practice and elsewhere and to give general recommendations even if these may need modification in specific cases. The examples in these chapters have been chosen to present standard methods of calculation or useful approaches to some of the common cases. Calculations are set out as systematically as possible, this being recommended as a way of limiting the chances for error. Chapter 12 is given over entirely to two design examples. These examples treat some other topics but are also intended to give some feel for orders of magnitude and to demonstrate the kind of assumption it is necessary to make for the purposes of a calculation. The examples show the effects of various loading conditions and how these alter the calculation procedure. Appendix A consists of background information and design data. Appendix B contains some typical details which are intended as a basis for working into drawings which will meet all the particular requirements of each case. The data in the appendices and the procedures given in earlier chapters may help to prevent points being overlooked, speed up the time spent on design and reduce errors. In doing calculations for strength, it is assumed that, normally, the necessary dimensions, mortar and type of unit, if not determined by other factors, will be arrived at by trial and error. However, where necessary, preliminary sizes for the calculation may be obtained by dimensioning a wall, for example, to have a slenderness ratio between about 15 and 18 – i.e. somewhat less than the maximum allowed by the Code – and to keep well within the limiting dimensions given in the Code too – see text. There are two points about terminology: references in this book to the 'Code' refer to BS 5628:Part 1 and 'imposed loads' are generally referred to as 'live loads', as this is still the term in common use. References to other sections or Appendices of this book are in bold type – e.g. **2.1**, **2.2**, **A1.1**, **A1.2**.

No Code of Practice can be comprehensive but, in large degree, it can cover those cases where normal conditions would apply and give starting points in cases where further thought or research is necessary. This book attempts to follow these goals too. In spite of the emphasis given in the book to practical design and systematic calculation procedures, the ultimate aim is to convey an understanding and some 'feel' for the current practices governing the use of this very well-known but not always well-understood method of construction.

Preface to second edition

The principal change to this second edition is the inclusion of new material concerning reinforced masonry given in Appendices A7, A13 and A14. However, the opportunity has been taken to rephrase script, update references, correct errors and to revise and expand Appendix A7 which explains methods of calculation for walls with openings.

Acknowledgements

Material from the Building Research Establishment, reproduced in Chapter 10 and Appendix A10, is Crown Copyright and reproduced by kind permission of the Controller of HM Stationery Office. Material from British Standards is reproduced by kind permission of the British Standards Institution, 2 Park Street, London W1A 2BS, from whom complete copies of the Codes of Practice may be obtained. Full titles are given in the reference section at the end of the book.

Extracts from BS 5628:Part 1 (1992) are to be found in this book as follows:

Tables 1, 2, 3	Chapter 2
Clauses 20.1; 22; 23.1.1; 25; 28; 32.2.1; 32.2.2	Chapter 5
Tables 4, 5, 7, Figures 2, 9 and Appendices B1, B2	Chapter 5
Clauses 36.3; 36.4.2; 36.4.3	Chapter 6
Clause 36.5.3	Chapter 7
Table 9	Appendix A6
Tables 6, 8	Appendix B1

and extracts from BS 5628:Part 2 (1985) are to be found in this book as follows:

Table 8	Appendix A13
Clause A2.3	Appendix A14

Thanks are also due to Richard Saunders and Chris Bailey of Harris and Sutherland for checking and commenting on the script.

List of symbols

A	horizontal cross-sectional area – m²
b	width of column or flange – mm
b_p	width of pier – m
E	modulus of elasticity – kN/mm²
E_n	nominal earth or water load
e_a	additional eccentricity due to deflection in walls
e_b	additional eccentricity due to bending moment
e_m	the larger of e_x and e_t
e_t	total design eccentricity in the mid-height region of a wall
e_x	eccentricity at top of a wall
e_x, e_y	distance between shear centre and resultant force on building in x- and y-directions – m
f_k	characteristic compressive strength of masonry – N/mm²
f_{kx}	characteristic flexural strength (tension) of masonry – N/mm²
f_{ka}	effective characteristic flexural strength of masonry (failure parallel to bed joints) – N/mm²
f_{kb}	effective characteristic flexural strength of masonry (failure perpendicular to bed joints) – N/mm²
f_v	characteristic shear strength of masonry – N/mm²
F_k	characteristic load – kN
f_{xr}	horizontal force on a shear wall – kN
F	horizontal force – kN
G_k	characteristic dead load
g_A	design vertical load per unit area – N/mm²
g_d	design vertical dead load per unit area – N/mm²
G	modulus of rigidity – kN/mm²
h	clear height of wall or column between lateral supports – m
h_L	clear height of wall to point of application of a lateral load – m

h_{ef}	effective height of wall or column – m
i	ratio of 'plastic' moment of resistance at supports to 'elastic' moment of resistance in span
I	moment of inertia (second moment of area) – m⁴
K	stiffness coefficient
k_x, k_y	radius of gyration about x- and y-axis – mm
L	length – m
l_{ef}	effective length of wall – m
n_w	design vertical load per unit length of wall – kN/m
P	vertical load – kN
Q_k	characteristic imposed (live) load
s	spacing of piers – m
t	overall thickness of a wall or column – mm
t_{ef}	effective thickness of a wall or column – mm
t_p	thickness of a pier (overall thickness) – mm
v_h	design shear stress – N/mm²
w	uniformly distributed (ud) load – kN/m²
W_k	characteristic wind load
W	total ud load on span – kN
x_c, y_c	distance to shear centre from reference point in x- and y-directions – m
y_1, y_2	distance from neutral axis to extremities of section – mm
Z	section modulus – m³
α	bending moment coefficient for laterally loaded panels
β	capacity reduction factor for walls allowing for effects of slenderness and eccentricity
γ_f	partial safety factor for load
γ_m	partial safety factor for material
γ_{mv}	partial safety factor material in shear
μ	orthogonal ratio = f_{ka}/f_{kb}

CHAPTER 1

Basic facts about the structural design of masonry

Introduction

Masonry is construction which uses brick or block units. Among other things the units are usually reasonably easy to handle, have good compressive strength and are able to make some kind of attachment to a mortar. Unreinforced masonry walls and columns, as built, are treated as integral elements by virtue of the alignment and pattern in which the units are laid on each other, the wall bond, and more crucially by the attachment, or bond, between the units. The properties of a masonry wall will depend on the type of mortar and masonry unit used and the quality of workmanship. Masonry units are usually pre-wetted or laid in a wet mortar so that the suction of water from the mortar by the masonry unit does not weaken the bond strength. This chapter sets out some of the important general ideas concerning this assemblage of mortar and units known as masonry. These ideas are related to topics such as safety, structural theory, construction practices and the properties of materials and are intended to provide a framework for the more detailed information given in later chapters. For those unfamiliar with elementary theory of structures, some terms used in this chapter are explained in Appendix A1. The chapter is written in the form of a series of notes in an attempt to be clear and as brief as possible. Nevertheless the chapter contains virtually all the theory that is necessary for understanding the design of masonry to codes of practice or local building regulations.

1.1 The theory of masonry design

(a) Masonry can take large compressive stresses but only very small tensile stresses (Fig. 1.1).

(b) Hence any bending moment, which, without pre-compression, causes tensile, as well as compressive, stresses is to be avoided (Fig. 1.2).

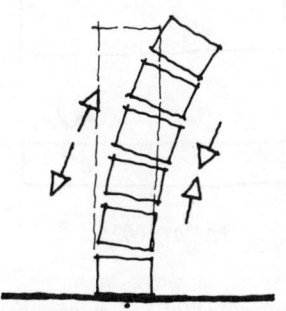

Fig. 1.2 Masonry element under bending.

(c) Forces which act through the centroid of any transverse section taken through a wall or column are said to be axial at that section. Forces which do not act through the centroid at any section are said to be eccentric at that section. In general the vertical load in a wall will have a different eccentricity at any of a series of horizontal sections taken between the top and bottom of the wall, the so-called running section of the wall. The centroid of the running section of the masonry element is at the point of intersection of the major and minor axes of bending and, for sections symmetrical about two axes, this is its central point (**A1.1** and **A1.2**).

Forces which are eccentric may be split up into two components, an axial force, P, and a bending moment, M, which is equal to P.e where e is the eccentricity at the section being considered (Fig. 1.3). For example, to consider a masonry

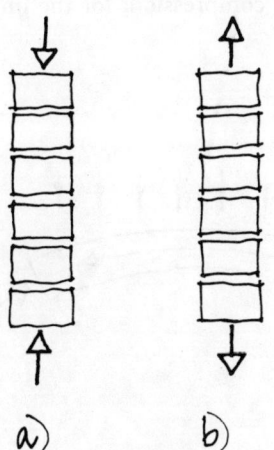

Fig. 1.1 Masonry element under (a) compression and (b) tension.

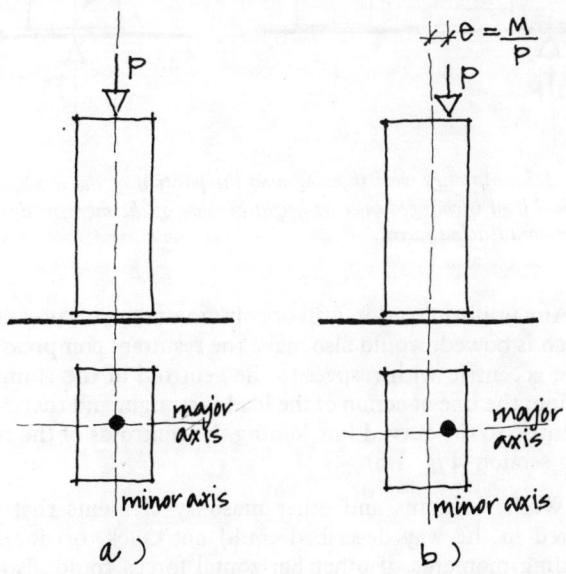

Fig. 1.3 Elevation and horizontal section of masonry element with (a) axial vertical load and (b) eccentric vertical load.

wall at any section as resisting a vertical axial load, P, and a bending moment, M, or as resisting a vertical load, P, at an eccentricity, M/P, are different ways of describing the same set of forces (**A1.5**). Both descriptions are used in masonry design.

(d) An eccentric force produces an increase in compressive stress over that produced by the same axial force. However, for a wall, vertical forces which are eccentric only about the major axis of bending do not usually produce any unacceptable increases of stress (Fig. 1.4).

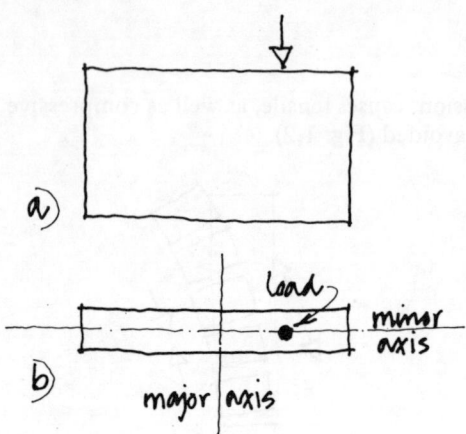

Fig. 1.4 (a) Elevation and (b) horizontal section of masonry wall with vertical load eccentric about major axis of bending only.

(e) A corollary of **1.1**c is that a force which initially is axial becomes eccentric when a bending moment, M, is applied to the masonry element. The eccentricity, *e*, at any point equals M/P (Fig. 1.5).

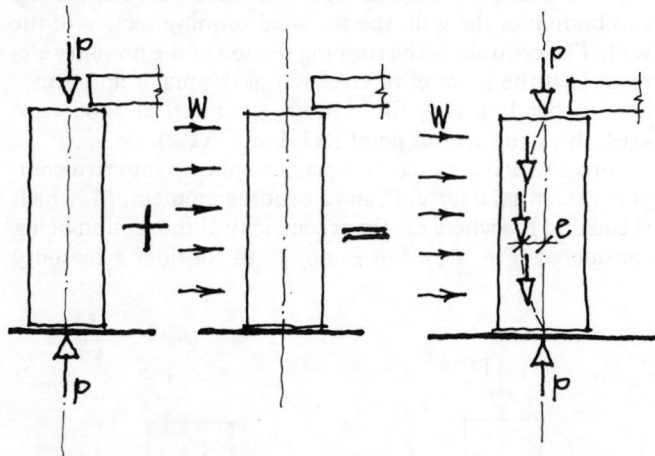

Fig. 1.5 Masonry wall showing how the position of the resultant vertical load is changed over its height because of the moment due to a horizontal wind force.

(f) Any uniform section wall or column under compression, which is bowed, would also make the resultant compression force eccentric with respect to the centroid of the running section; the line of action of the load is straight and therefore eccentric to the curved line joining the centroids of the running section (Fig. 1.6).

(g) Walls, columns and other masonry elements that are curved in the way described could not crack, or develop bending moments, if other horizontal forces could also be applied to the element to cancel out the tensile bending stresses that otherwise would occur (Fig. 1.7). Note,

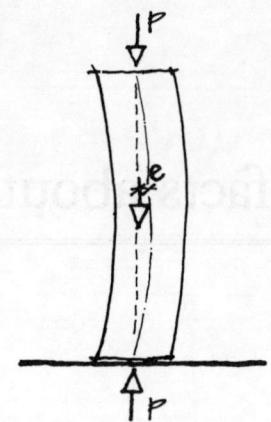

Fig. 1.6 Bowed masonry wall showing the eccentricity of the vertical load.

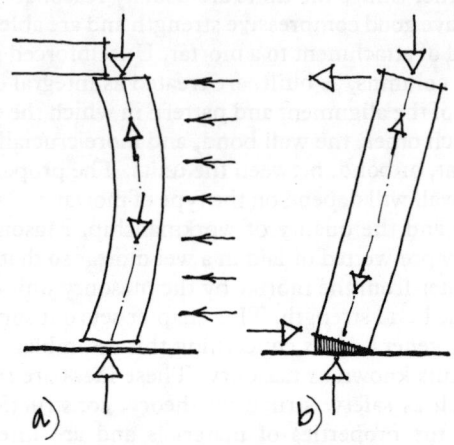

Fig. 1.7 (a) Bowed masonry wall with uniformly distributed horizontal load and (b) sloped masonry wall with point horizontal forces; bending moments could be eliminated in both these elements.

however, that if the horizontal forces reverse in direction then the cracks, or bending moments, become very much worse.

(h) A corollary of **1.1**g is that for any one known set of forces, for example self-weight, a masonry element in compression may be shaped to mobilise bending moments that will exactly balance those due to the applied loads. This is the principle of arch action. In general terms this shape is known as the funicular shape, the shape that will eliminate bending moment for any particular load case. Thus an efficient shape for masonry is that which makes the structure funicular, and in compression, for the important load case (Fig. 1.8).

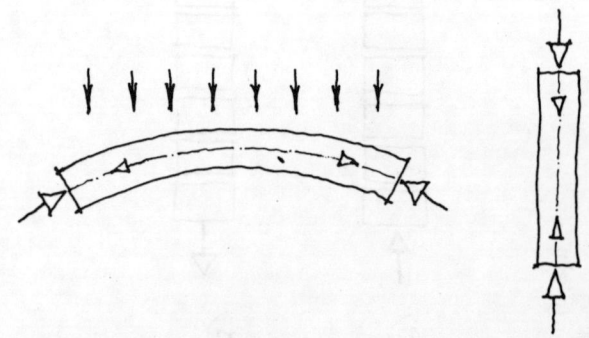

Fig. 1.8 Funicular arch and funicular vertical wall, both having axial load in the running section.

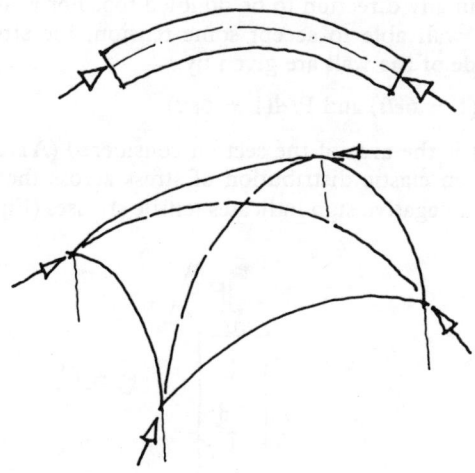

Fig. 1.9 Two- and three-dimensional funicular shapes.

(i) Both two- and three-dimensional shapes may be funicular (Fig. 1.9).

(j) The shape chosen is often that which is funicular for the dead load case. Any other loading case will normally cause some bending stresses in the elements, or cracking in other cases. The object of design is to adjust the shape, initial compression, or thickness so that other loading cases do not produce excessive bending stresses (Fig. 1.10) or excessive cracking (Fig. 1.11).

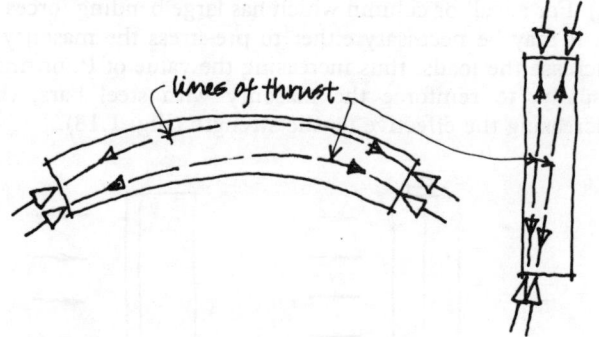

Fig 1.10 Arch and wall elements showing different positions for the lines of thrust within the arch or wall thickness.

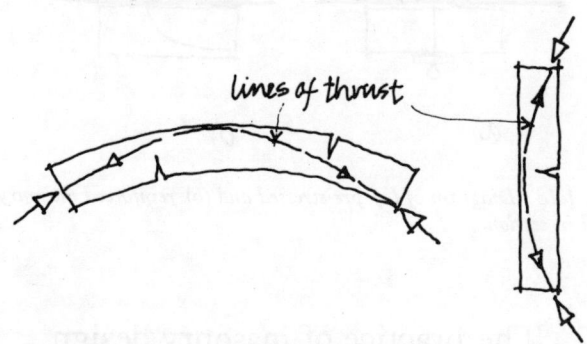

Fig. 1.11 Arch and wall elements having a line of thrust within the arch or wall thickness which causes cracking.

(k) Bending stresses do not produce a net tensile stress in the element until they are greater than the compressive stress due to the axial load. Until this point is reached the eccentricity of the resultant compression force will always fall within a central core area of the section (**A1.4**). For a rectangular section the core area is lozenge-shaped with the two diagonals having lengths one-third that of the sides of the section (Fig. 1.12); see Table A2.2. A general principle in

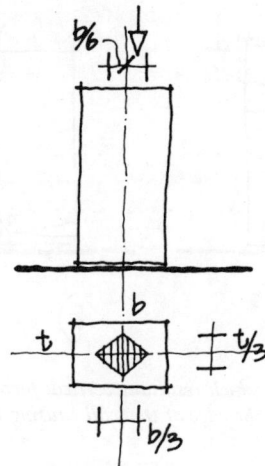

Fig. 1.12 Elevation and horizontal section on wall showing lozenge-shaped core area.

the design of masonry under vertical and horizontal load is that this core or kern area in the section be as large as possible. This is done by making the ratio of section modulus to area, Z/A, as high as possible about both the major and minor axes of bending (**A1.4**). For masonry under mainly horizontal load, by contrast, it is the section modulus, Z, which should be made high, although keeping the ratio Z/A high, too, will save material.

(l) If the resultant compression force does fall outside the core, that is the bending stresses exceed the axial compression stresses, then either tension stresses develop on one side of the section or cracks will occur in masonry which is unable to sustain such tensile stresses (**A1.5**). However, the structure is usually still quite safe but has adjusted itself to the new pattern of loading. The crack may develop almost fully across the section without necessarily causing failure (Fig. 1.13).

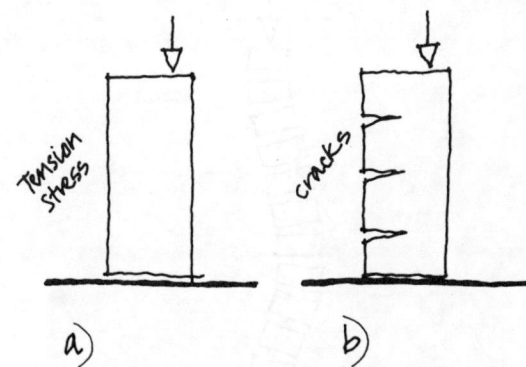

Fig. 1.13 Walls in which resultant vertical force falls outside core area causing (a) tension stresses or (b) cracking.

(m) If the resultant compressive force falls outside the thickness of the masonry, failure of the element is immediate for masonry unable to accept tension (**A1.5**). In addition if the resultant compressive force acts very near the edge of the masonry it is likely to cause crushing, as only a small area is then in compression (Fig. 1.14).

(n) Masonry elements in compression will eventually fail either by crushing of the masonry, or by buckling, or by a combination of these effects (Fig. 1.15).

(o) Buckling is only associated with slender elements having slenderness ratios above about six. Buckling failure is rapid and caused by a knock-on effect whereby the longitudinal

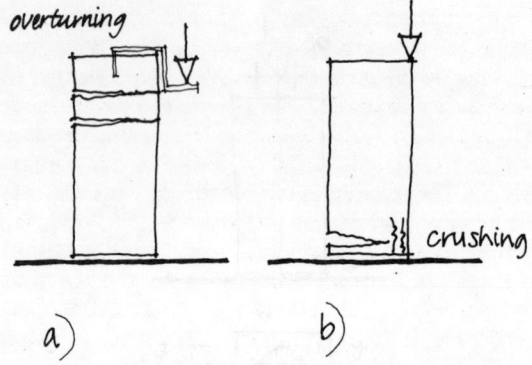

Fig. 1.14 Walls in which resultant vertical force falls (a) outside the wall or (b) near the edge of the wall causing failure.

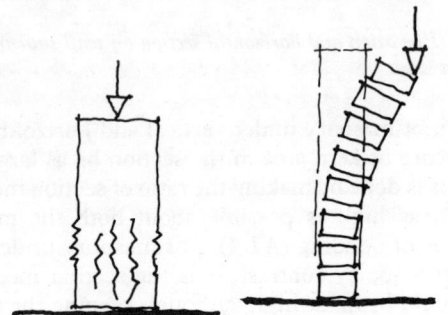

Fig. 1.15 Masonry elements fail by crushing or buckling.

forces cause the middle portion of the element to deflect laterally at an ever-increasing rate producing a large and uncontrollable bending moment. There is not necessarily any crushing of the masonry units. The buckling is usually initiated by lateral forces, an initial lack of straightness or, in some cases, uneven filling of the mortar joints which can cause axial forces to bow the element in one direction (Fig. 1.16).

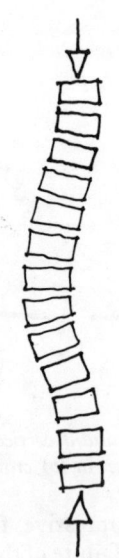

Fig. 1.16 Typical buckling failure of slender masonry element.

(p) Masonry elements in bending will usually fail by exceeding the tensile strength of the masonry. If the design criteria be that no tension or cracks develop in the masonry, then the ability of a solid, regular-shaped wall or column to accept bending is proportional to the axial compression and the depth of the section, t, in the direction of bending. For example, doubling either the axial compression or the plan depth of the cross-section would allow the bending moment

applied in any direction to be doubled too. For a solid rectangular wall able to accept some tension, the stresses at either side of the wall are given by

$$P/A(1 - 6e/t) \text{ and } P/A(1 + 6e/t)$$

where A is the area of the section considered (**A1.5**). This assumes an elastic distribution of stress across the section (**A1.3**). A negative sign indicates tensile stresses (Fig. 1.17).

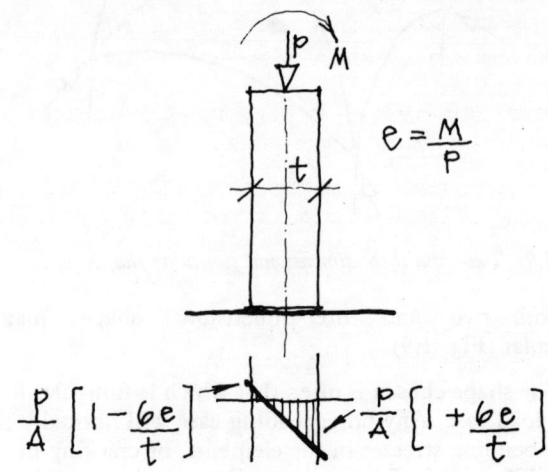

Fig. 1.17 Vertical section through solid masonry wall with diagram showing tensile and compressive stresses at section where axial force is P and bending moment is M.

(q) For a wall or column which has large bending forces on it, it may be necessary either to pre-stress the masonry or increase the loads, thus increasing the value of P, or, more usually, to reinforce the masonry with steel bars, thus increasing the effective tensile strength (Fig. 1.18).

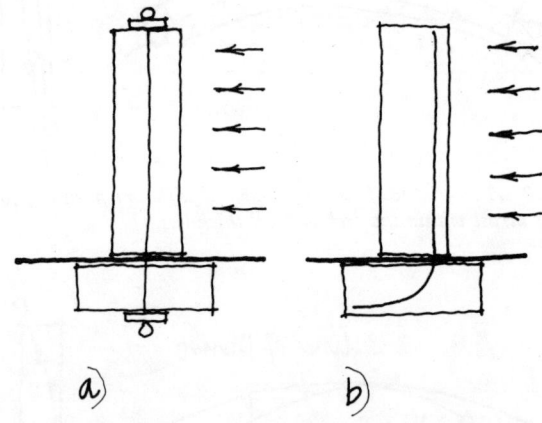

Fig. 1.18 Diagram of (a) pre-stressed and (b) reinforced masonry wall in section.

1.2 The practice of masonry design

(a) Most research work on unreinforced masonry has been done with clay brickwork. However, glass, natural stone, calcium silicate bricks, concrete bricks, aggregate concrete blocks and aerated concrete blocks are also in common use.

(b) Almost all recent test and research work has been on vertical masonry walls and columns.

(c) The compressive strength of a stout, vertical brickwork wall, that is a wall with a low slenderness ratio which fails by crushing of the masonry, depends for a given eccentricity of applied loading on:

- the cross-sectional area of the wall;
- the strength of the brick unit;
- the strength of the mortar;
- the thickness of the mortar bed;
- the height to width ratio of the brick unit;
- the type of bond;
- the number and size of unfilled voids in the bed joints;
- the degree of curvature of the wall, as built.

However, unfilled voids in the perpend joints have negligible effect on compressive strength.

(d) The compressive strength of a slender brickwork wall, which fails by buckling, depends, for a given applied eccentricity of loading, on how easily it bulges − its lateral deformability. Lateral deformability depends on:

- the cross-sectional area of the wall;
- the slenderness ratio of the wall;
- the elastic modulus of the brickwork;
- the number and size of gaps at the edges of the mortar joints (caused by shrinkage of the mortar and movement during laying of the brickwork);
- the thickness of the mortar joint;
- the lateral loading on the wall, from wind for example;
- the non-linear stress-strain relationship of the material in compression.

However, the axial compressive strength and the flexural strength of the brick wall, as such, do not affect the buckling strength of the wall.

(e) The maximum distance a laterally loaded vertical panel or element of brickwork will span depends on:

- the thickness of the panel or element;
- the number of supports along the edges of the panel or element and the extent to which these supports prevent rotation;
- the flexural strength of the brickwork in the two directions parallel and perpendicular to the bed joints;
- the lateral loading on the panel or element from wind or earthquake loads;
- the amount of vertical compressive stress where the bending stresses are greatest, usually near the middle and edges of the panel.

(f) Research results and rules of good practice are incorporated into codes of practice for use in design. This book uses a British Code of Practice, BS 5628:Part 1 (1992) referred to as 'the Code' elsewhere in this book.

(g) The Code of Practice, BS 5628:Part 1, uses limit state design. The object of this design procedure is to ensure that there is an adequate probability that a limit state will not be reached. Examples of limit states are failures due to strength, instability, excessive movement, excessive cracking and damage to secondary elements such as partitions or windows. The first two of these are ultimate limit states considering how the structure fails when it takes the design loads. The design loads include factors of safety. The rest are serviceability limit states considering how the structure behaves under the probable actual loads. A higher factor of safety used in the design gives a lower probability of failure.

(h) In the structural design of masonry, a calculation on the limit state of strength is necessary. The limit state of strength may be that in compression, tension, flexure (bending), or shear for example. The object is to make the design strength or design load resistance of a wall, for example, greater than the design load. The design load is obtained by multiplying the characteristic load, F_k, which may be thought of as the estimated actual loads, by safety factors, called the partial safety factors for loads, γ_f. The design vertical load resistance of a stout wall is obtained by dividing the characteristic compressive strength of the wall, f_k, which may be thought of as the strength at failure point, by a safety factor, called the partial safety factor for materials, γ_m, and then multiplying this by the area of the wall under consideration, A. Hence,

$$\gamma_f . F_k < A . f_k / \gamma_m.$$

(i) The Code of Practice BS 5628:Part 1 (App. B) assumes, for calculation purposes, that vertical load on a wall at the limit state causes a uniformly distributed (ud) compressive stress over part of the area of that wall (**A1.5**). However, in the design of shear walls or walls in flexure, the Code (Clauses 32.1 and 36.4.3) assumes an elastic distribution of stress (**A1.3** and **1.1**p).

(j) A wall may fail in other ways, that is to say, may reach other limit states, such as buckling or cracking. Increasing the strength of the wall would not necessarily improve the wall in these respects. Buckling largely depends on the slenderness ratio of the wall (**5.1.1**c). The Code of Practice, BS 5628:Part 1, treats buckling by limiting the allowable values of slenderness ratio and limiting the allowable design vertical load on slender elements. It is very often assumed in design work that if a wall satisfies the requirements for strength it will not crack excessively. The Code of Practice BS 5628:Part 1 (Clause 19) explicitly allows this assumption for axially applied loads. In practice cracking, or the effects of cracking, are very often reduced by good detail design.

(k) Other important design requirements are that the whole building and the individual elements in it be stable both in use and during construction. In addition the building should be able to accept some degree of accidental damage without this leading to extensive damage. This property of a building has been called its 'robustness'. In Britain for buildings classified as being five or more storeys high, and possibly for certain other types (**11.1**), there are specific requirements to be met if the building is to be considered as 'robust'; these provisions, as applied to masonry buildings, are set out in the Code of Practice, BS 5628:Part 1 (Clause 37).

(l) Elements of masonry designed in accordance with the Code of Practice, BS 5628:Part 1, will generally have an overall calculated factor of safety between about four and six, depending on the type of element. Hence loadings and material strengths used in calculations need only be known to a reasonable degree of accuracy, say 10% or 15%. For this reason too, calculations with masonry can be made simple, even at the expense of accuracy.

(m) The actual factor of safety on any element, or, put another way, the load which will cause failure, cannot usually be known with any accuracy, because of the inaccuracy of the design assumptions and inadequate control of all the other factors involved. In particular masonry as a form of construction is vulnerable to poor workmanship, for example that causing weak mortar, inaccurate construction or disturbance of newly laid masonry units. The Code of Practice BS 5628:Part 1 (Clause 27.3) recognises good site and manufacturing control by allowing the partial safety factor for materials, γ_m, to be reduced in such cases, if certain conditions are met.

(n) Reinforced masonry design is covered in BS 5628:Part 2 (1985) and general information about the use of masonry is given in BS 5628:Part 3 (1985). Reinforced masonry is used when greater strength, especially flexural strength, is required. It appears most convenient as a method of construction for vertical elements, although reinforced masonry beams are sometimes used, especially for lintels. Horizontal reinforcement can be used in masonry bed joints to help control cracking and improve flexural strength. The design of vertically reinforced elements in bending, only, is covered in Appendix A13; the design of horizontally reinforced walls in bending, by means of bedjoint reinforcement, is covered in Appendix A14. The use of light, galvanised or stainless steel wind posts built into the thickness of the wall is a convenient and theoretically superior alternative to bed-joint reinforcement, where conditions allow.

CHAPTER 2

The elements of masonry construction

2.1 Masonry units

Masonry units may be produced from a very large number of raw materials. In addition, only a relatively small investment in plant is needed to make the finished masonry units at moderate rates of production. This helps to account for the very widespread use of masonry as a structural material for buildings.

The properties required of a finished masonry unit, in general application, are some compressive strength, the ability to form a bond with mortar, durability, resistance to rain penetration, and good dimensional stability. The latter is the property in which only small changes in dimensions occur under load or with drying or because of changes in temperature or in the air moisture content.

To a lesser extent, masonry units generally require low thermal conductivity and sound transmission, fire resistance, and moderate tensile, shear and flexural strength. The units must also have a colour, texture and shape that is acceptable. Other important requirements of masonry units are that they be of a size and weight that is easy to handle and lay, and that they be reasonably convenient to cut and to fix to. The process of manufacture will also impose restrictions which, in practice, are as important as the finished properties in choosing suitable materials for masonry. Most commonly used masonry units are made from clay or shale mixtures which are fired, natural stone, or from various types of concrete. However, adobe (or unburnt clay), gypsum and glass have all been successfully used in masonry construction too. BS 6073 (1982) defines a brick as a masonry unit not exceeding 337.5 mm in length, 225 mm in thickness or 112.5 mm in height. A block is a masonry unit which is larger but does not exceed 650 mm in any of these dimensions.

2.1.1 Clay brickwork

Clay bricks are made by firing clay or shale deposits. Very many kinds of clay brick are produced. BS 3921 (1985) classifies clay bricks in terms of their type, compressive strength, water absorption, frost resistance, soluble salt content, and the degree to which they effloresce. The types of brick are solid, frogged, perforated and cellular bricks: solid bricks are those without any voids or depressions; frogged bricks those which have depressions on one or both bed faces which, however, do not exceed 20% of the gross volume of the brick; perforated bricks are those with holes not exceeding 25% of the gross volume of the brick, with no hole exceeding 10% of the gross volume and no vertical section through the brick containing less than 30% solid material; cellular bricks are those with voids which exceed 20% of the gross volume but are closed off at one end (Fig. 2.1). The structural properties of the brick unit of most interest are its

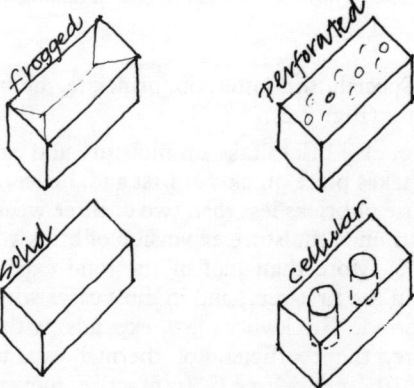

Fig. 2.1 Bricks defined as solid, frogged, perforated and cellular to BS 3921.

compressive strength and its water absorption, the water absorption being related to the flexural strength of the clay brickwork (Table 2.3). The quoted compressive strength of bricks with voids is the average over the gross area of the brick. Engineering bricks are called Class A if their compressive strength is at least 70 N/mm² and the water absorption no more than 4.5% and Class B if their compressive strength is at least 50 N/mm² and the water absorption no

Fig. 2.2 Snapheader and single bullnose Standard Special bricks.

more than 7%. For frost resistance, BS 3921 classifies bricks as either frost resistant (F), moderately frost resistant (M) or not frost resistant (O). For soluble salt content, bricks are classified either as having low (L) or normal (N) amounts of soluble salts. As both frost resistance and soluble salt content affect durability, the classifications for those two factors are combined to form six durability designations for clay bricks which are FL, FN, ML, MN, OL, ON. In general, frost resistant bricks with low salt content are most durable, the actual durability depending also on the degree of exposure (Table 2.4). Bricks are also classified as having nil, slight, moderate or heavy efflorescence. Bricks to BS 3921 are produced with nominal dimensions of 215 × 102.5 × 65mm, the so-called work size, and the coordinating size, used for dimensioning brickwork, is 225 × 112.5 × 75mm, which takes account of the mortar joint. The joint takes up errors in manufacture and construction (Fig. 2.3). Other Standards cover metric modular sized bricks and special shaped bricks. Two types of special shaped bricks are made, the Special and

7

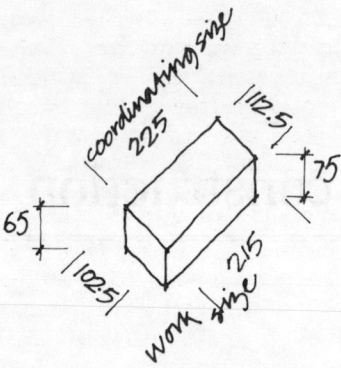

Fig. 2.3 Standard format brick showing work and coordinating sizes.

Standard Special, the latter, in principle, being in more common use (Fig. 2.2).

After firing, clay bricks take up moisture and expand. The expansion takes place quickly at first and, hence, it is best to avoid the use of bricks less than two or three weeks old. The total unrestrained moisture expansion of brickwork is up to about 0.1%. More than half of the total expansion takes place within the first year, and in most cases within a much shorter period. Brickwork also expands with a rise in temperature, the coefficient of thermal expansion being about 6×10^{-6} per degree C. In practice, movement joints in walls need to be provided about every 12 m length in a temperate climate or in any place where the wall is weakened by door or window openings and cracking would otherwise be likely to occur. Unlike longitudinal movement, vertical movement of load-bearing masonry is not restrained and, therefore, horizontal cracking is unlikely to be significant. The frost resistance of clay brickwork can cause problems. In general, high compressive strength indicates good frost resistance but low total water absorption, and the presence of voids after an initial period of water absorption indicate good resistance too. Normally, experience on the frost resistance of a brick enables it to be given a classification.

Clay bricks are subject to sulphate attack in wet conditions. The sulphates may be present in the bricks themselves or, for example, in soil against the brickwork or in smoke within a brick chimney. A chemical reaction can take place with the Portland cement in the mortar, causing spalling of the mortar and brick. The first priority, if possible, is to stop the brickwork getting soaked or staying wet for too long. In places where the brickwork is exposed, a strong mortar can reduce the amount of saturation by rain and, with the right choice of brick, attack can be prevented. Engineering bricks, which usually contain less than 0.5% of soluble sulphates, are much less liable to attack than other clay bricks some of which may have more than 3% of these salts. Clay bricks are also subject to efflorescence, that is the white crystallisation on the surface of the brick of salts which were previously in solution. Again, brickwork which is not allowed to become wet over a period of time will not effloresce. As noted, some kinds of brick are more liable to efflorescence than others. In many cases efflorescence occurs because the brickwork is not protected and is allowed to get wet during construction. Once the efflorescence appears, it can take several months to go, although it is not usually harmful.

2.1.2 Calcium silicate brickwork

Calcium silicate bricks are made by autoclaving lime, water and sand or flint, sometimes including other aggregates. Calcium silicate bricks are classified in BS 187 (1978) according to their compressive strength and drying

shrinkage. The types of brick are solid or cellular. They are produced to the same nominal sizes as clay bricks. After autoclaving, calcium silicate bricks shrink and this drying shrinkage will continue over a long period of time. Total drying shrinkage is up to about 0.03%. The coefficient of thermal expansion is about 12×10^{-6} per degree C, higher than for clay brickwork. Walls will need contraction joints about every 8 m. In long walls undergoing large variation in temperature, expansion joints will be necessary too. The frost resistance of calcium silicate bricks is related to their compressive strength and, generally, is good. Calcium silicate bricks do not normally contain soluble salts and, therefore, unlike clay brickwork, do not suffer from efflorescence after wetting.

2.1.3 Aggregate concrete blocks and autoclaved aerated concrete blocks

Concrete blocks are made by mixing cement, aggregates and water together. The blocks must be allowed to cure properly.

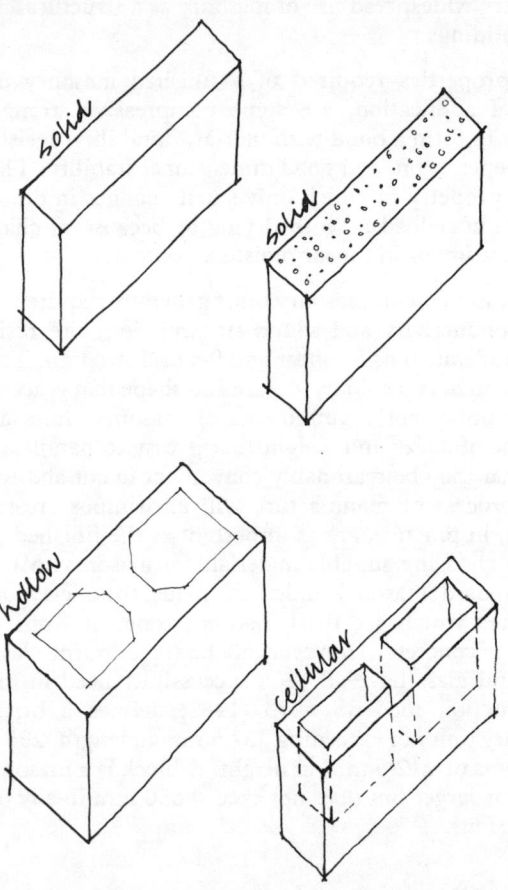

Fig. 2.4 Blocks defined as solid, hollow or cellular to BS 6073.

BS 6073 classifies blocks as being of a certain type, that is to say solid, hollow or cellular (Fig. 2.4). Solid blocks are those which have no holes or voids in them, except voids inherent in the material itself. For example, autoclaved aerated blocks are classed as solid blocks. Hollow blocks are those with voids. Cellular blocks are blocks with voids which are closed off at one end. Hollow blocks are often used with steel reinforcement in the voids. Special blocks are also available for sills and corners, for closing the cavity at the end of a wall and for making reinforced concrete lintels (Fig. 2.5). Higher thermal insulation may be obtained with lightweight concrete blocks or dense concrete hollow or cellular blocks with foam insulation in the voids (Fig. 2.6). As with bricks, the quoted compressive strength of blocks, whether solid or

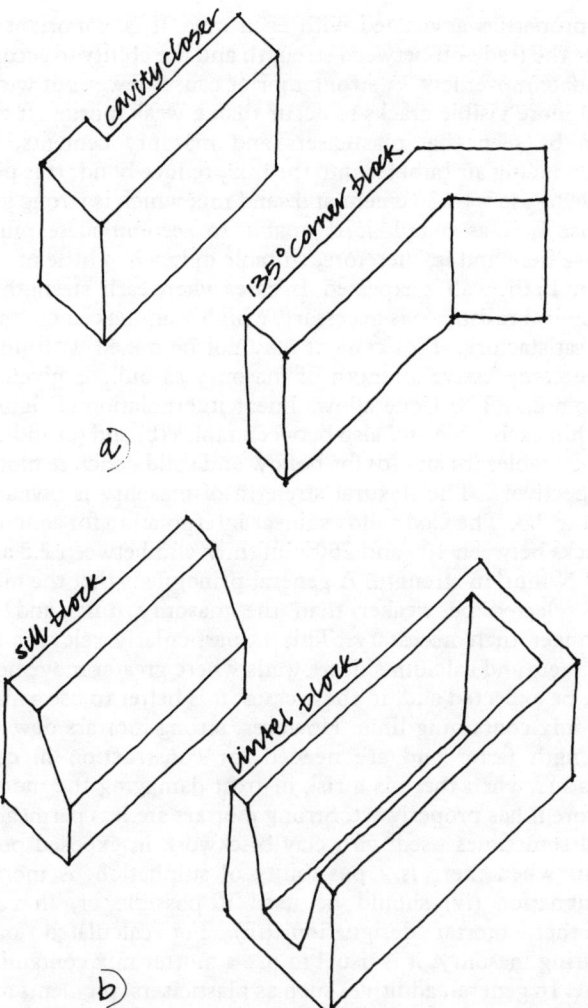

Fig. 2.5 (a) Cavity closer or corner block and 135° corner block; (b) sill and lintel blocks.

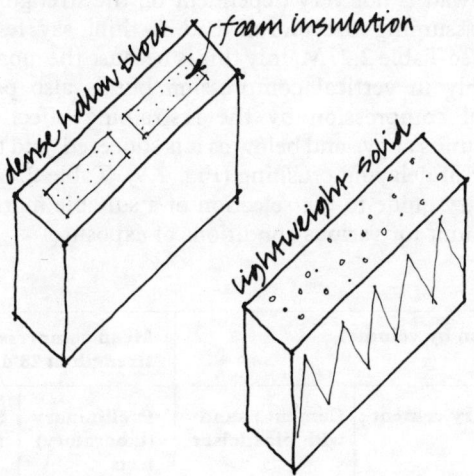

Fig. 2.6 Dense hollow block with insulation and lightweight solid block.

from lightweight, autoclaved, aerated concrete and those made from dense aggregate concrete. Hollow and cellular dense aggregate blocks may end up having an average compressive strength of the same order as solid lightweight blocks because of the effect of the voids (Fig. 2.6).

Blocks are manufactured with lengths of about 400 mm, heights of about 200 mm and in thicknesses between 75 mm and 225 mm. BS 6073 gives the height of a block as not exceeding its length or six times its thickness. The coordinating size of blockwork is obtained by adding 10 mm to the length and height of the work size. Common coordinating sizes are based on 400 × 200 or 450 × 225 modules (Fig. 2.7). After manufacture, concrete blocks lose moisture and shrink, and are also subject to temperature movements. The

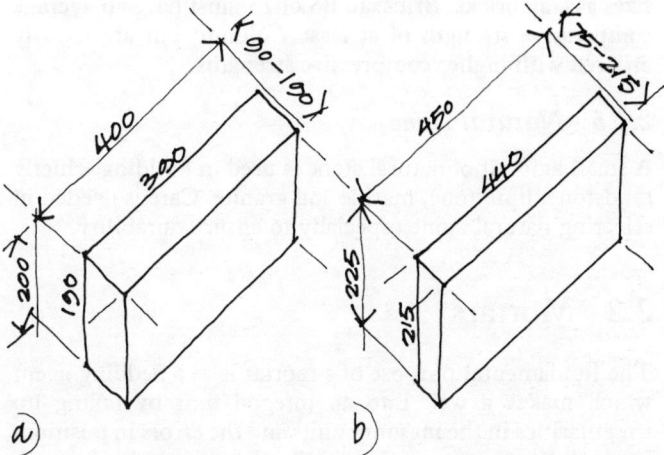

Fig. 2.7 Coordinating and work sizes for (a) 400 × 200 and (b) 450 × 225 concrete block units.

shrinkage is of the order of 0.06% and the coefficient of thermal expansion is about 8×10^{-6} per degree C. In practice, concrete block walls, especially those made from weaker blocks, are more likely to crack than clay brick walls and contraction joints in a wall would be provided about every 6 m in a temperate climate or where there are door and window openings. Occasionally long walls undergoing large variations in temperature will require expansion joints. It is important to remember that clay brickwork expands over a period of time whereas concrete blocks and *in situ* concrete will shrink. This effect may cause damage to an *in situ* concrete frame having brickwork infill panels without a soft joint at the top and in time will leave only the brick leaf of a brick/concrete cavity wall as a load-bearing member (Fig. 2.8). Generally, all available types of concrete block have extremely good frost resistance. It is also worth noting that

with voids, is based on an average strength over the gross area of the block. Thus, blocks with voids are made with material of higher compressive strength than the quoted figure for the block. This ties in with the method of calculation in BS 5628:Part 1 which treats masonry units with voids as solid but uses the average strength of the block, as quoted (**5.1.1**g, **5.1.2**e). Concrete blocks to BS 6073 (75 mm thick or less) must have an average compressive strength of at least 2.8 N/mm² but are usually ordered with higher compressive strengths. In practice blocks divide into those made

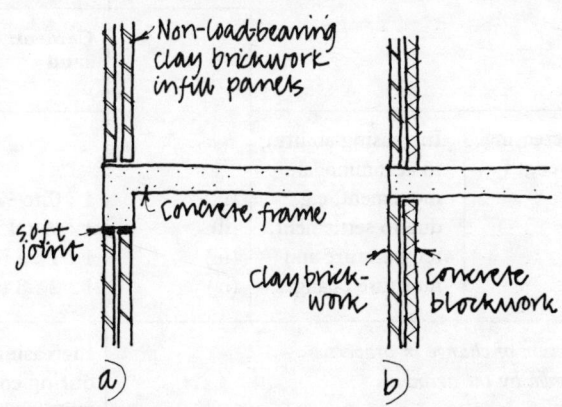

Fig. 2.8 Expansion of clay bricks and shrinkage of concrete causing (a) need for soft joint at top of infill panels and (b) loss of load-bearing capacity on concrete blockwork inner leaf.

blockwork is very much quicker to lay than brickwork, assuming that the block used is easy to handle. However, a corollary of this is that complicated work, unless specifically dimensioned for blockwork, is more conveniently done on a smaller brick module; this saves cutting and ensures a better standard of work. In general concrete blocks are less likely to soak than common clay bricks because even if they absorb moisture on their face, their permeability is lower. However, in general, they weather less well than bricks.

2.1.4 Concrete bricks

Concrete bricks are made in a similar way to dense aggregate blocks and have similar properties. The standard for concrete bricks is also BS 6073. The bricks are made in the same sizes as clay bricks. Bricks to BS 6073 must have an average compressive strength of at least 7 N/mm^2 but are usually ordered with higher compressive strengths.

2.1.5 Natural stone

A small amount of natural stone is used in building, chiefly sandstone, limestone, marble and granite. Care is needed in selecting natural stone especially to ensure durability.

2.2 Mortars

The fundamental purpose of a mortar is as a bedding agent which makes a wall into an integral unit by taking up irregularities in the masonry units and the errors in positioning and thus evens out the distribution of load in the wall. Many old masonry buildings with thick walls have mortars of only limited strength which are, nevertheless, perfectly satisfactory in this role. Generally, the mortar is also required to help prevent ingress of water from the outside of the building. Increasingly, mortar is also being relied on to give the finished masonry flexural as well as compressive strength. Most mortars consist of approximately 3 parts of sand mixed with 1 part of binder, the binder filling approximately the total volume of the voids existing in the sand before it was mixed. The binder usually consists of cement and plasticiser or cement and lime. Plasticiser and lime both increase the workability of the mix. Masonry cement is also in use and consists of Portland Cement mixed with fine filler and air-entraining agents, which are both plasticisers. Table 2.1 shows the standard mortar mixes in use and the changes

in properties associated with each mix. It is important to note the trade-off between strength and the ability to accommodate movement. A strong mortar causes fewer but wider and more visible cracks to occur than a weak mortar. It can also be seen that plasticisers and masonry cements, by introducing air bubbles into the mix, reduce bond. It is possible to use a 1 : 3 cement and sand mix which is strong and dense but, as noted, it is unable to accommodate much movement and is, therefore, suitable only when little movement in the wall is expected. In cases where early strength or chemical resistance is necessary, a high alumina cement may be satisfactory. This cement may not be mixed with lime. The compressive strength of masonry as built is given in Table 2.2. The Code allows linear interpolation of figures within each table and also between tables (b) and (c) and between tables (b) and (d) for hollow and solid concrete blocks respectively. The flexural strength of masonry is given in Table 2.3. The Code allows linear interpolation for concrete blocks between 100 and 250 mm thick and between 2.8 and 7.0 N/mm^2 in strength. A general principle is that the mortar selected be weaker than the masonry units and no stronger than necessary. This is particularly relevant for concrete and calcium silicate walls where greater movement can be expected and, in these cases, it is better to use a mortar mix containing lime. However, strong mortars develop strength faster and are needed for construction in cold weather when there is a risk of frost damaging the mortar before it has properly set. Strong mortars are less permeable and sometimes used with clay brickwork in exposed positions when there is a possibility of sulphation. A mortar designation (iv) should be used if possible or, in cold weather, mortar designation (iii). For calculated load-bearing masonry, it is usual to use a mortar mix containing lime. In general, additives such as plasticisers, pigments and anti-freeze admixtures should be accurately measured out or the strength and (especially) the durability of the mortar may be affected. Note that the compressive strength of the finished masonry wall is not very dependent on the strength of the mortar, assuming the mortar bed is thin, say less than 12 mm; see Table 2.2. Mainly this is because the mortar bed is not only in vertical compression but is also put into horizontal compression by the restraining effect of the masonry units above and below as it is squeezed, and this has the effect of delaying crushing (Fig. 2.9). Table 2.4 gives a preliminary guide to the selection of a suitable mortar and masonry unit for various conditions of exposure.

Table 2.1 Mortar designations (after Table 1 of BS 5628:Part 1)

		Mortar designation	Type of mortar (proportion by volume)			Mean compressive strength at 28 days	
			Cement: lime : sand	Masonry cement : sand	Cement : sand with plasticiser	Preliminary (laboratory) tests	Site tests
↑ Increasing strength	Increasing ability to accommodate movement, e.g. due to settlement, temperature and moisture changes ↓	(i) (ii) (iii) (iv)	1 : 0 to ¼ : 3 1 : ½ : 4 to 4½ 1 : 1 : 5 to 6 1 : 2 : 8 to 9	— 1 : 2½ to 3½ 1 : 4 to 5 1 : 5½ to 6½	— 1 : 3 to 4 1 : 5 to 6 1 : 7 to 8	N/mm^2 16.0 6.5 3.5 1.5	N/mm^2 11.0 4.5 2.5 1.0
Direction of change in properties is shown by the arrows			Increasing resistance to frost attack during construction → Improvement in bond and consequent resistance to rain penetration ←				

Table 2.2 *Characteristic compressive strength of masonry, f_k, in N/mm² (after Table 2 of BS 5628:Part 1)*

(a) Constructed with standard format bricks

Mortar designation	Compressive strength of unit (N/mm²)								
	5	10	15	20	27.5	35	50	70	100
(i)	2.5	4.4	6.0	7.4	9.2	11.4	15.0	19.2	24.0
(ii)	2.5	4.2	5.3	6.4	7.9	9.4	12.2	15.1	18.2
(iii)	2.5	4.1	5.0	5.8	7.1	8.5	10.6	13.1	15.5
(iv)	2.2	3.5	4.4	5.2	6.2	7.3	9.0	10.8	12.7

(b) Constructed with blocks having a ratio of height to least horizontal dimension of 0.6

Mortar designation	Compressive strength of unit (N/mm²)							
	2.8	3.5	5.0	7.0	10	15	20	35 or greater
(i)	1.4	1.7	2.5	3.4	4.4	6.0	7.4	11.4
(ii)	1.4	1.7	2.5	3.2	4.2	5.3	6.4	9.4
(iii)	1.4	1.7	2.5	3.2	4.1	5.0	5.8	8.5
(iv)	1.4	1.7	2.2	2.8	3.5	4.4	5.2	7.3

(c) Constructed with hollow blocks having a ratio of height to least horizontal dimension of between 2.0 and 4.0

Mortar designation	Compressive strength of unit (N/mm²)							
	2.8	3.5	5.0	7.0	10	15	20	35 or greater
(i)	2.8	3.5	5.0	5.7	6.1	6.8	7.5	11.4
(ii)	2.8	3.5	5.0	5.5	5.7	6.1	6.5	9.4
(iii)	2.8	3.5	5.0	5.4	5.5	5.7	5.9	8.5
(iv)	2.8	3.5	4.4	4.8	4.9	5.1	5.3	7.3

(d) Constructed from solid concrete blocks having a ratio of height to least horizontal dimension of between 2.0 and 4.0

Mortar designation	Compressive strength of unit (N/mm²)							
	2.8	3.5	5.0	7.0	10	15	20	35 or greater
(i)	2.8	3.5	5.0	6.8	8.8	12.0	14.8	22.8
(ii)	2.8	3.5	5.0	6.4	8.4	10.6	12.8	18.8
(iii)	2.8	3.5	5.0	6.4	8.2	10.0	11.6	17.0
(iv)	2.8	3.5	4.4	5.6	7.0	8.8	10.4	14.6

Table 2.3 *Characteristic flexural strength of masonry, f_{kx}, in N/mm² (after Table 3 of BS 5628:Part 1)*

	Plane of failure parallel to bed joints			Plane of failure perpendicular to bed joints		
Mortar designation	(i)	(ii) and (iii)	(iv)	(i)	(ii) and (iii)	(iv)
Clay bricks having a water absorption						
less than 7%	0.7	0.5	0.4	2.0	1.5	1.2
between 7% and 12%	0.5	0.4	0.35	1.5	1.1	1.0
over 12%	0.4	0.3	0.25	1.1	0.9	0.8
Calcium silicate bricks	0.3		0.2	0.9		0.6
Concrete bricks	0.3		0.2	0.9		0.6
Concrete blocks (solid or hollow) of compressive strength in N/mm²:						
2.8				0.40		0.4
3.5 used in walls of thickness* up to 100 mm	0.25		0.2	0.45		0.4
7.0				0.60		0.5
2.8				0.25		0.2
3.5 used in walls of thickness* 250 mm	0.15		0.1	0.25		0.2
7.0				0.35		0.3
10.5				0.75		0.6
14.0 used in walls of any thickness*	0.25		0.2	0.90†		0.7†
and over						

*The thickness should be taken to be the thickness of the wall, for a single-leaf wall, or the thickness of the leaf, for a cavity wall.

†When used with flexural strength in parallel direction, assume the orthogonal ratio $\mu = 0.3$.

Table 2.4 *General recommendations for selection of a durable brick, block and mortar in unrendered walls*

	Clay brickwork		Calcium silicate brickwork		Concrete blockwork		Concrete brickwork	
	Min. durability designation	Mortar designation	Min. class of brick	Mortar designation	Min. strength of block unit (N/mm^2)	Mortar designation	Min. strength of brick (N/mm^2)	Mortar designation
Internal walls (including the inner leaf of cavity walls)	ON	(iv)	Class 2	(iv)	any strength	(iv)	15	(iv)
	MN	(iii)	Class 2	(iii)	any strength	(iii)	15	(iii)
	*if risk of frost during construction**							
External walls (including the outer leaf of cavity walls) above dpc level	MN or FN, if exposed†	(iii) or (ii) if exposed	Class 2	(iv)	any strength	(iv)	15	(iv)
	as above	as above	as above	(iii)	as above	(iii)	as above	(iii)
	*if risk of frost during construction**							
External walls (including the outer leaf of cavity walls) below dpc level	MN or FL, if waterlogged site†	(iii) or (i), if waterlogged site	Class 3	(iii)	3.5 (dense concrete) 7.0 (lightweight concrete)	(iii)	20	(iii)
Chimneys, parapet walls and freestanding walls	MN or FL, if exposed†	(iii) or (ii), if exposed	Class 3	(iii)	3.5 (dense concrete) 7.0 (lightweight concrete	(iii)	20	(iii)
Sills and copings	FL	(i)	Class 5	(ii)	high strength	(ii)	30	(ii)
	or use purpose-made precast concrete products							

*During construction in cold weather there could be frost damage to a weak mortar, until it has hardened; clay brickwork which is not frost resistant and has become wet is also liable to frost damage.

†To avoid frost damage, bricks known to have frost resistance are used in exposed positions or on waterlogged sites.

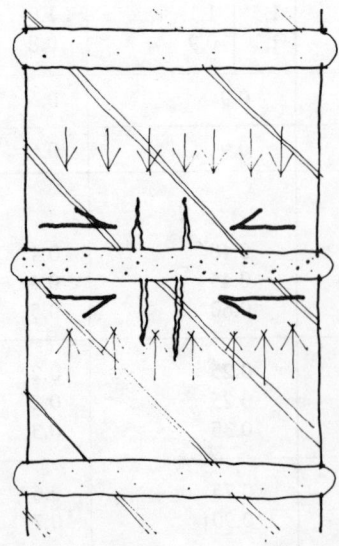

Fig. 2.9 Thin mortar layer restrained horizontally by friction of bricks above and below.

2.3 Wall bonds

Masonry units are bonded together in a wall, so that it becomes an integral element, by using a mortar which gives a good bond or attachment and, to a lesser extent, by choosing a suitable stacking arrangement of the masonry units. This latter property is the wall bond. A bond, in a wall, is usually said to exist if the cross joints in adjacent layers of the wall are staggered by not less than a quarter of the length of the masonry unit. For example, in a single-leaf wall with English or Flemish bond, the cross joints are both staggered by 25% of the length of the masonry unit (Fig. 2.10) whereas in a single-leaf wall with the common form of stretcher bonding, the cross joints are staggered by 50% of the length of the masonry units (Fig. 2.11). Another criterion for 225 mm thick walls of two skins is the amount of material which is put across the centre line joint. A collar-jointed wall has none except metal ties (Fig. 2.18) whereas an English bond wall and a Flemish bond wall have 50% and 33% respectively of their material crossing the central joint and are stronger in compression than the collar-jointed wall (Fig. 2.10). The Code (Clause 23.1.2) allows a higher compressive

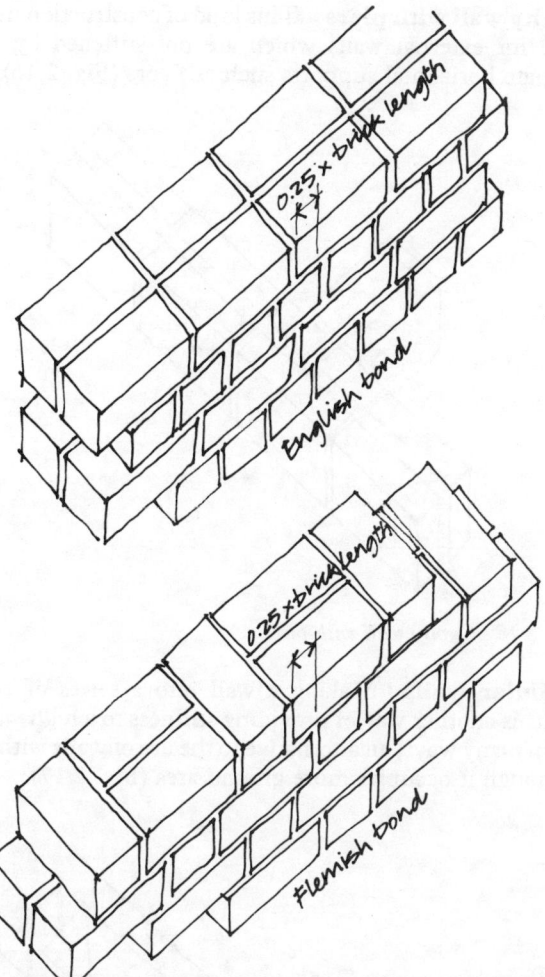

Fig. 2.10 Wall bonds in single-leaf brick walls (225 mm thick).

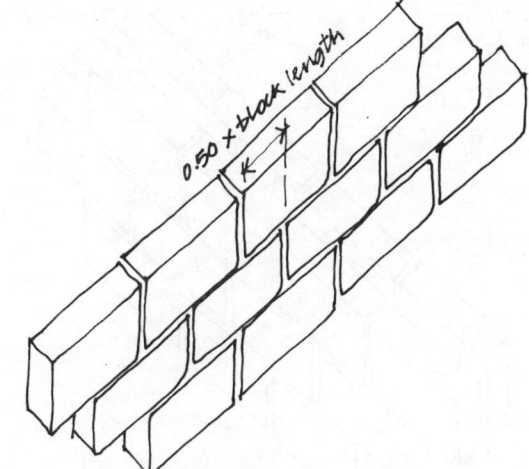

Fig. 2.11 Stretcher bond in concrete block wall.

stress to be taken on a brick wall with the same thickness as a standard brick than that on the same wall but 225 mm thick which is weakened by the centre line joint between the two skins. There are a very large number of different bonds in use for brickwork. Concrete blockwork, by contrast, is almost always used in common stretcher bond.

2.4 Types of masonry element

Single-leaf wall This is the basic form, a wall bonded so as to behave as one unit (Fig. 2.12). It is defined by BS 5628:Part 1 as a wall of brick or blocks, solidly set in mortar, which overlap in one or more directions. It can be built in a variety of thicknesses. Walls referred to as being of half, single or one and a half brick thickness, have thicknesses of approximately 102, 225 and 327 mm respectively, these being proportions of the brick length. Even with reasonable care in manufacture of the masonry units, it can be difficult to obtain a good finish on both sides of a single-leaf wall because of variations in the dimensions of the masonry units.

Single-leaf wall with piers Piers are added to the wall where it is required to stiffen the wall and make it stronger or where extra bearing area is required for beams and trusses (Fig. 2.13).

Cavity wall This is used mainly as an external wall to prevent rain water from penetrating to the inside and, thus, also preventing a loss of thermal insulation (Fig. 2.14). Thermal insulation boards can be placed in the cavity. This type of wall could also be used as an acoustic barrier, for example, by filling the cavity with sand. Cavity walls built

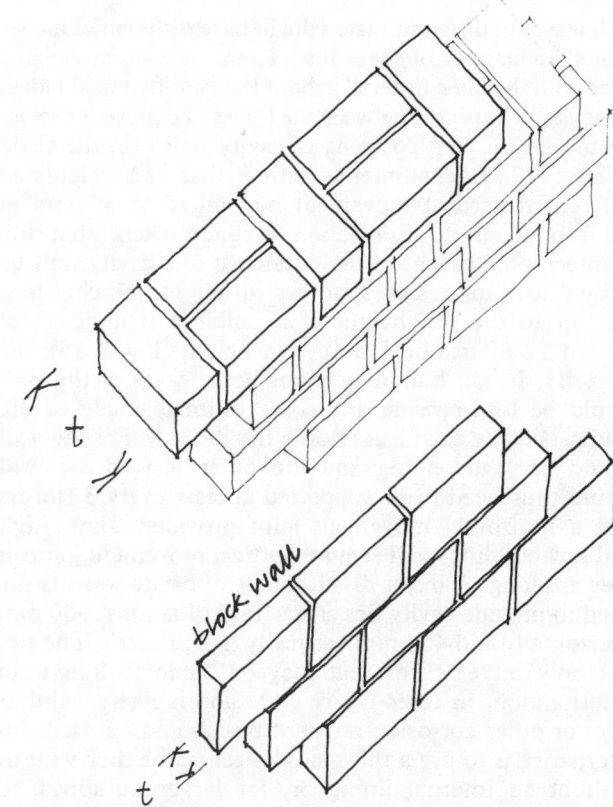

Fig. 2.12 Single-leaf walls in brick and block.

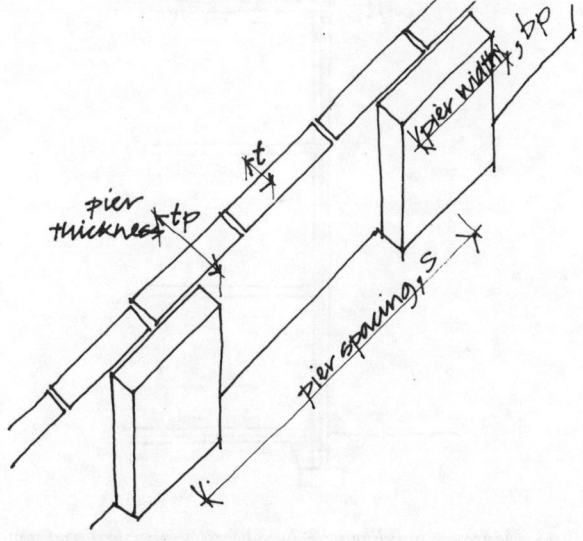

Fig. 2.13 Single-leaf wall with piers, showing pier dimensions.

13

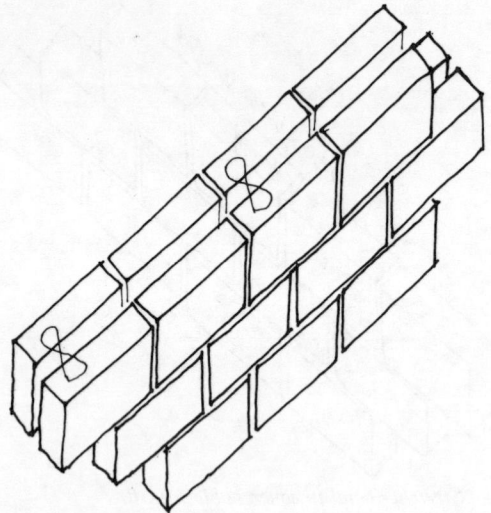

Fig. 2.14   Cavity wall.

with leaves of different material will have differential movements, sometimes of more than 2 mm in a storey height. Even with the same material in both leaves, differential movement can be expected between the leaves. To prevent excessive movement and loosening of cavity wall ties, the Code (Clause 29.2) recommends either that the calculated differential vertical movement be limited to 30 mm at the top or, if no calculation is undertaken, that the uninterrupted height of the outer leaf of a cavity wall be limited to 9 metres or 3 storeys in height, whichever is less, or to the full height of a building if it does not exceed 12 metres or 4 storeys in height (Fig. 2.15); see App. **B5**. In tall buildings either both leaves of the wall would be load-bearing and some estimate made of the load carried by each leaf or only the inner leaf of the wall would be load-bearing and the outer leaf of the wall would thus need to be supported at least every 3 storeys and a horizontal movement joint provided. Both high and low buildings will require vertical movement joints if they are long. A major disadvantage of cavity walls is the need to provide cavity ties at spacings of at most 900 mm horizontally and 450 mm vertically (Fig. B1.26). The ties not only increase cost but may be liable to long-term deterioration. In cases where corrosion is likely stainless steel or other corrosion-resistant ties should be used. An alternative is to use a thick single-leaf wall either with or without an internal lining or, for larger buildings, to have a cellular wall construction.

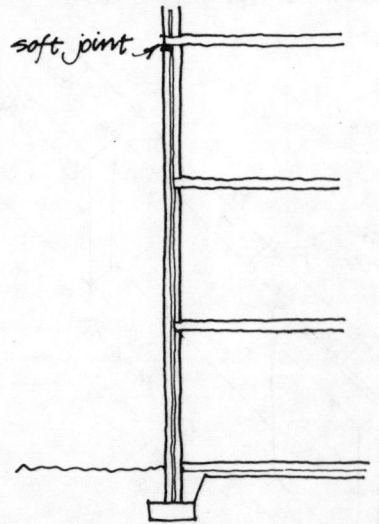

Fig. 2.15   *Masonry building with load-bearing inner leaf and non-load-bearing outer leaf.*

Cavity wall with piers   This kind of construction may be used for external walls which are not stiffened by intermediate horizontal supports such as floors (Fig. 2.16).

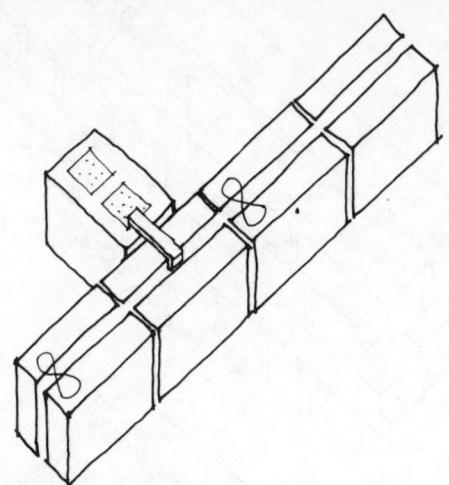

Fig. 2.16   *Cavity wall with piers.*

Cellular walls   Making a wall into a series of cellular units is another way of providing stiffness to a high wall and is, in many ways, an alternative to the use of walls with piers, although it occupies more ground area (Fig. 2.17).

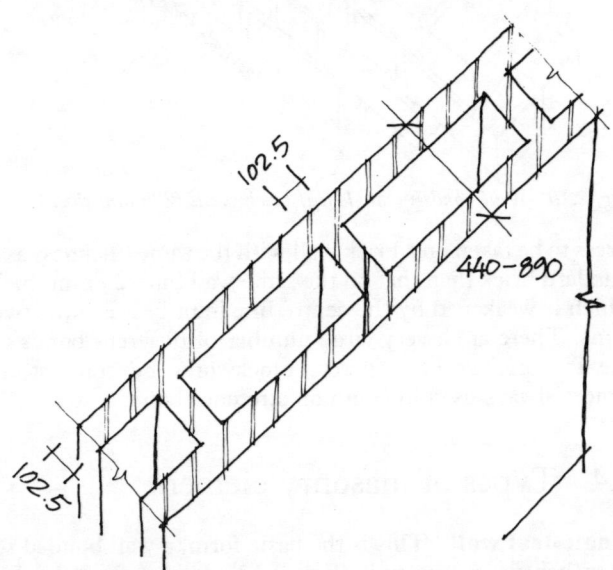

Fig. 2.17   *Cellular wall.*

Collar-jointed walls   These walls are defined by the Code as double-leaf walls separately constructed with a vertical collar-joint not exceeding 25 mm in width (Fig. 2.18). They are treated, structurally, as a cavity wall or as a single-leaf wall provided certain conditions are met (Clause 29.5 of the Code). This construction is often used where a fair faced finish is required on both faces of a wall about 225 mm thick.

Faced walls   These walls are defined by the Code as walls constructed of two different types of masonry unit which are, nevertheless, compatible (Fig. 2.19). They are treated, structurally, as a single-leaf wall made of the weaker material or as a veneered wall.

Veneered walls   These walls are defined by the Code as walls in which the veneer has no structural effect, except that of its own weight (Fig. 2.20).

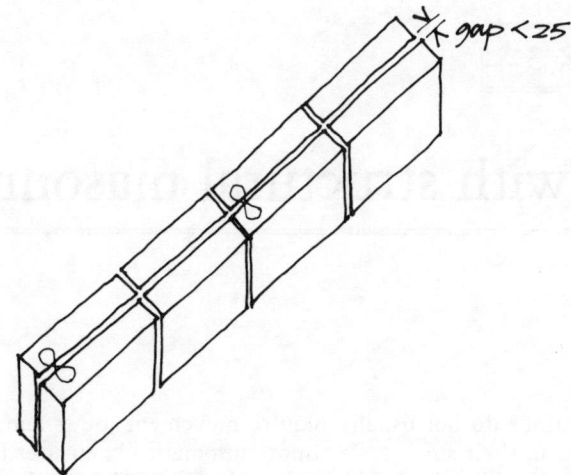

Fig. 2.18 *Collar-jointed wall.*

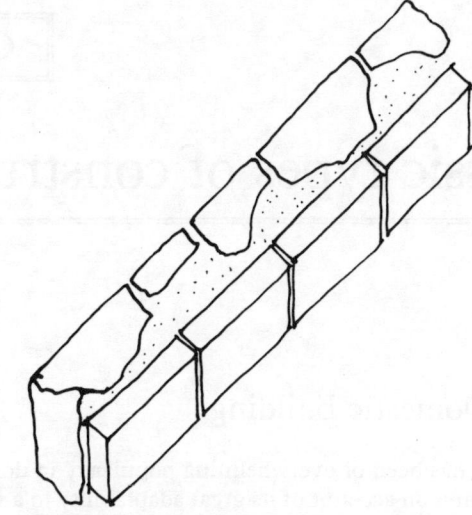

Fig. 2.19 *Wall faced with stone.*

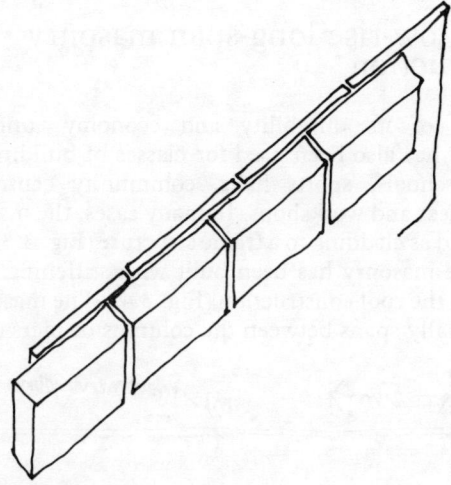

Fig. 2.20 *Veneered wall.*

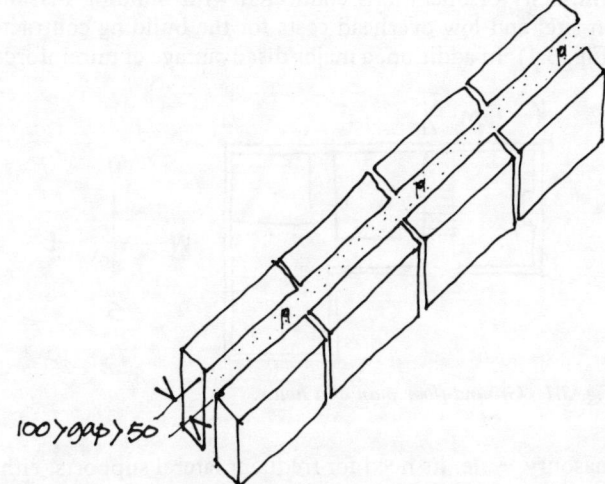

Fig. 2.21 *Grouted cavity wall with vertical reinforcement.*

Grouted cavity walls These walls are defined by the Code as walls of two leaves separated by a space of between 50 and 100 mm which is filled with concrete and which may be treated as a single-leaf wall for structural purposes, provided certain conditions are met (Clause 29.7 of the Code); see Fig. 2.21. This method of construction is a convenient way of giving additional strength to a cavity wall. The concrete infilling is often reinforced with steel bars. An alternative to this is the use of hollow masonry filled with concrete and this may also be reinforced. Reinforced masonry is covered in BS 5628:Part 2.

Columns According to the Code, a column is an isolated vertical element of masonry whose width is not greater than four times its thickness (Fig. 2.22). In practice, most columns are rectangular in plan but other vertical elements of different plan shapes but similar proportions should also be treated as columns (**5.2**).

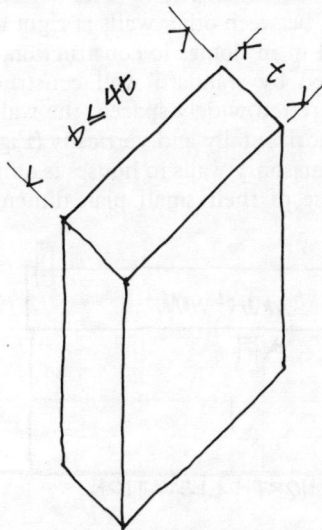

Fig. 2.22 *Dimensions defining a rectangular column to BS 5628: Part 1.*

CHAPTER 3

Classic types of construction with structural masonry

3.1 Domestic building

Masonry has been of overwhelming popularity in domestic construction on account of its great adaptability to a variety of plan and elevational treatments; the relatively low cost of the mass-produced masonry unit; the small number and simplicity of operations connected with building masonry on site; and low overhead costs for the building contractor (Fig. 3.1). In addition, a major disadvantage of unreinforced

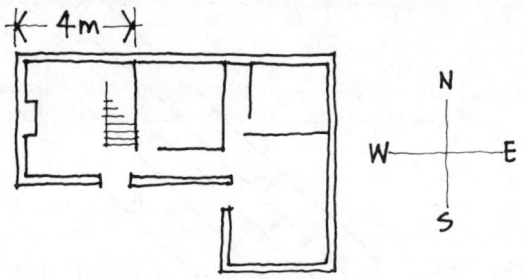

Fig. 3.1 Ground-floor plan of a house.

masonry − i.e. its need for frequent lateral supports, either vertical or horizontal, to avoid instability and collapse − is overcome by the frequency of return walls and floors common in domestic construction. The horizontal span for a masonry wall, between other walls at right angles to it, is of the order of 4 m in domestic construction, a distance that can be bridged by standard wall constructions. If these return walls are too widely spaced, the wall may still span vertically or horizontally and vertically (Fig. 3.2). The vertical span for masonry walls in houses is only of the order of 2.5 m. Because of their small plan dimensions, domestic

buildings do not usually require movement joints and, in general, their small scale almost automatically ensures that such a building is strong and durable. On good ground, strip foundations are almost exclusively used for walls in a domestic building.

3.2 Low-rise long-span masonry construction

Because of its durability and economy, unreinforced masonry has also been used for classes of building such as shops, schools, sports halls, community centres, small warehouses and workshops. In many cases, the masonry has been used as cladding to a frame structure (Fig. 3.3). In other cases the masonry has been built with stiffening piers and tied into the roof construction (Fig. 3.4). The masonry cladding usually spans between the columns or piers (Fig. 3.5).

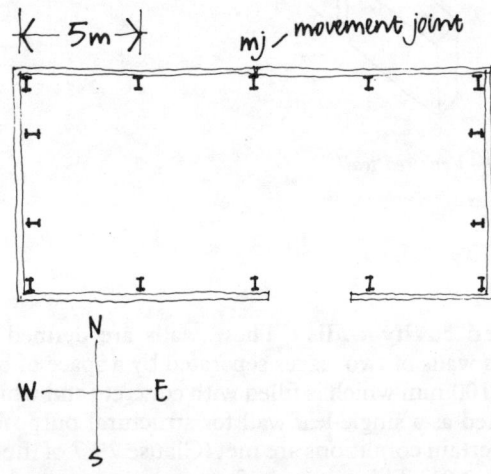

Fig. 3.3 Plan of hall with rigid steel frame and masonry cladding.

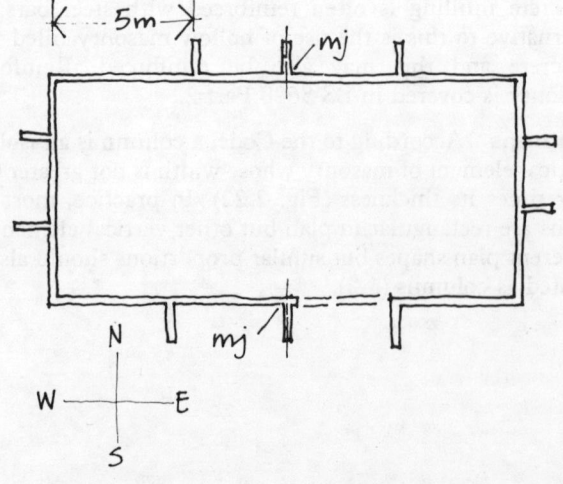

Fig. 3.4 Plan of hall with cavity wall stabilised by masonry piers.

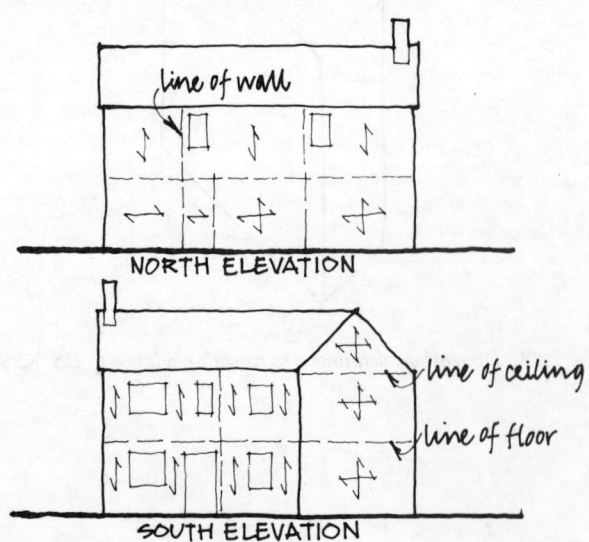

Fig. 3.2 Elevations of house in Fig. 3.1 showing directions of wall spans.

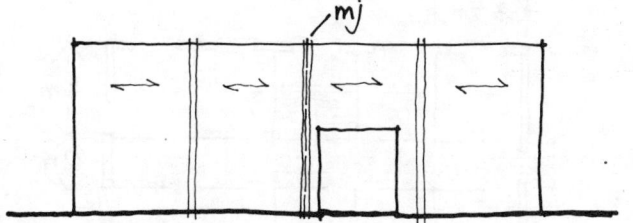

Fig. 3.5 Elevation of hall shown in Fig. 3.4.

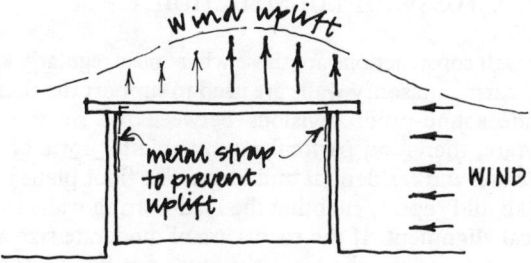

Fig. 3.7 Uplift on lightweight roof.

Both kinds of construction have been used for buildings such as halls and assembly areas. For schools or community centres, load-bearing masonry is also very common, the rooms in them being built to be of sufficient size for classes or meetings (Fig. 3.6). An important feature of a load-bearing masonry building of this type is its larger scale compared to domestic building. The necessary strength and stability are more difficult to achieve than in a house because the required internal spaces are larger and the masonry walls must therefore span larger distances. In addition failure of an element of the building, such as a roof truss, is more likely to lead to a progressive collapse than it would in a smaller scale building such as a house (**11.1**). Nevertheless masonry buildings of this type are usually constructed by methods traditionally used for domestic construction and have been aptly referred to as being of 'jumbo-domestic' construction (Sutherland 1978).

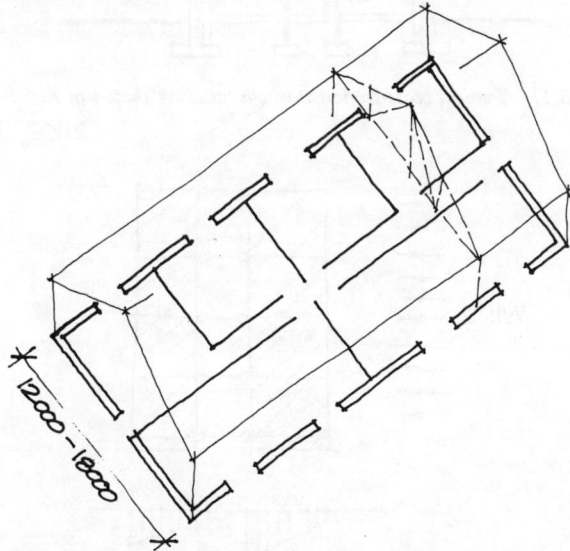

Fig. 3.6 Plan on a single-storey classroom block.

The stability of load-bearing masonry buildings or masonry clad buildings has two aspects: (a) the overall stiffness, strength and stability of the building-functions which may be taken by a rigid frame; and (b) the local strength of all the masonry elements such as the masonry cladding connected to a rigid frame structure. There is further discussion of this in Chapter 8. In practice, masonry wall panels, in buildings of this size, are very often required to span up to their safe limit and, most often, difficulties are experienced in conveniently introducing lateral supports for these walls; they may either be horizontal supports, such as floors and roofs, or vertical supports, such as columns, intersecting walls and piers (Figs 3.3, 3.4, 3.6). Difficulties may also occur in connecting the wall or roof elements to their lateral supports. The supports must be strong enough and stiff enough so that they do support the wall or other element in the way assumed in the design. For example, wind uplift on lightweight roofs means that a connection between the roof and the wall must

ensure the design uplift is balanced out by the dead weight of the wall (Fig. 3.7). Similarly if the roof construction is relied on to support the top edge of a wall, then it must be stiff enough to do this properly (Fig. 3.8). In the examples illustrated in Figs 3.3 and 3.4 the roof construction is connected to all four outside walls and acts like a plate, or diaphragm, and would ensure that wind forces on the north or south elevations are transferred to ground through the east and west gable walls (Fig. 3.9). In the example illustrated in Fig. 3.6, the roof could also be connected to internal walls to provide extra lateral support against horizontal loads. In general flexible roof constructions, such as those using timber, need to have a length to width ratio of about four or less to be stiff enough to prevent damage to the masonry walls. For the kinds of buildings discussed in this section, strip foundations provide adequate support to the walls on good ground.

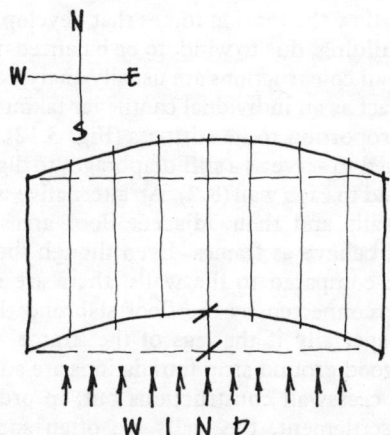

Fig. 3.8 Plan on a roof with insufficient stiffness to support tops of north and south walls.

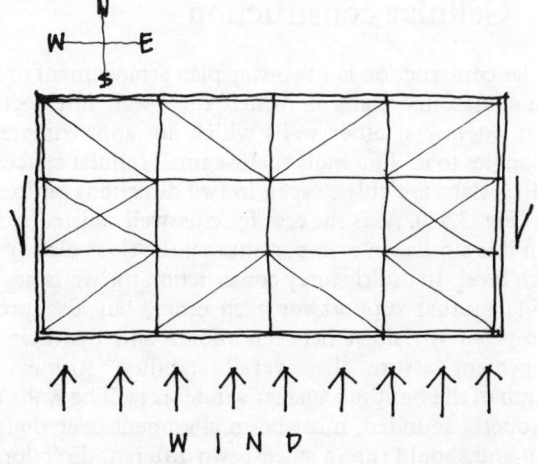

Fig. 3.9 Plan on roof of hall in Fig. 3.4 showing the steel members which make the roof rigid.

3.3 Crosswall construction

Crosswall constructions are those where near regularly spaced load-bearing masonry walls are used to support the floor and provide sound-proof divisions between the room spaces. They are, therefore, particularly suitable for rows of shops or in hotel and residential buildings. The floor plans at each level should repeat, enabling the load-bearing walls to be in vertical alignment. If the rooms are of moderate size and of similar dimensions the floor structure can be an economic one. The floor slabs span one way on to the crosswalls, although the direction of span may change between one part of the building and another (Fig. 3.10). A typical spacing for the walls is about 3 m, but even buildings with double this span may be economic. Non-load-bearing walls may be placed almost anywhere on the floor slab if it is of reasonable strength and stiffness. A reinforced concrete slab is most suitable and, in addition, gives good sound insulation between floors. The overall stability of the building against lateral forces, such as wind, is assured by properly founded crosswalls and other walls or bracing running approximately at right angles to the crosswall (Figs 3.11 and 3.12). A small number of bracing walls at right angles to the crosswalls combined with poor connections to a flexible or discontinuous floor slab could make crosswall construction liable to horizontal progressive collapse (**11.1**). Crosswall structures are well suited to multi-storey construction and in this case the crosswalls in the lower storeys are able to take high lateral forces by virtue of the vertical load on them. Wind forces can be the critical factor in higher multi-storey buildings. As with almost any masonry construction, the plan should allow the tension forces that develop at the bottom of the building, due to wind, to be balanced out by dead load. Crosswall constructions are usually analysed assuming each wall to act as an individual cantilever taking a horizontal load in proportion to its stiffness (Fig. 3.12). The floor slab is assumed to serve as a stiff diaphragm to distribute the horizontal load to each wall (**8.2**). An alternative is to assume that some walls and their adjacent floor areas can work together and behave as frames. Even though the floors are very flexible compared to the walls, there are many such 'frames' interconnected by the floor slab and the effect is significant especially if the legs of the 'frame' are closely spaced. On good ground strip foundations are adequate, for multi-storey crosswall constructions but, in order to limit differential settlement, the walls are often supported by ground beams on piles or by a thick raft of about 600 mm depth, depending on soil conditions.

3.4 Cellular construction

Cellular construction is a two-way plan arrangement of load-bearing masonry walls in which each wall intersects, or almost intersects, other walls which are approximately at right angles to it. The walls enclose small cellular spaces and the floor slabs are able to span in two directions on to these walls (Fig. 3.13). As is the case for crosswall construction, to which it is similar, it is important that the floor plans repeat at each level. In multi-storey construction the walls not only provide mutual support for each other, but also provide sound-proof divisions between rooms and by their plan arrangement assure the overall stability, stiffness and strength of the building against wind forces. The walls must be properly founded, must be in alignment over their full height and should run in at least two different directions. In principle the walls are spaced closely enough so that, unlike crosswall construction, the whole building can behave as a

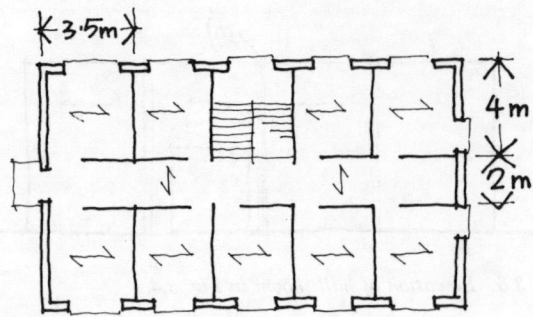

Fig. 3.10 Typical floor plan of crosswall load-bearing masonry building showing load-bearing walls and directions of floor span.

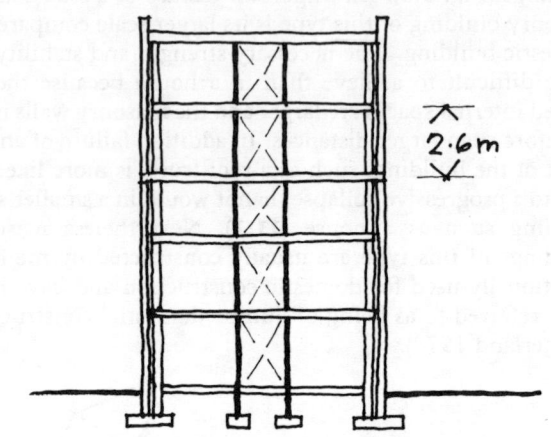

Fig. 3.11 Typical cross-section through building shown in Fig. 3.10.

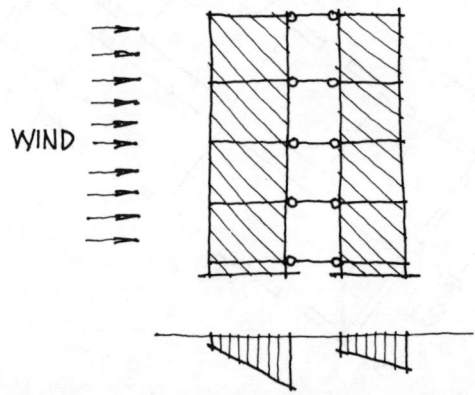

Fig. 3.12 Structural model and typical compressive stress distribution in crosswalls, due to vertical and horizontal loads.

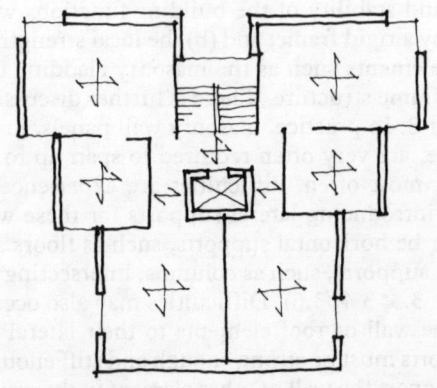

Fig. 3.13 Typical floor plan of cellular load-bearing masonry building, showing load-bearing walls and directions of floor span.

box structure under horizontal loads deflecting somewhat like a single cantilever, rather than the two connected cantilevers shown in Fig. 3.12. In practice the behaviour of the building under horizontal load is likely to be something between these two extremes and it may be most accurate to assume that the building behaves as a series of frames interconnected by the floor slab (3.3). Vertical loads may be evenly distributed between the walls enabling them to be relatively thin, and the floor construction can usually be an economic one if it is only required to span over short distances. These last two factors can make cellular construction economic over heights of up to about twenty storeys, for residential buildings and the like, where only small room areas are required. Below twenty storeys, even for narrow buildings, a good plan arrangement will ensure that no tension occurs in

the bottom load-bearing walls when the building is under wind load. The object, as with crosswall construction, is to arrange the walls in plan so as to balance out tension forces with dead load; in some cases this may mean that the floor spans have to be increased. An accurate method of calculation may show that the tensile stresses predicted by a simpler method do not in fact occur and this can bring significant savings in the steel and reinforced material that would otherwise be required. Bearing in mind the close spacing of walls and their weight, a raft of about 600 mm depth is often the most economic way of supporting a building of cellular construction even on good ground. However, strip foundations and ground beams on piles are also used, depending on soil conditions.

CHAPTER 4

Foundations for masonry buildings

4.1 Introduction

Foundation design starts with the gathering of information on the proposed site for the building. Information is required on the type of soil at each level, on the groundwater level and possible variations in this level, and on the possibility of ground movement – for example, that due to subsidence at low levels. The state of existing buildings, especially masonry buildings, could give an indication of poor ground. In compressive soils the trial holes or borings for the site investigation would go to a depth at least 50% greater than the width of the widest foundation or the effective loaded width of the soil; for example, for a building in which the piles or foundations are so closely spaced as to load the whole area under the building, the boring would be taken to a depth one and a half times the width of the building. For a square or circular foundation the net increase in vertical stress at this depth is about 20% of that at foundation level. Compression of strata below this level will only have a small effect on the total differential settlement, unless there are very soft layers present, such as peat.

Knowing the kind of building proposed, an estimate can then be made of the loads on the foundations and the amount of settlement that the building can accept. Usually, it will then be possible to decide on the type of foundation, the foundation level and the allowable bearing pressure at that level. Vertical loads on the building are usually the ones that determine the type and size of the foundation but horizontal loads, due to wind or retained earth, may sometimes have an influence too. The Allowable Bearing Pressure (**4.2**) on the soil must be two or three times less than the Ultimate Bearing Pressure, the one which would cause shear failure in the soil, and low enough not to cause unacceptable absolute or differential settlement, this last factor usually being the critical one for masonry buildings which are extremely sensitive to differential movements (Fig. 4.1). The foundations must be placed at a sufficiently low level so that they will not be significantly affected by frost heave (Fig. 4.2) or by swelling and shrinkage of the soil at foundation level due to seasonal changes in the weather. Frost heave can occur in chalk and chalky soils, silts, fine sand and fine sandy soils. In the UK, foundations on soils susceptible to frost will not heave if founded at depths greater than about 450 mm to 600 mm. Swelling and shrinkage can occur in clay where there are changes in the moisture content of the clay, especially in the over-consolidated fat clays. Foundations on shrinkable clays need to have a minimum depth of 900 mm (App. **B2**).

The proposed foundation should not be put at a level where it may undermine an existing building foundation (Fig. 4.3) unless the existing foundation is to be underpinned, nor should it be positioned where its effect on the soil beneath the existing foundation causes settlement which the existing building may not be able to tolerate (Fig. 4.4). It is possible

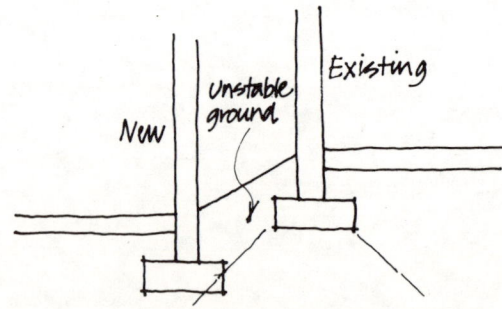

Fig. 4.3 Existing foundation undermined by new one.

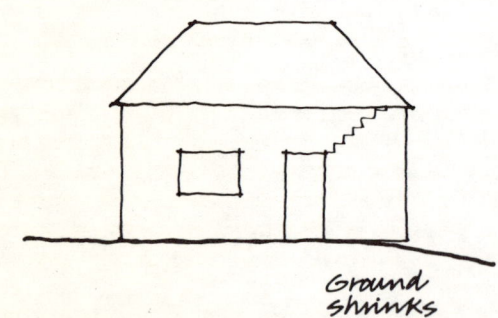

Fig. 4.1 Cracking of masonry due to ground settlement.

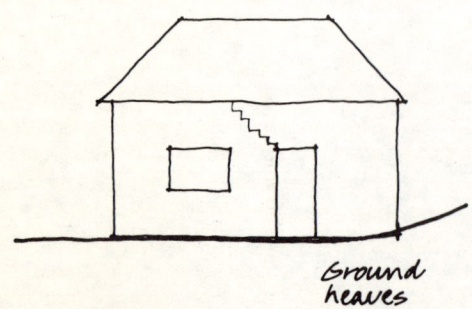

Fig. 4.2 Cracking of masonry due to ground heave.

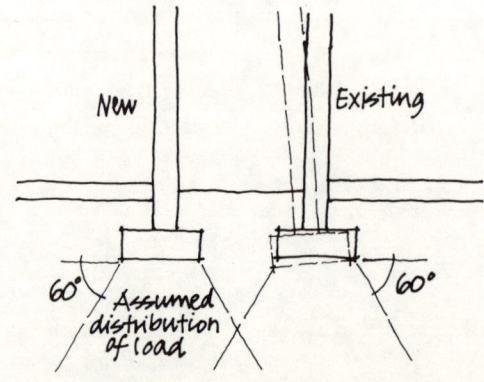

Fig. 4.4 Existing foundation made to rotate by extra pressure on ground from new foundation.

for general landslips to occur, for example those due to the sliding and rotation of a clay slope. Movement of soil at the foundation level may also be caused by other factors such as the softening of clay and silt soils from the presence of water or the washing out of the fine particles in silt and clay mixtures, especially with boulder clay, or in sand and gravel mixtures, causing local collapse of the ground. Both these effects could be produced by groundwater flows or water from broken soil pipes, for example. Movements can also occur as a result of any significant movements in the level of the groundwater. For example, an immediate loss of bearing capacity and settlement may occur due to lowering of the groundwater level and the consequent decrease in pore water pressure. A 'hard spot' in the ground below the building, for example that due to the presence of an existing foundation or wall at a lower level, or the presence of a soft peat layer at depth may also cause differential movement leading to cracks in masonry buildings. Some typical foundation details are given later in this chapter and in App. **B2**.

4.2 Allowable soil bearing pressure

The figure chosen for the allowable bearing pressure should be less than the ultimate bearing pressure divided by a factor of safety – this is the requirement for strength – and less than that which could cause excessive movement in the building, see **4.3**. The allowable bearing pressure, when quoted, usually means the net allowable bearing pressure, that is, the pressure that the soil can be allowed to support on top of the overburden of soil it already carries, if any. This is the meaning adopted here. This pressure is then compared with the net increase in soil pressure under the foundation due to the new weight of the building and, if present, the effect of other new buildings or loads on the surface. The weight of any soil overburden is not included, however (Fig. 4.5).

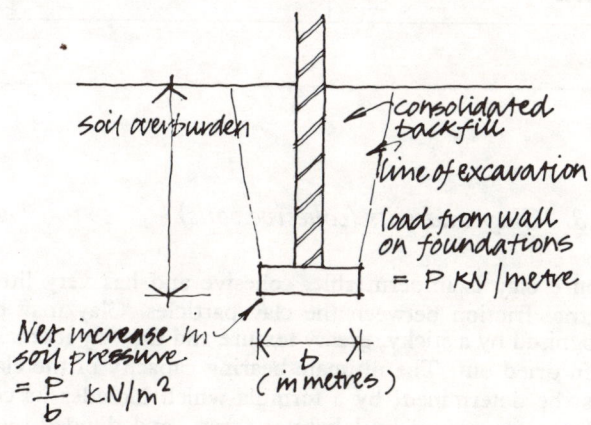

Fig. 4.5   Net increase in soil pressure under foundation.

The Actual Bearing Pressure beneath a foundation, by contrast, includes all the loads on that foundation including the soil overburden. If for any reason the soil at foundation level cannot be said to be confined by the intergranular forces of the overburden, then the quoted allowable bearing pressure must be compared to the soil pressure due to the total weight of the new building and soil overburden. Sand and gravel overburdens which have been stabilised provide these intergranular forces, whereas this is not completely true for a clay, except one that is fully consolidated, although it is often assumed to be true for design purposes. A case where the actual bearing pressure beneath the foundation should

be calculated, is for a foundation overlain by an unconsolidated and recent fill (Fig. 4.6); in time a selected fill will stabilise and consolidate and would then be able to confine the soil at lower levels properly. In practice the fill is often compacted as it is put in so that the problem need not arise. Preliminary values for allowable bearing pressure, the Presumed Bearing Pressure, are given in Table 4.1. These values may be checked against or improved upon by local knowledge and site inspection. Further information on soil identification is given in App. **A10**. BS 5930 (1981) gives advice on site investigations, including the description and identification of soil. Brief details of the various soils follow.

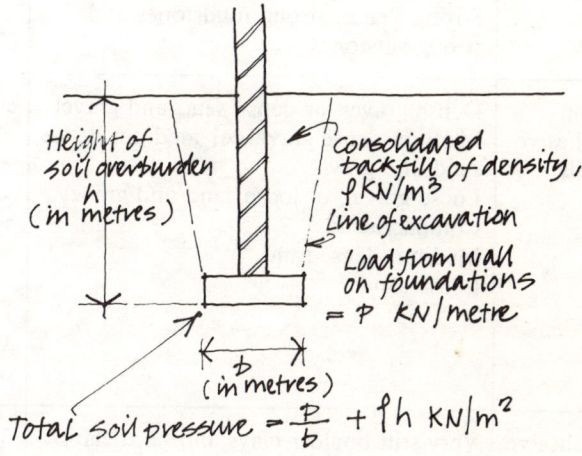

Fig. 4.6   Total soil pressure under foundation.

4.2.1   Rocks

Allowable bearing pressures for rock are mostly quite high and, therefore, it is not usually necessary to investigate in detail. However, some weathered rocks may only have poor bearing capacity. In cases where an investigation is necessary, it may be difficult to establish how good the condition of the rock is below the surface. A commonly found rock with unusual and very variable properties is chalk which breaks and softens easily when wetted, and, like some other formations, may contain swallow holes.

4.2.2   Sands and gravels (non-cohesive soils)

The allowable bearing pressure of non-cohesive soils depends on the compaction and the grading. A well-graded soil has material of many different sizes in balanced proportions and, thus, can be compacted to a dense mixture. There is an optimum moisture content at which the mixture compacts best. The degree of compaction can be measured on site but often a visual inspection is sufficient and, sometimes, this is a better guide. For sands, however, the Standard Penetration Test provides a measure of compaction which is correlated to well-established values of allowable bearing pressure based on limiting total settlement to 25 mm.

The aim is to make all foundations have settlements of similar amounts. For the equal settlement of two foundations of different size, or width if they are strip foundations, the larger foundation will need to have a lower actual bearing pressure than the smaller foundation. This is simply because the larger foundation has greater effect on the lower soil levels than the smaller one, compresses a greater depth of soil and would cause more settlement if the actual bearing pressures were equal (**4.1**). It is worth noting that raising of the groundwater level in sand does not markedly increase

Table 4.1 *Presumed bearing pressure under vertical static loading (adapted from Table 1 of BS 8004: 1986)*

NOTE. These values are for preliminary design purposes only, and may need alteration upwards or downwards. No addition has been made for the depth of embedment of the foundation.

Category	Types of rocks and soils	Presumed allowable bearing value		Remarks
		kN/m²*	kgf/cm²* tonf/ft²	
Rocks	Strong igneous and gneissic rocks in sound condition Strong limestones and strong sandstones Schists and slates Strong shales, strong mudstones and strong siltstones	10 000 4 000 3 000 2 000	100 40 30 20	These values are based on the assumption that the foundations are taken down to unweathered rock.
Non-cohesive soils	Dense gravel, or dense sand and gravel Medium dense gravel, or medium dense sand and gravel Loose gravel, or loose sand and gravel Compact sand Medium dense sand Loose sand	> 600 < 200 to 600 < 200 > 300 100 to 300 < 100 Value depending on degree of looseness	> 6 < 2 to 6 < 2 > 3 1 to 3 < 1	Width of foundation not less than 1 m. Groundwater level assumed to be a depth not less than below the base of the foundation.
Cohesive soils	Very stiff boulder clays and hard clays Stiff clays Firm clays Soft clays and silts	300 to 600 150 to 300 75 to 150 < 75	3 to 6 1.5 to 3 0.75 to 1.5 < 0.75	Group 3 is susceptible to long-term consolidation settlement
	Very soft clays and silts	Not applicable		
Peat and organic soils		Not applicable		
Made ground or fill		Not applicable		
*107.25 kN/m² = 1.094 kgf/cm² = 1 tonf/ft².				

settlement and, therefore, does not affect the allowable bearing pressure if this is fixed, as is normal, by considerations of the likely settlement. Nevertheless, the ultimate bearing capacity of the submerged sand is less than the same sand when it was dry. In these conditions an immediate re-lowering of the water table would not, in general, cause significant settlement such as could be the case for a poorly consolidated soil.

Settlement is usually the critical factor in setting a figure for the allowable bearing pressure on sands and gravels but with shallow and narrow foundations on loose sand the allowable bearing pressure is set by the need for strength. Loose sand likely to be subject to vibration needs special care. Settlement of sands and gravels is usually small and takes place almost instantaneously. If settlement is the governing factor, an increase in the allowable bearing pressure can be made due to embedment, the effect of a reduction in settlement as foundations go deeper into the founding layer. It should be remembered that the use of an allowable bearing pressure for non-cohesive soils is usually simply a convenient alternative to doing a calculation of the settlement. In simple cases, quoted figures for the allowable bearing pressure are sufficiently accurate in design.

4.2.3 *Clay and silts (cohesive soils)*

A pure clay is impermeable, cohesive and has very little internal friction between the clay particles. Clay may be recognised by a sticky, greasy texture and the way it cracks when dried out. The ultimate bearing capacity of the clay must be determined, by a formula which includes its cohesion as measured by laboratory tests, and divided by a factor of safety. Settlement calculations are then done in order to establish an allowable pressure which would limit settlement to acceptable amounts. The lower of these two figures then becomes the allowable bearing pressure on the soil. For simpler cases, a quoted figure, for example from Building Regulations or from a Code of Practice, may be used (Table 4.1). Clays soften and rapidly lose strength when wetted; for this reason clay when excavated for foundations should be protected as soon as possible by a layer of concrete blinding. As explained previously, the confining effect of a clay overburden is doubtful and should not be relied upon unless the clay is fully consolidated. Silt is partly cohesive, like clay, and partly granular, like sand, although its internal friction is low. It may be distinguished from clay by its gritty texture and the way it disintegrates when wetted. The allowable bearing pressure is calculated in the same way as for clay. However, silts can be tricky soils to assess, bearing in mind their sometimes sudden loss in strength when wet. As is the case for non-cohesive soils, a large foundation on clay or silt will settle more than a small foundation having equal actual bearing pressures. This will be shown up by a calculation of settlement but must be taken into account if this calculation is not done.

When the water content of clay is reduced, the clay shrinks. Some clays have particularly high shrinkage when they dry out; in general these are the fat clays – i.e. those with a high clay fraction. In the UK the shrinkable clays generally only occur in the south-east of England although carboniferous shales and glacial deposits can behave in similar ways to the shrinkable clays. Shrinkage is made very much more severe by the fine roots from trees or shrubs which increases the moisture lost over that due to surface evaporation; this is a common cause of damage to foundations even those at some depth. Damage is more likely to be caused by trees that have a large root spread or have a large moisture demand such as poplar, elms and willows. Other factors include the groundwater conditions and the soil condition. See BS 5837, Chapter 4 of NHBC Standards and BRE Digests 240 and 298 for more details.

The aim of the foundation design in a shrinkable clay is to position the foundation at a depth which avoids significant shrinkage and swelling. Research at the Building Research Establishment (BRE) indicates that in the UK significant movement only occurs in the top metre of a shrinkable clay, except near large trees and shrubs. A practical minimum depth has been established as 900 mm in these conditions. Near trees or shrubs or in a minority of particularly shrinkable clays, the foundations will need to be taken deeper, usually 1.5 or 2 m but occasionally up to 4 or 5 m in depth. Trench fill foundations are in common use or, for greater depths, short bored or driven piles; an alternative is the use of a pier in mass concrete or a pad with stem (Figs 4.16 to 4.29). Some clay sites recently cleared of trees and shrubs can experience significant heave over periods of time up to ten years because of re-swelling of the clay previously dried out by the trees and vegetation. As heave can be very damaging to masonry buildings, BRE recommends the use of bored pile foundations with the whole ground floor suspended from beams, not just the masonry walls. There may be uplift on the pile caused by the upward moving clay near the surface. In these cases the piles may be sleeved at the top to allow-sliding or, better, taken to a greater depth with extra reinforcement provided near the top of the pile. Note that foundations put down on clay in very dry conditions may also be liable to uplift when the clay re-swells with the rain.

Soft clay soils frequently experience long-term settlement as a result of the extra weight of the building. Again, foundations taken to depth either with trench fill, pier or bored pile foundations can greatly reduce the settlement. In such cases a floating ground floor slab is often adequate. Alluvial clays form a crust at the top, usually at least one metre thick, and this, in general, overlies a much softer clay layer. If possible, in these clays, the foundations should be kept well above the softer layer to avoid overstressing it.

4.2.4 Peat and organic soils

Peat and organic soils have a very low allowable bearing pressure, although these may be improved by pre-loading with a surcharge, usually a fill material. Without pre-loading, it might still be possible to support a lightly loaded building which could accept some differential settlement, by using an imported granular fill under the foundations. However, in general, for masonry buildings piling will be necessary with the walls supported by suspended ground-floor beams resting on the piles. Alternatives are a stem foundation (Fig. 4.18) or, if there is a stiffer layer over the peat, a raft of medium or high stiffness (Figs 4.21 and 4.25).

4.2.5 Fill (made ground)

There is an increasing need to be able to found on fill material. Ideally, the fill would be of a selected material such as a graded mixture of gravel, sand and clay in balanced proportions. The surface on which the fill is placed should be cleared of debris and organic matter and rolled. The fill material would be placed in thin layers, about 250 mm deep, and at a moisture content which enabled each layer to be densely compacted. In such conditions an allowable bearing pressure comparable with soil in its natural state can be obtained. By contrast, fill material such as household waste, even if compacted in layers, could experience about twenty times the settlement of a compacted, properly-graded, granular fill. Normally foundations cannot be placed at all on fill containing household waste. Nor can foundations be placed on fill material tipped in an uncontrolled way unless the material can compact sufficiently under its own weight and has been allowed to do so over a period of time of about ten years. A quicker method is some kind of ground improvement such as excavation of the fill and the replacing of it in thin compacted layers or pre-loading with a temporary surcharge such as a rock fill to effect consolidation. After pre-loading, further settlement can still be expected in deep fill due to the self-weight of the fill or due to variations in groundwater level. However, load-bearing masonry, if placed on a stiff raft foundation, could be completely satisfactory on such ground. Alternatively, piling can be used and the masonry walls supported by a suspended ground-floor construction.

4.3   Settlement and heave

Settlement is the downward movement and heave is the upward movement of a slab or a foundation. If all the foundations of a building were to settle or heave by the same amount, or otherwise move like a rigid body, no damage would be caused to the building by small movements, although incoming services may need flexible connections to the buildings. It is the differential movement between the parts of a building block that causes damage and, therefore, this is usually the most relevant measure of movement. Two types of movement may be distinguished, 'shear' and 'bending' deformation (Figs 4.7 and 4.8). In a framed building, a masonry building mainly of crosswall construction or a masonry building with large window openings, the differential movement due to local 'shear' deformation of the building is much larger than overall 'bending' deformation of the building block (Fig. 4.9). 'Shear' deformation is measured by the Angular Distortion, that is, the relative rotation of two adjacent foundations divided by the distance between them, S/L. For a load-bearing wall without openings or only small openings 'shear' deflection is much smaller than the overall 'bending' deflection. The 'bending' deformation is measured by the Deflection Ratio, that is, the maximum relative deflection of a wall from a line joining the ends of a section of wall divided by the length of this section, d/L (Fig. 4.10). At a certain point, differential movement – as measured by either of these two parameters – will cause cracking. For load-bearing walls, the length to height ratio of

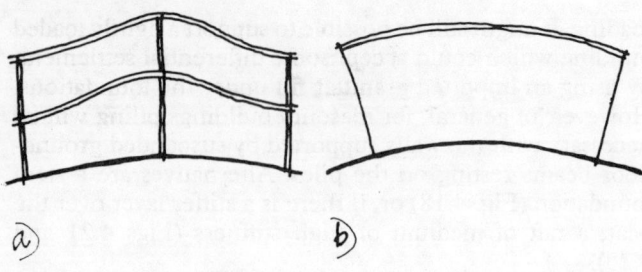
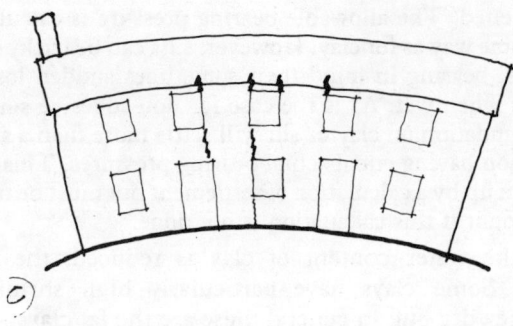

Fig. 4.7 *(a) Frame undergoing 'shear' deformation and (b) similar sized solid wall undergoing 'bending' deformation.*

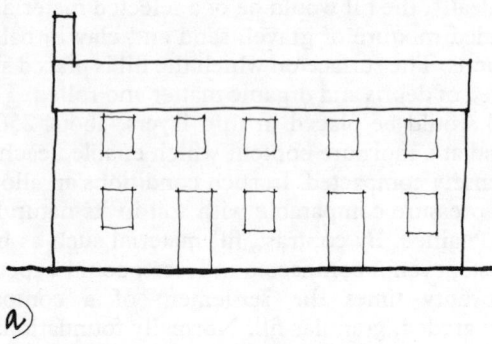
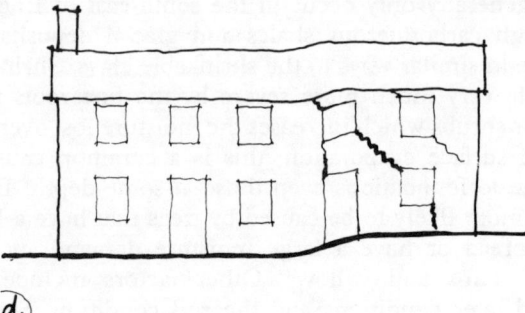
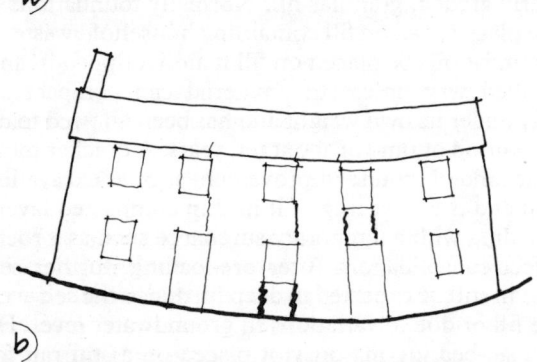
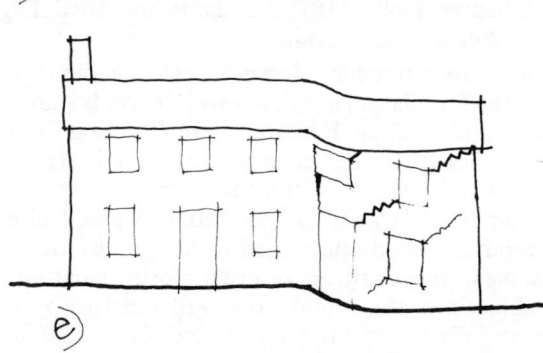

Fig. 4.8 *Masonry building (a) undeformed, (b) bending in sagging mode, (c) bending in hogging mode, (d) and (e) deformed in shear at one end.*

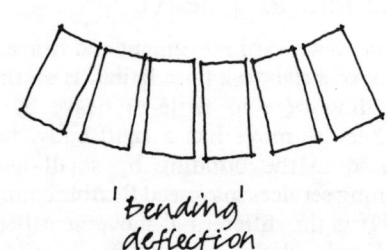

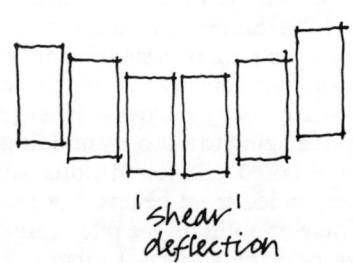

'bending' deflection

'shear' deflection

Total deflection = bending deflection + shear deflection

Fig. 4.9 *Models of bending and shear deflections which may occur during settlement of a masonry wall with openings.*

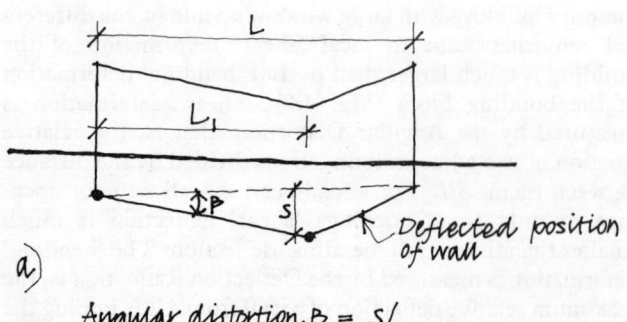
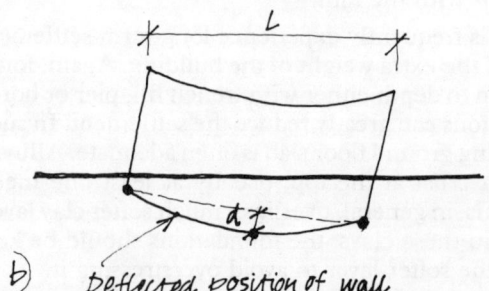

Angular distortion, $\beta = s/L_1$

Deflection ratio = d/L

Fig. 4.10 *Definitions of (a) angular distortion of two points relative to one another (b) of deflection ratio of a wall, based on the point with maximum deflection relative to the whole wall.*

the wall, *L/H*, is usually significant when assessing the likelihood of cracking.

Burland and Wroth (1975) suggest that the onset of visible cracking for unreinforced, load-bearing walls occurs when the deflection ratio is 1/2500 with *L/H* = 1 or 1/1250 with *L/H* = 5, assuming the wall to deflect in a sagging shape. They suggest deflection ratios of half these figures for walls in a hogging shape because, generally, cracks in hogging walls can start and propagate much more easily. Design values of deflection ratio would be about half to a third of all these values, depending on the soil and the type of building (Fig. 4.11). For the wall of a building in which the local 'shear' deformation is likely to be significant, the angular distortion may be used as a measure of differential movement. This should not exceed about 1/600 and may need to be less to avoid visible cracking.

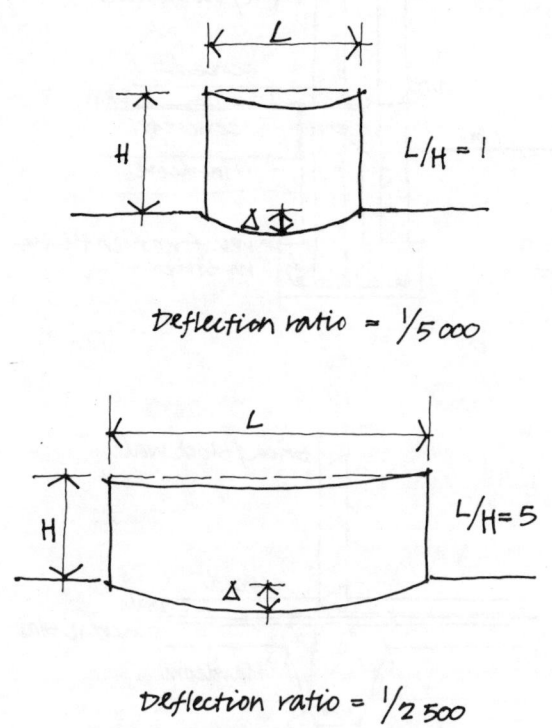

Fig. 4.11 *Suggested design values of deflection ratio for load-bearing wall in soft clay.*

In practice, the actual differential movements are extremely difficult to predict with reasonable accuracy and a calculated figure for differential movement, to compare with those considered to be just acceptable, will often be based on taking a proportion of the calculated total settlement. This approach would be suitable for a routine building where it is required to check the order of magnitude of settlement. In other cases, if heave or settlement is expected and it is decided to use a raft, a more sophisticated calculation may be required which takes account of the soil characteristics as well as the interaction of these with the building foundation and possibly the interaction with the building structure. Certainly, the design should take some account of the effect the masonry has in stiffening the foundation system and, thus, preventing visible cracking.

Load-bearing walls provide considerable stiffening to the foundations if they deflect in a sagging shape and this accounts for the fact that most masonry buildings experience very little relative deflection (Fig. 4.12). The problem is to identify conditions where cracking may be initiated and where it would be unacceptable. In existing buildings BRE

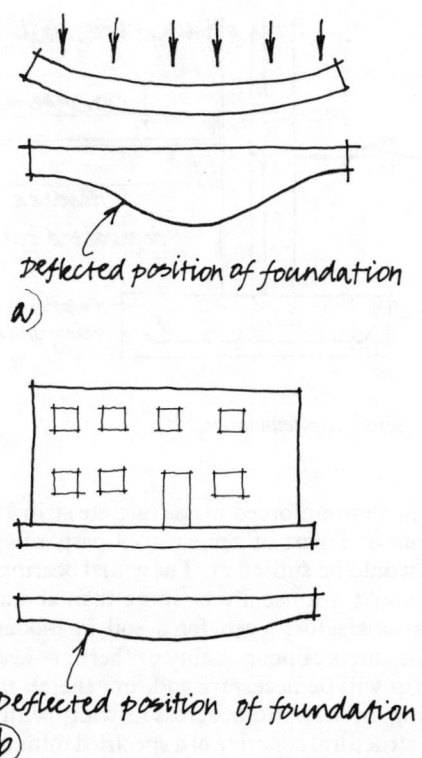

Fig. 4.12 *Deflection of (a) flexible foundation, and (b) foundation supporting similar load but with rigid wall above.*

Digest 251 (1981) classifies damage due to crack widths up to 5 mm as 'slight'. The only major concern with cracks of this size is to ensure that they are not a sign of progressive movement leading to much wider cracks, out-of-plumb walls (**12.5**) and, in an extreme case, instability. Cracking is less likely in a wall under high and uniform gravity load. Where a weak, flexible mortar has been used, the cracking may be so evenly spread out that it is not easily visible and is otherwise not significant. If cracking is a problem, reinforcement in the bed joints at the bottom of a wall if the wall deflects in a sagging shape − or at the top of a wall if it deflects in a hogging shape − may prevent any visible cracks occurring at all.

4.4 Types of foundation

4.4.1 Strip foundations

This is the most popular foundation type for load-bearing masonry. A traditional strip foundation is shown in Fig. 4.13. For lightly loaded walls such as those of a two-storey

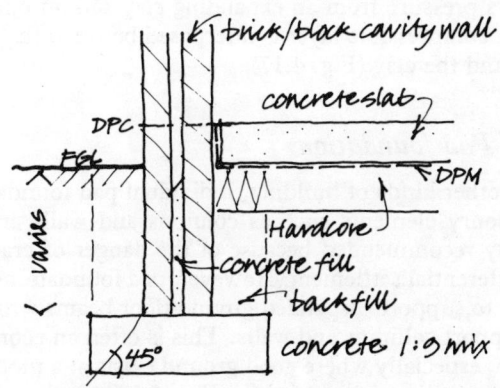

Fig. 4.13 *Traditional strip foundation.*

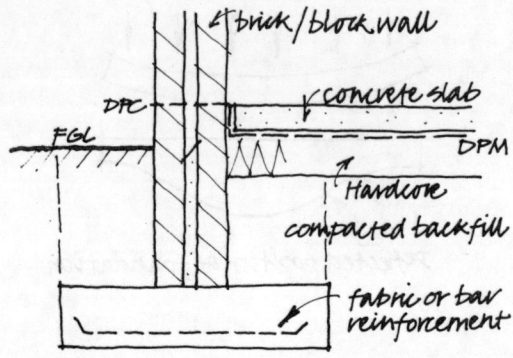

Fig. 4.14 Wide strip foundation.

building, an unreinforced mass concrete strip 150 mm deep with a mix of 1 part of cement to 9 parts of aggregate, by volume, would be sufficient. The actual bearing pressure on the soil would not usually be more than about 100 kN/m² which is satisfactory even for a soil of moderate quality. Where the soil is of poor quality or there are heavier loads, a wider strip will be necessary and, in general, this will need reinforcing, in the bottom, across its width with steel bars or mesh. A structural concrete of a specified minimum strength of 20 N/mm² should be used (Fig. 4.14). As is always the case, foundations subject to chemical attack, for example from sulphates in the soil, should use a dense, strong concrete mix and the use of a sulphate-resisting cement should be considered. Strip foundations can be made deep and reinforced, top and bottom, in the longitudinal direction as well as across the width, so that they are able to span over soft spots in the ground (Fig. 4.15). Unreinforced strip foundations which need to step on sloping ground should not go up in one step by more than the slab thickness and should have a lap length of the same amount, or 300 mm if this is greater.

4.4.2 Trench fill foundations

An alternative to the use of a strip foundation is the trench fill foundation which has a narrower width but is much stiffer in the longitudinal direction than a traditional strip foundation (Fig. 4.16). It may be reinforced in the longitudinal direction if soft spots are thought to exist in the soil. Usually, an unreinforced 1 to 9 mass concrete mix is used. This type of foundation is likely to be economic where the trench can be excavated by machine and has stable sides, such as a clay would have, and where the foundations need to be deep for any reason, for example in order to reach a level at which there is good bearing capacity or to found at sufficient depth in a shrinkable clay. It is thought that trench fill foundations, especially deep foundations, are susceptible to sideways pressure from an expanding clay soil on one side unless a compressible layer is interposed between the foundation and the clay (Fig. 4.17).

4.4.3 Pad foundations

Unlike other kinds of building, individual pad foundations for masonry elements such as columns and walls are not generally recommended because of the danger of cracking with differential settlement. However, pad foundations may be used to support suspended ground-floor beams which in turn support columns and walls. This is often an economic solution, especially where good ground exists at a moderate depth and piling is unsuitable (Fig. 4.18). A foundation using a mass concrete pier is similar in principle (Fig. 4.19).

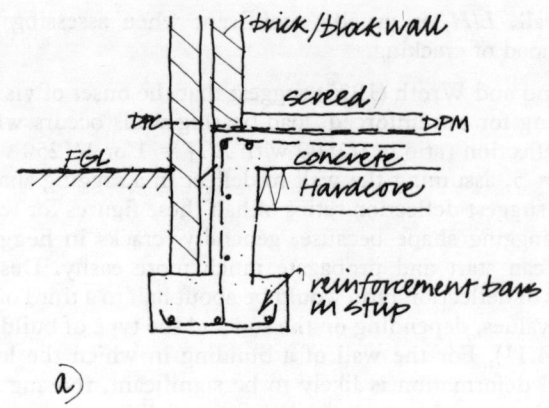

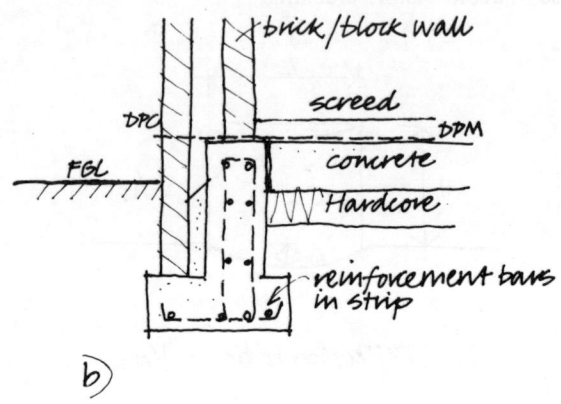

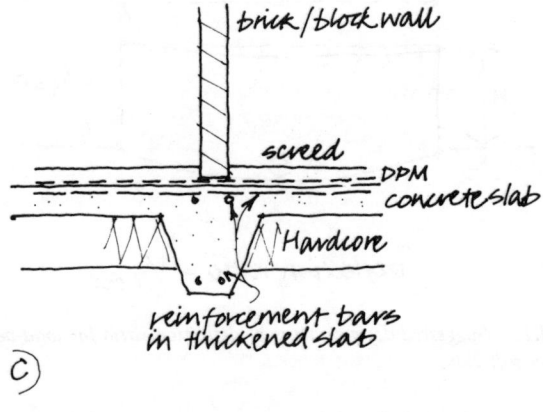

Fig. 4.15 Deepened strip footing: (a) and (b) for external wall, (c) for internal wall.

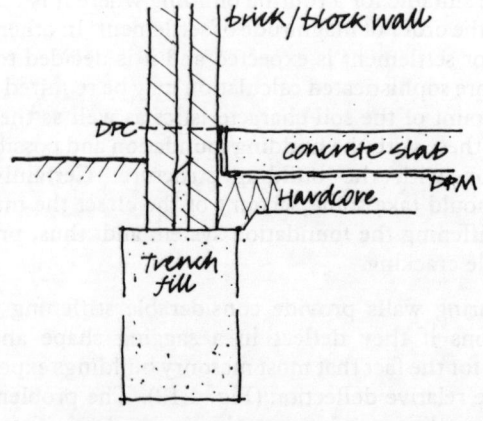

Fig. 4.16 Trench fill foundation with floating slab.

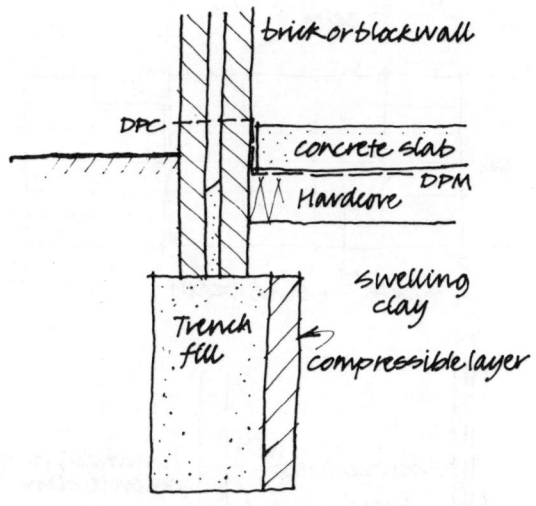

Fig. 4.17 Trench fill foundation with compressible layer on one side.

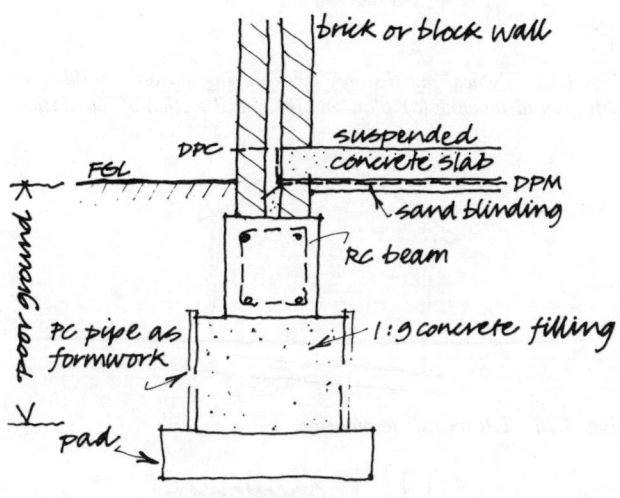

Fig. 4.18 Stem foundation to support wall and floor using concrete pipe as formwork.

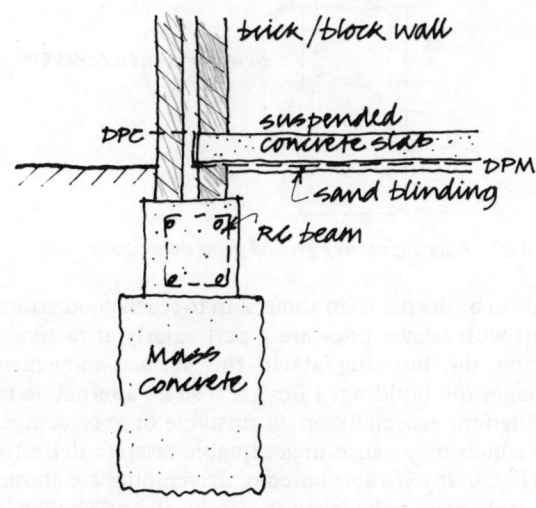

Fig. 4.19 Mass concrete foundation to support wall and floor.

4.4.4 Rafts

Rafts are used on compressible soils to reduce the actual bearing pressure on the soil and to reduce differential settlement, and on unstable ground, in general, where it is necessary to ensure that the building block is tied together at foundation level so as to act as one unit. An example in this last category would be a building which uses the raft to bridge across depressions in the ground caused by soft spots or swallow holes in chalk (Fig. 4.20). Rafts are often used where there is a great depth of poor soil and piling would be uneconomic. In the particular case of masonry buildings, the function of the raft is mainly to prevent cracks, the high-bending stiffness of the raft counteracting the differential settlement. However, the stiffness of the raft is only a matter of degree, rafts being anything from about 150 mm to 2 m deep. The required depth of the raft depends on the compressibility of the soil and the stiffness of the masonry superstructure, although it is not always possible to rely on the stiffness of masonry, as it may crack. Compared with a high building, a low building of a similar type has a more flexible superstructure and therefore can accept greater differential settlement; it also puts smaller loads on the ground thereby giving rise to smaller differential settlements, in spite of its flexibility. Rafts for low buildings may be built with relatively shallow slabs, therefore.

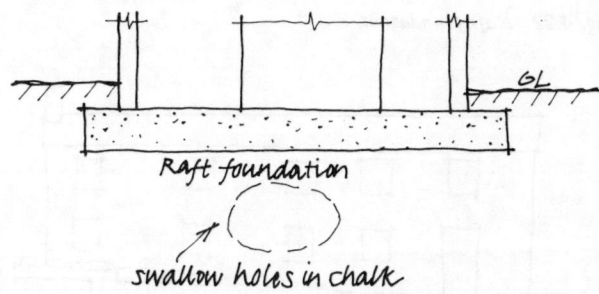

Fig. 4.20 Raft used to bridge across voids in ground.

For low-rise buildings, Tomlinson, Driscoll and Burland (1978) recommend a raft with a downstand reinforced edge beam on highly compressible soils, such as soft clays or fill. However, flat raft slabs are also suitable (Fig. 4.21 and Fig. 4.22). The flat raft slab may require stepping on an uneven site (Fig. 4.23). A raft with edge beams, by contrast, can be stepped at these edges without the necessity of altering the ground-floor levels. In the UK, the authors suggest that founding depths are 250 mm to 300 mm on soils not susceptible to frost, and 450 mm on soils susceptible to frost. On sloping sites, rafts may be stepped or the ground levelled (Fig. 4.24). Multi-storey masonry buildings are often built with rafts on soft ground. As already stated, a greater overall depth of raft would be required than for a low-rise building (Fig. 4.25). To prevent the occurrence of large bending moments in the raft, it is usually necessary to split up long buildings into blocks or separate rafts. In designs of this kind, careful analysis gives economy. Even on reasonably good ground a solid raft may still be more convenient to build and better able to prevent differential settlement than other types of foundation.

4.4.5 Piled foundations

Piles can be more economic on poor soil than raft or strip foundations, even for low-rise buildings, if there is good ground at a reasonable depth. For example, piles would be cheaper than a trench fill foundation where the trench fill

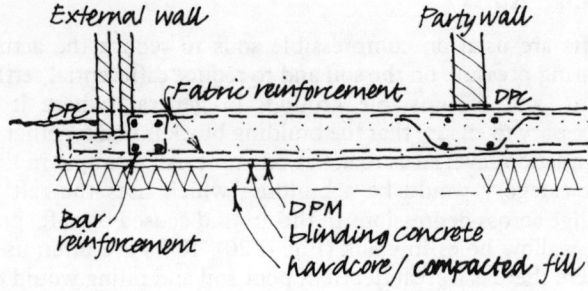

Fig. 4.21 *Raft foundation for soil of high compressibility.*

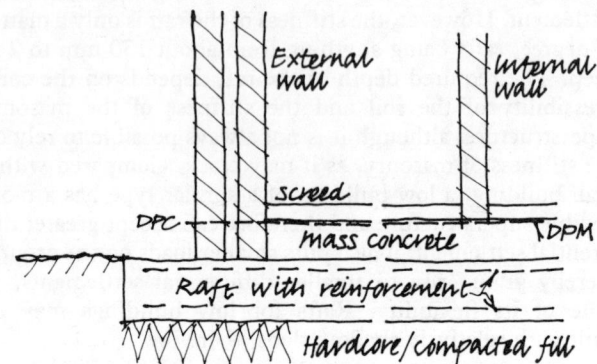

Fig. 4.22 *Raft foundation.*

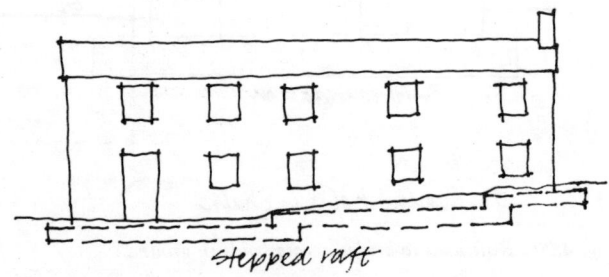

Fig. 4.23 *Stepped raft on uneven site.*

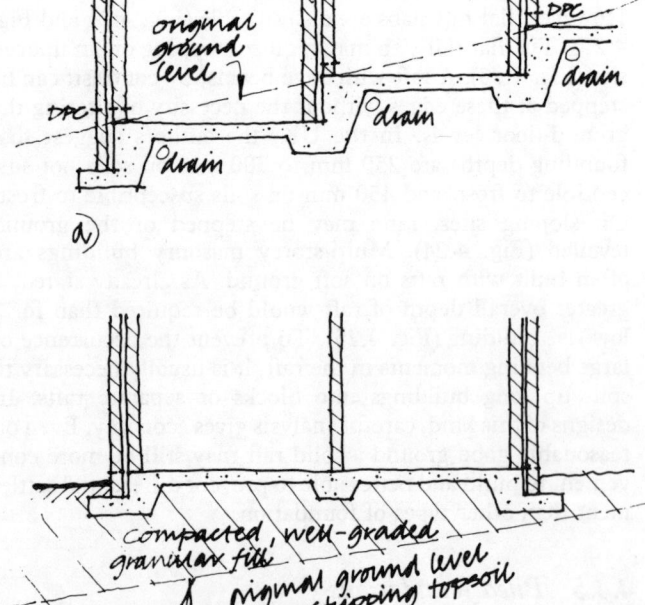

Fig. 4.24 *Raft foundation on sloping site (a) with stepped foundation and (b) after site has been levelled with compacted fill.*

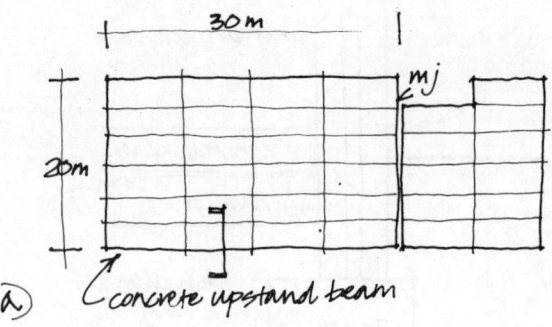

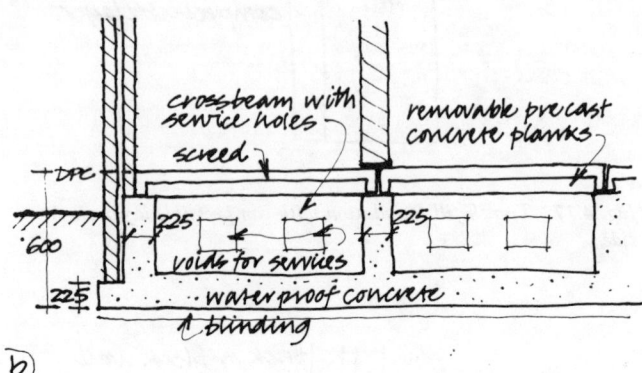

Fig. 4.25 *Typical multi-storey load-bearing masonry building on soft ground showing (a) plan and (b) detail section of foundation.*

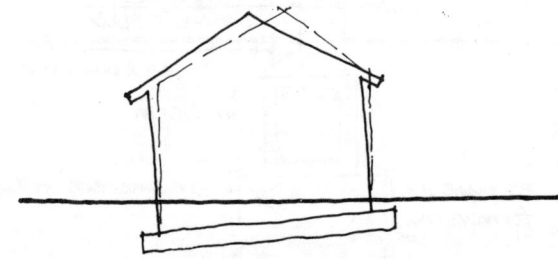

Fig. 4.26 *Tilt on raft foundation.*

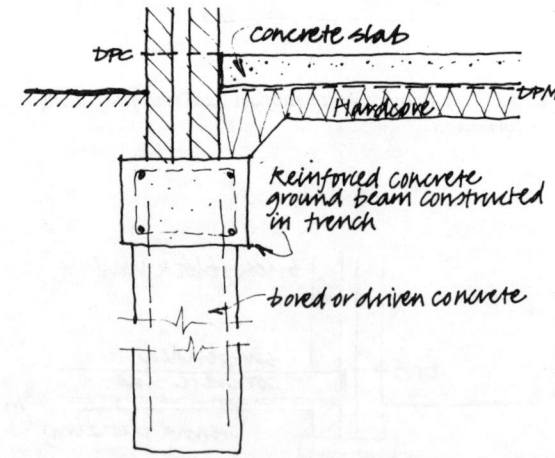

Fig. 4.27 *Piles supporting ground beams and walls.*

needs to be deeper than about 2 m to reach good ground. For a soil with heave, piles are a particularly attractive option, placing the building above the surface movement that damages the building. Piles are also an alternative to a raft foundation, especially on an unstable or very compressible soil which may cause unacceptable relative deflection and tilt (Fig. 4.26). If short bored or driven piles are chosen, they may only need to be taken to depths of 2 or 3 m for low-rise buildings, depending on the soil present (Table A10.4). The ground-floor beams supporting the load-bearing walls can rest directly on the piles (Fig. 4.27). A typical layout of piles

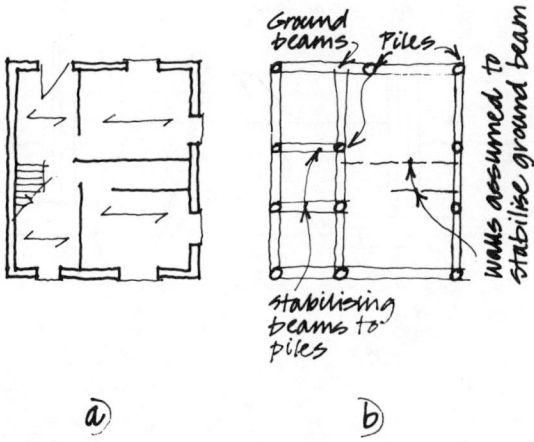

Fig. 4.28 (a) Ground floor of house and (b) typical layout of piles and ground beams.

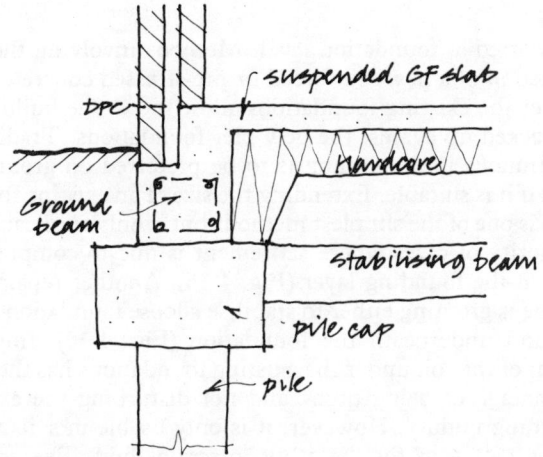

Fig. 4.29 Section through pile foundation at edge of multi-storey building.

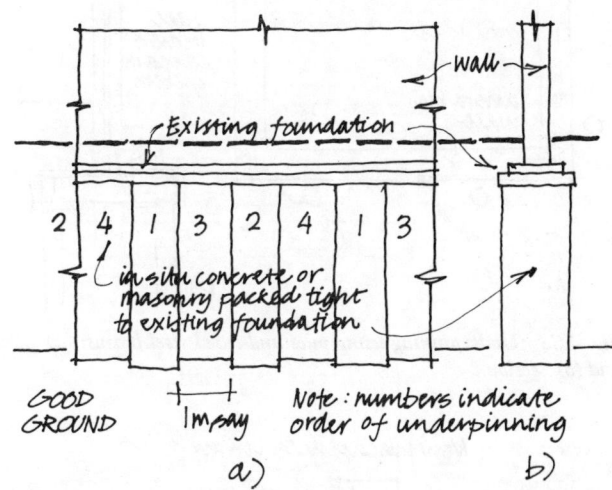

Fig. 4.30 Traditional underpinning: (a) elevation and (b) section.

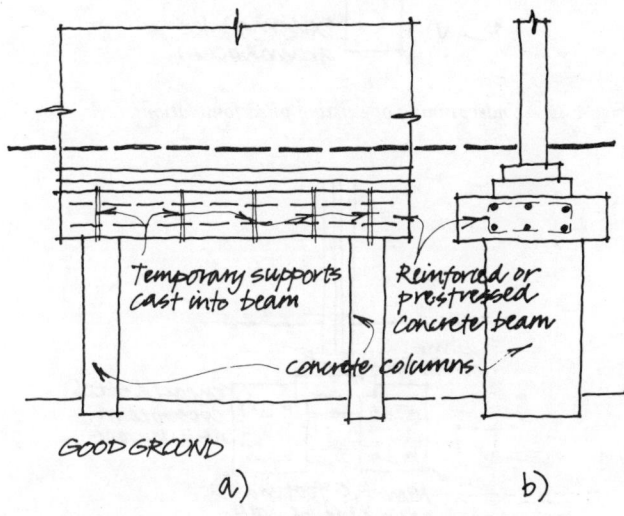

Fig. 4.31 Underpinning using new beams and columns: (a) elevation and (b) section.

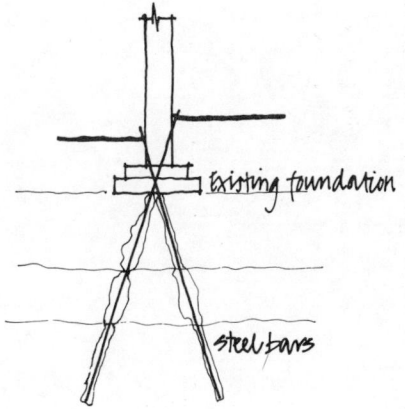

Fig. 4.32 Micropiles, reinforced with steel bars, connecting masonry walls to soil strata below.

and ground beams for a house is shown in Fig. 4.28. Brick walls resting on ground beams may act compositely (**10.1**). If piling is chosen for multi-storey building, bored or driven piles would be used but, compared with low-rise building, pile lengths would be longer and, in general, it would be necessary to use pile caps to support the suspended ground-floor beams (Fig. 4.29). On sites where the water table is high it may also be more economic to install piles than to excavate foundations in waterlogged ground whatever its allowable bearing pressure.

4.5 Underpinning of masonry buildings

Sometimes it is necessary to increase the load-bearing capacity of an existing foundation because the building is to be increased in height or because the existing foundations are found to be on the point of failure. Either the foundations may be taken down to better ground or increased in size or the soil below the foundations may be improved. In most cases, but depending on the particular circumstances, it will be better to extend the foundations down to good ground. The traditional method is by continuous underpinning of the existing foundation (Fig. 4.30). An alternative is to use the existing foundation, or a new reinforced or prestressed concrete beam built below the existing foundation, to span between piers taken to good ground (Fig. 4.31). Micropiles, formed of a cement grout injected at high

pressures and reinforced with steel bars, are sometimes convenient (Fig. 4.32); the piles so formed work mainly in friction. Piling is also used in conjunction with beams, taken through the wall and supporting it (Fig. 4.33). For buildings already founded on piles, it will be necessary to install more piles and another pile cap (Fig. 4.34). With all these methods, cracking is most likely to occur during installation of the underpinning when the building may only be partially

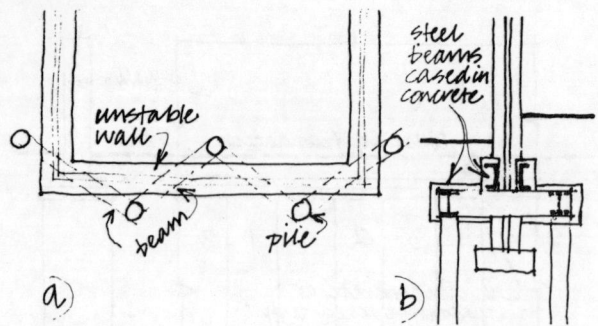

Fig. 4.33 Underpinning using piles and cased steel beams: (a) plan and (b) section.

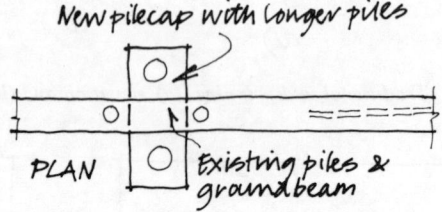

Fig. 4.34 Underpinning of existing piled foundations.

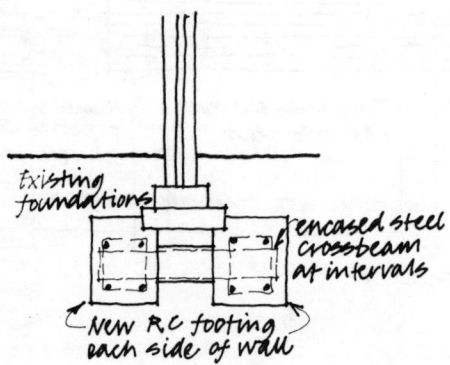

Fig. 4.35 Widening an existing foundation.

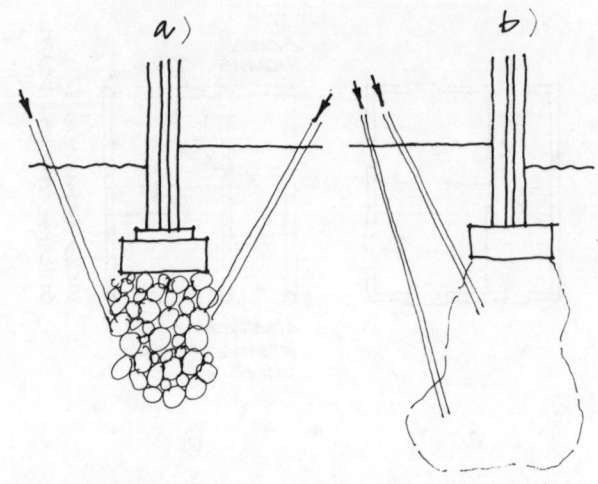

Fig. 4.36 Grout injection used to improve: (a) loose stone foundation and (b) ground.

supported at foundation level. Methods involving the construction of a new reinforced or pre-stressed concrete beam under the existing foundations allow the whole building to be jacked up against the new pier foundations. Traditional continuous underpinning is to be preferred on grounds of cost if it is suitable. Extending the size of an existing foundation is one of the simplest methods but is only applicable in a minority of cases where settlement is due to compressible soil in the founding layer (Fig. 4.35). Another repair technique is grouting either to stabilise a loose foundation or the ground underneath the foundation (Fig. 4.36). Improvement of the soil under the existing foundations has the great advantage of being quick and not disturbing the existing building unduly. However, it is only usable in soils which allow pathways for the grout injection fluid. The grout is usually a cement mortar, or a silicate for fine-grained soils. A similar technique may be used for filling up large voids in the ground if this is the cause of problems. Other underpinning details are given in App. **B6**.

CHAPTER 5

Walls and columns under vertical and horizontal load

Introduction

Structural elements must be strong enough to resist forces applied to them from any direction. For convenience these forces can be split up into vertical and horizontal components. This chapter is concerned with the calculation of load-bearing masonry walls and columns which have to take vertical forces, mainly those due to gravity loads, and horizontal forces, mainly those due to wind loads but which in earthquake areas will also include those due to ground movements. The Code requires particular combinations of these forces to be considered in the design. The wall or column, in general, has only two ways of resisting these loads, either with its compressive strength or its flexural strength, although in some cases the wall may be able to take direct tensile stresses while in other cases it may have very little flexural strength at all. However, axial forces will increase the effective flexural strength of the wall. Thus under a horizontal load combined with a high axial load, the initial flexural strength of a wall may not be relevant to the calculation of the wall's effectiveness in resisting horizontal load. Note that under vertical load too, because masonry is usually very strong in compression, the calculation of the vertical strength of the wall is very often a check not so much on the compressive strength of the masonry units but on the wall's slenderness ratio, out-of-plumbness, eccentricity of load or the kind of lateral support the wall receives. Happily, in most cases, as discussed in the text, the loads which cause maximum compressive stress or the maximum flexural stress to occur can each be associated with one particular load combination so that in practice it is not usually necessary to calculate all possible combinations of load in order to find the worst case. It is assumed in this chapter that the wall or column is laterally supported and that these supports form part of a building which, overall, is stable. Lateral stability is discussed in Chapter 8. More information about walls under horizontal loads is provided in Chapter 6 which, to some extent, repeats information given in this chapter. Excerpts from the Code of Practice concerned with walls and columns are included in this chapter but the excerpts given in Chapter 6 may be relevant too. Other design information is given in App. **A** and **B**. This chapter is set out in the form of a series of instructions, as if for someone needing to assess a masonry building and undertake calculations. The format helps to separate the main points and provides a logical sequence in which the calculations for walls and columns may be done. The design of columns is similar to that for walls and the section concerned with columns only sets out the differences in approach. It is hoped that the examples at the end will clarify the text in the main part of this chapter.

5.1 Procedure for checking walls under vertical and horizontal loads

5.1.1 Vertical loads

(a) By inspection of the drawings, or by calculation if necessary, decide which of the walls in the building is most critical and at what level.

(b) Work out the characteristic loadings for floors, roofs and walls, keeping dead, live and wind loads separate.

(c) Check the Slenderness Ratio (SR) of the wall. The slenderness ratio is calculated from the effective thickness and the effective height or length, whichever is the lesser, and should not exceed the allowable slenderness ratio of 27, or 20 for walls less than 90 mm thick in buildings of more than two storeys (Clause 28.1 of the Code (**5.3**)); see Figs 5.11, 5.15, 5.16. The effective height or length is reduced if there is enhanced resistance to lateral movement (Clauses 28.2.2; 28.2.3; 28.3 of the Code (**5.3**)). See Figs 5.13 and 5.14 and also App. **B4** and **B5** which give typical details providing simple or enhanced resistance to lateral movement.

Table 5.1   Partial safety factors for material strength, γ_m (after Table 4 of BS 5628:Part 1)

		Category of construction control	
		Special	Normal
Category of manufacturing control of structural units	Special	2.5	3.1
	Normal	2.8	3.5

The clear height of a wall, h, is taken between points able to provide a lateral support to the wall (Fig. 5.1). Enhanced resistance to lateral movement can be assumed at the foundation level, or dpc at ground level, if no tensile stresses or cracks can develop there under any of the design loads; see App. **B4**.

Walls with piers For walls with piers either the slenderness ratio of the pier and the slenderness ratio of the wall panel between piers are checked separately or, for walls with

$$6 \leqslant s/b_p \leqslant 20 \text{ and } 1 \leqslant t_p/t \leqslant 3,$$

the slenderness ratio of the walls and piers acting together is checked instead (Fig. 5.2). In the latter case, Tables 5.2 and 5.3 give the effective thickness of the wall. Even in this latter case the slenderness ratio of the pier alone may still be required, for example when the pier takes an additional point

Table 5.2   Effective thickness of wall and columns, t_{ef}, (after Figure 3 of BS 5628:Part 1)

Column	Single-leaf wall	Cavity wall	Walls stiffened by piers	
			Single-leaf	**Cavity**
Plan shapes				
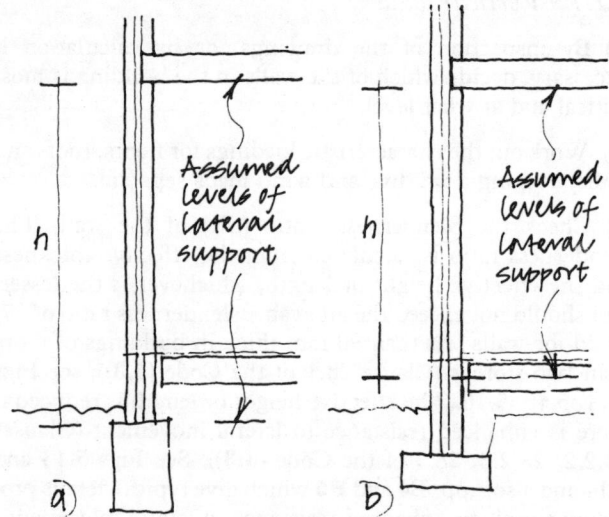				
Effective thickness				
t or b, depending on direction of bending	t	the greatest of (a) $2/3\,(t_1 + t_2)$ or (b) t_1 or (c) t_2	$t \times K$	the greatest of (a) $2/3\,(t_1 + Kt_2)$ or (b) t_1 or (c) Kt_2 where K is the stiffness coefficient from Table 5.3

Fig. 5.1   Clear height of wall, h, taken between: (a) floor and top of foundation and (b) floor and dpc level.

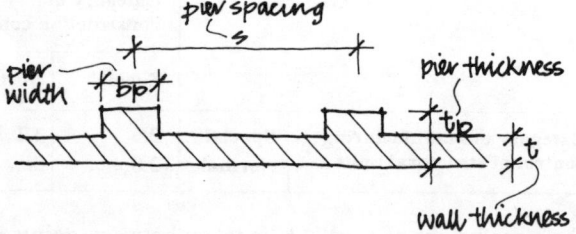

Fig. 5.2   Plan on wall with piers.

load not on the rest of the wall. In calculating the slenderness ratio of a pier, the effective thickness is taken as the actual thickness, t_p, and Clause 28.3.1.4 of the Code (**5.3**) establishes whether the pier be treated as a wall or a column for ·effective height considerations (Clauses 28.3.1.1; 28.3.1.2 of the Code (**5.3**)); see Fig. 5.18. Except for piers which are large relative to the wall, it is not necessary to calculate the slenderness ratio of the pier in the plane parallel to the wall.

Cavity walls   For cavity walls an effective thickness of the wall is used; see Table 5.2.

(d) Consider the vertical loading, which would cause the maximum compressive stress.

Table 5.3   Stiffness coefficients, K, for walls stiffened by piers (after Table 5 of BS 5628:Part 1)

Ratio of pier spacing (centre to centre) to pier width	Ratio t_p/t of pier thickness to actual thickness of wall to which it is bonded		
	1	**2**	**3**
6	1.0	1.4	2.0
10	1.0	1.2	1.4
20	1.0	1.0	1.0

NOTE. Linear interpolation between the values given in Table 5.3 is permissible, but not extrapolation outside the limits given.

(e) Work out the characteristic vertical loads on the particular part which is being checked, keeping dead, live and wind loads separate.

(f) Load combinations. Select either Case (a), (b) or (c) from Clause 22 of the Code (**5.3**) whichever gives the worst combination of loads. For vertical loading in low-rise buildings almost invariably Case (a) and with one of the combinations illustrated in Fig. 5.3 will give the worst case. See Haseltine and Moore (1981, p.57) for more details. Two combinations are considered for a wall loaded from two sides

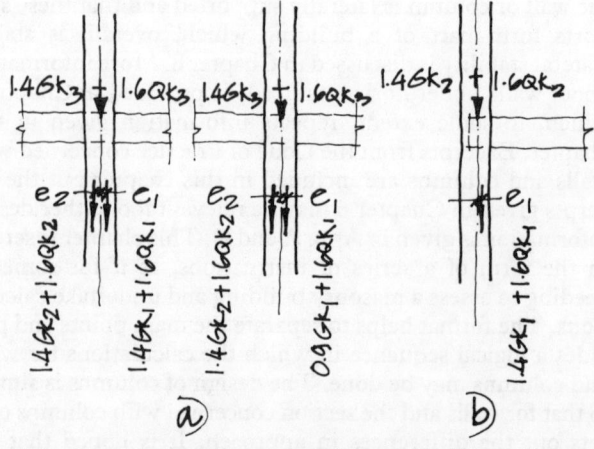

Fig. 5.3   Usual load combinations to be considered for wall supporting vertical load from: (a) two sides and (b) one side.

and one combination for a wall loaded from one side only. Other combinations of loads for Case (a) may give greater resultant eccentricities at the top of the wall, for example if the axial load from higher levels is reduced; however, overall the combinations shown are more severe for the vast majority of practical cases in which the spans are not grossly dissimilar.

(g) Design vertical loads and eccentricity. For the load combination selected, calculate the design vertical load and the resultant eccentricity of the loads at the top of the wall, e_x, and hence knowing the slenderness ratio find the capacity reduction factor, β, from Table 5.4 (Table 7 of the Code). Note that Table 5.4, in general, applies only to solid rectangular walls and columns; see Appendix B of the Code (**5.3**) and Chapter 12 for the assumptions used in preparing Table 7 of the Code. For information about eccentricity see App. **A11**.

Table 5.4 Capacity reduction factor, β (after Table 7 of BS 5628:Part 1)

Slenderness ratio h_{ef}/t_{ef}	Eccentricity at top of wall, e_x			
	Up to 0.05t*	0.1t	0.2t	0.3t
0	1.00	0.88	0.66	0.44
6	1.00	0.88	0.66	0.44
8	1.00	0.88	0.66	0.44
10	0.97	0.88	0.66	0.44
12	0.93	0.87	0.66	0.44
14	0.89	0.83	0.66	0.44
16	0.83	0.77	0.64	0.44
18	0.77	0.70	0.57	0.44
20	0.70	0.64	0.51	0.37
22	0.62	0.56	0.43	0.30
24	0.53	0.47	0.34	
26	0.45	0.38		
27	0.40	0.33		

*It is not necessary to consider the effects of eccentricities up to and including 0.05t.

NOTE 1. Linear interpolation between eccentricities and slenderness ratios is permitted.

NOTE 2. The derivation of β is given in Appendix B of the Code (**5.3**).

Hollow blocks   Hollow concrete blocks may be designed as solid blocks having the same compressive strength as the average compressive strength of the hollow block (**2.1.3**). See Appendices B3 and B4 of the Code for more details.

Walls with piers or bowed walls   Where a wall with piers is used – i.e. a wall behaving as a T-section – or where, in a small number of cases, there are bending moments in the middle of a wall due to bowing of the wall, it will be necessary to calculate the eccentricity in the middle of the wall as well. For these cases the use of Table 5.4 is no longer valid and the factor, β, and the design vertical load resistance of the wall may then be calculated, using Appendix B of the Code (**5.3**); see **12.5**.

Note that lateral wind forces on a wall will also cause bending moments in the middle of the wall and Appendix B of

the Code may then be used in precisely the same way to check that the wall is safe under the worst combination of wind loads, usually Case (b) from Clause 22 of the Code (Example 5.6). However, this method will only be appropriate for thick walls or walls in which the axial load is high relative to the bending moment produced by wind forces. If wind forces are resisted by the flexural strength of the wall, see **5.1.2**.

(h) Design vertical load resistance. Calculate the design vertical load resistance of the wall which must be greater than the design vertical load. For a straight wall, the design vertical load resistance (Clause 32.2.1 of the Code (**5.3**)) is equal to

$$(\beta \cdot t \cdot f_k)/\gamma_m$$

where γ_m is the partial safety factor for material strength (Table 5.1) and
f_k is the characteristic compressive strength of the masonry (Table 2.2).

For a wall with piers, the design vertical load resistance depends on the area in compression, but it may often be more conveniently checked by ignoring the projecting area of the piers or, alternatively, only considering the rectangular area of the pier. For a cavity wall only the area of the load-bearing leaf is considered to take vertical load (Clause 29.1.1 of the Code).

Do not forget Clauses 23.1.1; 23.1.2; 23.1.3 of the Code. The first requires f_k to be multiplied by a reduction factor of $(0.70 + 1.5\,A)$ where the area of wall, A, is less than 0.2 m^2 (**5.3**). The second allows an increase in f_k for a brick wall where the thickness is equal to the width of the brick. The third allows increases for modular bricks.

When use is made of Appendix B of the Code (**5.3**) to calculate the design vertical load resistance, it will be desirable in thin walls to limit the design eccentricity, e_t, to say $0.35\,t$ when this is due to a permanent vertical load acting alone, even if the design vertical load resistance is found to be adequate. This is because of the rapid fall off in the load resistance at large eccentricities. Large eccentricities caused by wind forces can be safely taken in flexure using the values from Table 2.3 (Table 3 of the Code); see **5.1.2**. However, it is not certain that large eccentricities due to permanent loads can also rely on the same values for flexural strength given in Table 2.3; see **12.5**.

5.1.2 Horizontal (lateral) loads

(a) Consider the horizontal loading which would cause the maximum tensile or compressive stresses.

(b) Check that the limiting dimensions of the wall or wall panel as set out in Clause 36.3 of the Code (**6.2**) are not exceeded. The wall panel is the portion of the wall between vertical and horizontal lateral supports (Fig. 5.4).

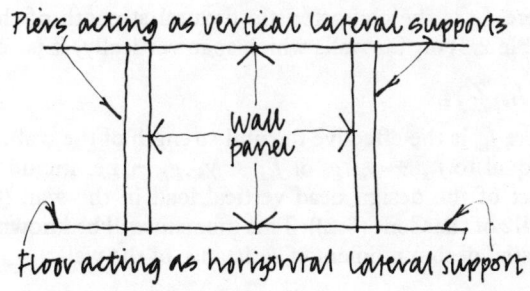

Fig. 5.4 Elevation on wall with piers.

(c) Select Case (a), (b) or (c) from Clause 22 of the Code (**5.3**) whichever gives the worst combination of loads. For horizontal loading on a building usually only Cases (b) and (c) will be relevant. For the design of elements in low-rise buildings, where tensile stresses are more likely to be critical than compressive stresses, almost invariably Case (b) with the combination $0.9G_k + 1.4W_k$ will give the worst case.

(d) Work out the characteristic vertical loads on the wall, or each leaf of that wall in the case of a cavity wall, and the total horizontal load on the wall, keeping dead, live and wind loads separate.

Minimum horizontal load According to Clause 20.1 of the Code (**5.3**) a building should be able to resist, from any direction in plan, a minimum design horizontal force which is 1.5% of the characteristic dead load above any level. This provision is one of the conditions to be considered in Cases (b) and (c) of Clause 22 of the Code (**5.3**), although the wind forces are usually greater than this minimum force, except for a wind force blowing on the short side of a long building (Fig. 5.5).

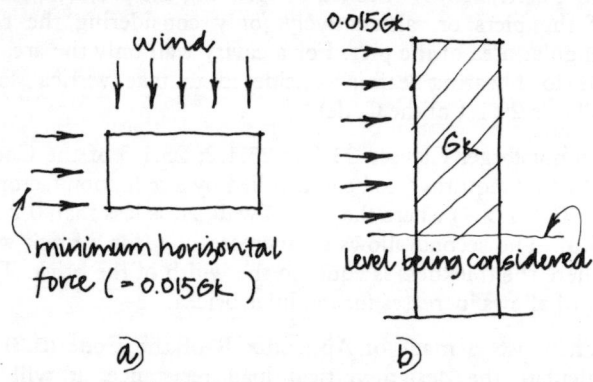

Fig. 5.5 (a) Minimum horizontal force may be greater than the wind force on the short side of a long building; (b) minimum horizontal force to be applied above any level.

(e) Design moment of resistance. Decide how the wall will span between its supports – i.e. with a horizontal or vertical one-way span or a two-way span. Work out the characteristic flexural strength of the wall in the direction in which the wall spans. For two-way spans, by convention, the Code considers the horizontal moments in the wall, rather than the vertical; with a yield line method, it is immaterial which is chosen. For walls which span vertically it will first be necessary to calculate the design vertical load at the point under consideration. The design moment of resistance of the wall may then be calculated and this should be greater than the design moment.

The design moment of resistance of the wall (Clause 36.4.3 of the Code (**6.2**)) is equal to

$$f_{kx} \cdot Z/\gamma_m$$

where f_{kx} is the characteristic flexural strength of the wall (Table 2.3) or, for walls which span vertically, it is equal to

$$f_{ka} \cdot Z/\gamma_m$$

where f_{ka} is the effective flexural strength of the wall, which is equal to $f_{kx} + \gamma_m \cdot g_d$ or $f_{kx} + \gamma_m \cdot g_A$ – i.e. including the effect of the design dead vertical load in the wall (Clause 36.4.2 of the Code (**6.2**)). This moment will be known as the 'elastic' design moment of resistance of the wall.

It is usual to assume that the design moment of resistance of hollow blocks is the same as that of solid blocks having the

same average compressive strength (**2.1.3**). See Roberts et al. (1983, p.102) for further discussion.

For thick walls or walls with high axial load it may not be necessary to rely on the flexural strength of the wall if any; see **5.1.1** g and h.

(f) Design moment. The design moment on the wall should be calculated for the load combination selected. The design moment may be calculated using Tables A4.1 and A4.2 for one-way spans, or Table A6.1 for two-way spans. The design moment is of the general form:

$W_k \cdot \gamma_f \cdot L^2 \times$ (bending moment coefficient) per unit height for a wall spanning horizontally or in two directions and

$W_k \cdot \gamma_f \cdot h^2 \times$ (bending moment coefficient) per unit width for a wall spanning vertically (Clause 36.4.2 of the Code (**6.2**)).

Continuity may be assumed at the lateral supports at the top and bottom of each storey-height of wall if the design loads do not cause any cracks to develop there, or any flexural stresses to develop there that exceed the design flexural strength of the wall, f_{kx}/γ_m or f_{ka}/γ_m. There may be continuity across a horizontal dpc, even if it has no tensile strength, because of the effect of axial load in preventing tensile flexural stresses developing there. See App. **B5**.

One-way spans The elastic bending moment patterns shown in Table A4.1 assume that the supports to the wall provide either: (a) geometrical fixity – i.e. the wall is not able to rotate at this edge – or (b) simple support – i.e. the wall is free to rotate at this edge. Over the intermediate support the wall is assumed to be fully continuous. If there are intermediate supports, in some cases, Table A5.1 may be used in conjunction with Table A4.1 to give a reduction in the bending moment due to the width of the supports. In other cases the appropriate bending moment coefficients may be obtained using different assumptions; see **6.1**g.

Two-way spans For walls which span in two directions it will first be necessary to calculate the design vertical load at the point under consideration and hence the orthogonal ratio, μ, before using Table A6.1 (Table 9 of the Code). For a definition of the orthogonal ratio, see **6.1**h. See **6.1**g for further discussion on bending moment coefficients.

(g) For cavity walls, the vertical load and the design moment of resistance is calculated separately for each leaf. In general the design horizontal load can be assumed to be shared between the leaves in proportion to their design moment of resistance, as long as the wall ties are strong enough (**6.1**h). For each leaf of the wall the design moment of resistance must then be greater than the design moment. However the Code (Clause 36.4.5) allows an alternative and more convenient assumption, which is suitable in the majority of cases. This is that the design lateral strength of a cavity wall, that is the maximum design load (e.g. wind load) per unit area on the cavity wall, may be taken as the sum of the design lateral strengths of each individual leaf of the cavity wall (Example 12.6, p. 87). See **6.1**h for a definition of the orthogonal ratio, μ.

(h) Check shear strength. In some exceptional cases, it may be necessary to check that the design shear strength of the masonry, f_v/γ_{mv} (Clause 25 of the Code (**5.3**)), is greater than the design shear stress, v_h, calculated for the load case from Clause 22 of the Code (**5.3**) which gives the highest value of design shear stress. The design shear stress, v_h, is the horizontal design load divided by the horizontal area of wall taking the load (Clause 33 of the Code). The partial safety factor for masonry strength in shear, γ_{mv}, equals 2.5 for any of the mortars listed in Table 2.1 (Clause 27.4 of the Code). Note that the design shear strength is likely to be low at flexible damp-proof courses (dpc).

(i) Check connections. Connections between masonry walls and lateral supports are essential to the stability of the wall. According to Clause 28.2.1 of the Code (**5.3**), the lateral support must be capable of taking the applied static design horizontal force plus 2.5% of the design vertical load at that point in the wall (Fig. 5.12). Connections taking forces either in the horizontal or vertical directions must be calculated using the load from Clause 22 of the Code (**5.3**) which gives the highest value of the design connection force in the direction considered. The design strength of the connection must be greater than the design load. If wall ties are used then their characteristic strengths may be found by referring to Table B1.2. The design strength of a wall tie is the characteristic strength divided by the partial safety factor for materials, γ_m, which equals 3.0 for a wall tie (Clause 27.5 of the Code). The design strength of a steel strap is $f_y . A/\gamma_m$ where A is the area of the strap, f_y is the quoted characteristic strength of the strap, which is equal to about 250 N/mm² for mild steel, and γ_m is equal to 1.15 (Appendix C of the Code). Consideration should be given to increasing the number of cavity wall ties at supports. However, in many cases a calculation is not very useful, for example because no information is available on the performance of the proposed joint. Appendix C of the Code gives details which are assumed to provide sufficient restraint and stability for walls in typical cases. See App. **B4**, **B5** and Chapter 8.

5.2 Procedure for checking columns under vertical and horizontal loads

5.2.1 Vertical loads

(a) A column is defined for the purposes of the Code as a portion of masonry whose width, b, is less than or equal to 4 times its thickness or, it is assumed here, its effective thickness (Fig. 5.6).

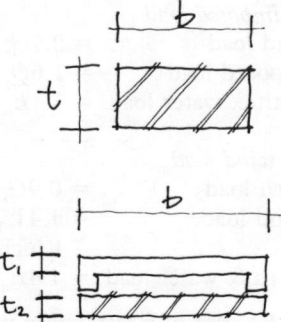

Fig. 5.6 Portion of masonry, shown in plan, is a 'column' if $b \leqslant 4t$ or $b \leqslant 4t_{ef}$.

(b) The calculations proceed in general as those for walls, except that account must now be taken of bending, or buckling, about the major as well as the minor axis of bending (Fig. 5.7). Note that calculations on columns usually consider the whole area of the column rather than a unit length as is usually the case for walls.

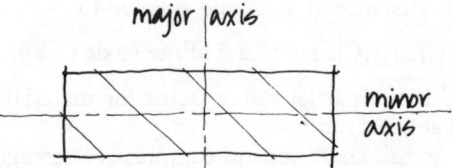

Fig. 5.7 Plan on rectangular column showing major and minor axes.

(c) Slenderness ratio of column. The slenderness ratio (SR) is calculated, from the effective thickness and the effective height of the column, about the major or minor axes of bending whichever gives the greater value of slenderness ratio. The effective height of an isolated column is never less than the distance between lateral supports and is greater than the actual height of the column in a direction for which no lateral support is available (Clause 28.3.1.2 of the Code (**5.3**)); see Fig. 5.8. For columns which are formed because two openings in a wall are close together other rules apply (Clause 28.3.1.3 of the Code (**5.3**)); see Fig. 5.17. Where a column does not have a rectangular shape in plan, the effective thicknesses about the major and minor axes of bending may be assumed to be those of a rectangular column having the same radius of gyration (Table A2.1) about these axes as the actual column (Fig. 5.9).

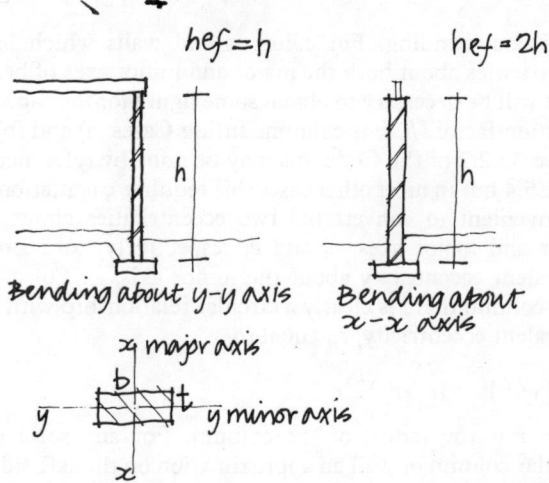

Fig. 5.8 Plan and elevations on a rectangular column supported at the top in one direction only.

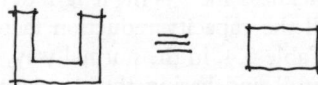

Fig. 5.9 Plan on U-shaped column and equivalent rectangular column having the same radii of gyration.

(d) Vertical loading. Consider the vertical loading which would cause the maximum compressive stress. The calculation proceeds as that for a wall except that the capacity reduction factor, β, now depends on the eccentricities of the load about both the major and minor axes of bending. Four cases are distinguished in Clause 32.2.2 of the Code (Fig. 5.10). The factor, β, may be worked out from Table 5.4 in all cases except Case (d) when it must be calculated using Appendix B of the Code (**5.3**). As is the case for walls, where there are large moments in the middle of the column it will be necessary to calculate the eccentricity in the middle of the column as well as at the top. The factor, β, in this case is also calculated using Appendix B of the Code. The design vertical load resistance of a column is equal to

$$\beta \cdot b \cdot t \cdot f_k / \gamma_m \quad \text{(Clause 32.2.2 of the Code (5.3))}$$

where γ_m is the partial safety factor for material strength (Table 5.1) and

f_k is the characteristic compressive strength of the masonry (Table 2.2).

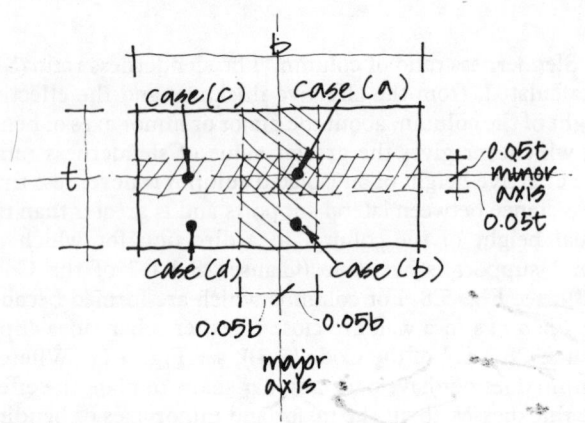

Fig. 5.10 *Plan of column showing different cases used in BS 5628: Part 1 for calculating β, dependent on position of resultant axial load (5.3).*

(e) Bi-axial bending. For columns and walls which have eccentricities about both the major and minor axes of bending, it will be necessary to obtain some figure for the capacity reduction factor, β. For columns fitting Cases (a) and (b) of Clause 32.2.2 of the Code this may be done by reference to Table 5.4 but in most other cases this requires calculation. It is convenient to convert the two eccentricities about the minor and major axes, e_1 and e_2 respectively, to a single equivalent eccentricity about the minor axis, e_r. For a circular column there is clearly a circular relationship with the equivalent eccentricity, e_r, equal to

$$[(e_1/r)^2 + (e_2/r)^2]^{1/2} r$$

where r is the radius of the column. For any solid rectangular column or wall an approximation on the safe side is to take the equivalent eccentricity about the minor axis, e_r, as equal to

$$1.1 [(e_1/t)^2 + (e_2/b)^2]^{1/2} t$$

where t is the thickness and b is the length of the wall or column. For a wall the capacity reduction factor, β, may be obtained from Table 5.4, in the normal way, using the e_r at the top of the wall and basing the slenderness ratio on t appropriate to the minor axis. For columns included in Cases (c) and (d) of Clause 23.2.2, the capacity reduction factor, β, should be calculated using Appendix B of the Code. Note that when calculating e_r in the middle of the column, e_1 and e_2 must both include the additional effects due to slenderness (Clause 23.2.2 (5.3)). In all cases the design vertical load resistance, as normal, is equal to

$$(\beta \cdot b \cdot t \cdot f_k) / \gamma_m$$

5.2.2 Horizontal (lateral) loads

(a) Horizontal loading. The calculation proceeds as that for a wall.

(b) Shear. Check shear, if necessary. The calculation proceeds as that for a wall.

(c) Connections. The calculation proceeds as that for a wall.

5.3 Extracts from BS 5628:Part 1

20 Stability

20.1 General considerations   The designer responsible for the overall stability of the structure should ensure the compatibility of the design and details of parts and components. There should be no doubt of this responsibility for overall stability when some or all of the design and details are not made by the same designer.

To ensure a robust and stable design it will be necessary to consider the layout of structure on plan, returns at the ends of walls, interaction between intersecting walls and the interaction between masonry walls and the other parts of the structure.

The design recommendations in section four assume that all the lateral forces acting on the whole structure are resisted by walls in planes parallel to these forces, or by suitable bracing.

As well as the above general considerations, attention should be given to the following recommendations:

(a) Buildings should be designed to be capable of resisting a uniformly distributed horizontal load equal to 1.5% of the total characteristic dead load above any level (see clause 22(b) and (c));

(b) connections of the type indicated in appendix C should be provided as appropriate at floors and roofs.

. . . . . .

22 Design loads: partial safety factor, γ_f

When using the design relationship for the ultimate limit state given in sections four and five, the design load should be taken as the sum of the products of the component characteristic loads multiplied by the appropriate partial safety factor, as shown below. Where alternative values are shown, that producing the more severe conditions should be selected.

(a) *Dead and imposed load*
design dead load $= 0.9G_k$ or $1.4G_k$
design imposed load $= 1.6Q_k$
design earth & water load $= 1.4E_n$

(b) *Dead and wind load*
design dead load $= 0.9G_k$ or $1.4G_k$
design wind load $= 1.4W_k$ or $0.015G_k$ whichever is the larger
design earth & water load $= 1.4E_n$

In the particular case of freestanding walls and laterally loaded wall panels, whose removal would in no way affect the stability of the remaining structure, γ_f applied on the wind load may be taken as 1.2.

(c) *Dead, imposed and wind load*
design dead load $= 1.2G_k$
design imposed load $= 1.2Q_k$
design wind load $= 1.2W_k$ or $0.015G_k$ whichever is the larger
design earth & water load $= 1.2E_n$

(d) *Accidental damage (see clause 37)*

design dead load	$= 0.95G_k$ or $1.05G_k$
design imposed load	$= 0.35Q_k$ except that, in the case of buildings used predominantly for storage, or where the imposed load is of a permanent nature, $1.05Q_k$ should be used.
design wind load	$= 0.35W_k$

where

G_k is the characteristic dead load,

Q_k is the characteristic imposed load,

W_k is the characteristic wind load,

and the numerical values are the appropriate γ_f factors. In design, each of the load combinations (a) to (d) should be considered and that giving the most severe conditions should be adopted.

In certain circumstances other values of γ_f may be appropriate e.g. in farm buildings. Reference should be made to the relevant British Standards, e.g. BS 5502:Part 1: Section 1.2.

. . . . . .

23.1.1 *Walls or columns of small plan area.*

Where the horizontal cross-sectional area of a loaded wall or column is less than 0.2 m², the characteristic compressive strength should be multiplied by the factor:

$$(0.70 + 1.5A)$$

where

    A is the horizontal loaded cross-sectional area of the wall or column (m²).

. . . . . .

25 *Characteristic shear strength of masonry, f_v*

The characteristic shear strength of masonry, f_v, in the horizontal direction of the horizontal plane (see figure 2) may be taken as $0.35 + 0.6g_A$ N/mm² with a maximum of 1.75 N/mm² for walls built in mortar designations (i) and (ii) or $0.15 + 0.6g_A$ N/mm² with a maximum of 1.4 N/mm² for walls built in mortar designations (iii) and (iv), where

    g_A is the design vertical load per unit area of wall cross-section due to the vertical loads calculated from the appropriate loading condition specified in clause 22.

. . . . . .

28 *Consideration of slenderness of walls and columns*

28.1 **Slenderness ratio**   The slenderness ratio should not exceed 27, except in the case of walls less than 90 mm thick, in buildings of more than two storeys, where it should not exceed 20.

28.2 **Lateral support**   A lateral support may be provided along either a horizontal or a vertical line, depending on whether the slenderness ratio is based on a vertical or horizontal dimension.

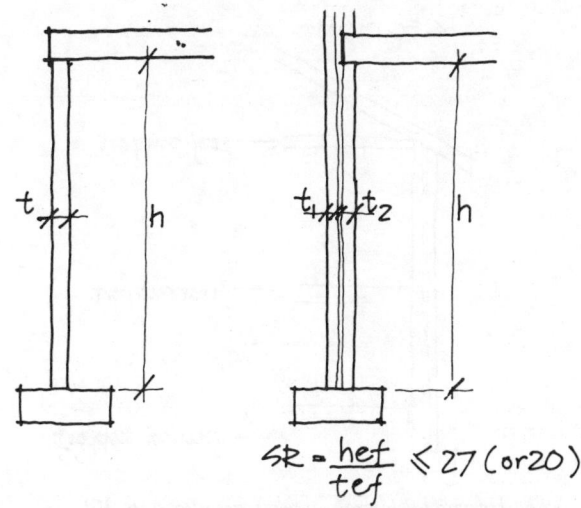

$$SR = \frac{hef}{tef} \leqslant 27 \ (or\ 20)$$

Fig. 5.11   Slenderness ratio to be not more than 27 (or 20).

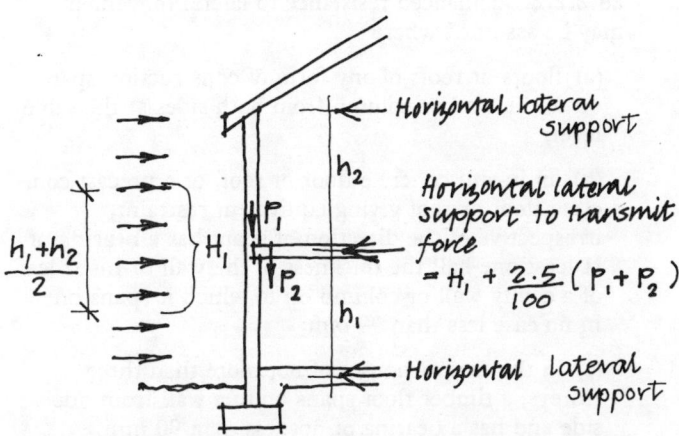

Fig. 5.12   Horizontal force to be taken by a floor element.

28.2.1 *Horizontal or vertical lateral supports*

Horizontal or vertical lateral supports should be capable of transmitting to the elements of construction that provide lateral stability to the structure as a whole, the sum of the following design lateral forces:

   (a) the simple static reactions to the total applied design horizontal forces at the line of lateral support, and

   (b) 2.5% of the total design vertical load that the wall or column is designed to carry at the line of lateral support; the elements of construction that provide lateral stability to the structure as a whole need not be designed to support this force.

However, the designer should satisfy himself that loads applied to lateral supports will be transmitted to the elements of construction providing stability, e.g. by the floors or roofs acting as horizontal girders.

28.2.2 *Horizontal lateral supports*

28.2.2.1   Simple resistance to lateral movement may be assumed in the case of houses of not more than three storeys where timber floor members, spaced apart at a distance of not more than 1.2 m, are connected by suitable joist hangers effectively fixed to the joist, as indicated in appendix C. In all other cases, including buildings of more than three storeys, a connection capable of providing simple resistance to lateral movement may be assumed where connections are of the form illustrated in Appendix C.

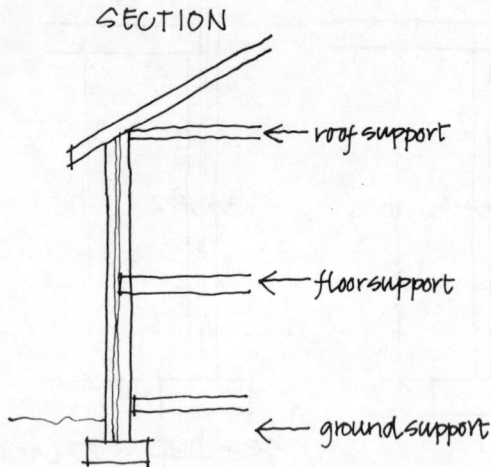

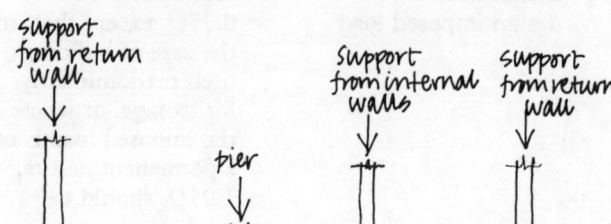

Fig. 5.13 *Horizontal lateral supports; see also App.* **B4**.

Fig. 5.14 *Vertical lateral supports; see also App.* **B5**.

28.2.2.2 Enhanced resistance to lateral movement may be assumed where:

(a) floors or roofs of any form of construction span on to the wall or column from both sides at the same level;

(b) an in-situ concrete floor or roof, or a precast concrete floor or roof giving equivalent restraint, irrespective of the direction of span, has a bearing of at least one-half the thickness of the wall or inner leaf of a cavity wall or column on to which it spans but in no case less than 90 mm;

(c) in the case of houses of not more than three storeys, a timber floor spans on to a wall from one side and has a bearing of not less than 90 mm.

Preferably, columns should be provided with lateral support in both horizontal directions.

28.2.3 *Vertical lateral supports*

28.2.3.1 Simple resistance to lateral movement may be assumed where an intersecting or return wall not less than the thickness of the supported wall or load-bearing leaf of a cavity wall extends from the intersection at least ten times the thickness of the supported wall or load-bearing leaf and is connected to it by metal anchors calculated in accordance with 28.2.1 and evenly distributed throughout the height at not more than 300 mm centres.

28.2.3.2 Enhanced resistance to lateral movement may be assumed where an intersecting or return wall as

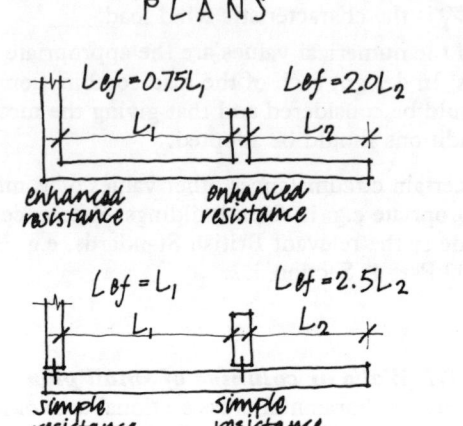

Fig. 5.15 *Effective length of wall depends on whether vertical supports provide simple or enhanced resistance to lateral movement.*

described in 28.2.3.1 is properly bonded to the supported wall or load-bearing leaf of a cavity wall.

28.2.3.3 In all other cases of vertical lateral support, simple or enhanced resistance to lateral movement may be established by calculation.

28.3 Effective height or length The effective height or length of a load-bearing wall or column should be assessed taking account of the relative stiffness of the elements of structure connected to the wall or column and the efficiency of the connections. In the absence of detailed calculations, the designer may take the effective height or length from 28.3.1 or 28.3.2.

28.3.1 *Effective height*

28.3.1.1 *Walls* The effective height of a wall may

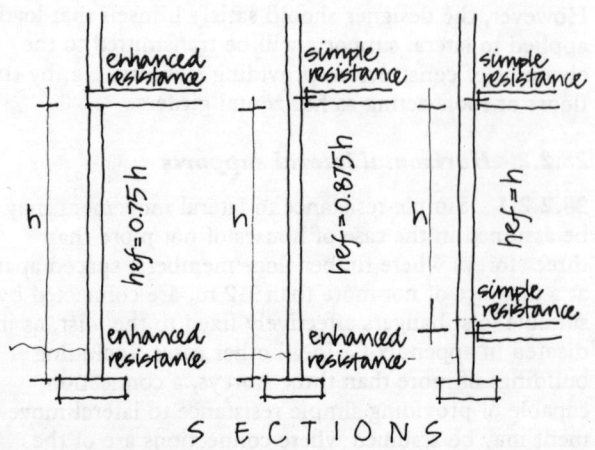

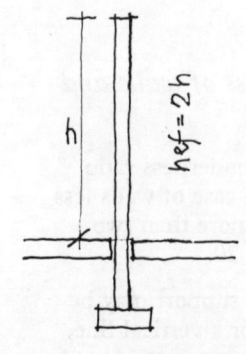

NB. This construction, without vertical piers, is only suitable for parapets or low garden walls

Fig. 5.16 *Effective height of wall depends on whether the horizontal supports provide simple or enhanced resistance to lateral movement.*

be taken as:

(a) 0.75 times the clear distance between lateral supports which provide enhanced resistance to lateral movement, or

(b) the clear distance between lateral supports which provide simple resistance to lateral movement.

28.3.1.2 Columns
The effective height of a column should be taken as the distance between lateral supports or twice the height of the column in respect of a direction in which lateral support is not provided.

28.3.1.3 Columns formed by adjacent openings in walls
Where openings occur in a wall such that the masonry between any two openings is, by definition, a column, the effective height of the column should be taken as follows:

(a) Where an enhanced resistance to lateral movement of the wall containing the column is provided, the effective height should be taken as 0.75 times the distance between the supports plus 0.25 times the height of the taller of the two openings.

(b) Where a simple resistance to lateral movement of the wall containing the column is provided, the effective height should be taken as the distance between the supports.

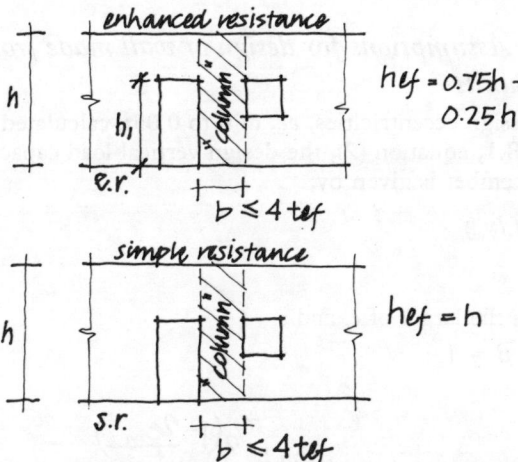

Fig. 5.17 Effective height of a 'column' formed by adjacent openings depending on whether there is enhanced or simple resistance to lateral movement at the supports.

28.3.1.4 Piers
Where the thickness of a pier is not greater than 1.5 times the thickness of the wall of which it forms a part, it may be treated as a wall for effective height consideration; otherwise the pier should be treated as a column in the plane at right angles to the wall.

NOTE The thickness of a pier, t_p, is the overall thickness including the thickness of the wall or, when bonded into one leaf of a cavity wall, the thickness obtained by treating this leaf as an independent wall.

28.3.2 Effective length
The effective length of a wall may be taken as:

(a) 0.75 times the clear distance between vertical lateral supports or twice the distance between a support and a free edge, where lateral supports provide enhanced resistance to lateral movement;

(b) the clear distance between lateral supports or 2.5 times the distance between a support and a free edge where lateral supports provide simple resistance to lateral movement.

28.4 Effective thickness
The effective thickness of a wall, column or pier is given in 28.4.1 and 28.4.2 and is illustrated in figure 3.

28.4.1 Walls and columns not stiffened by piers or intersecting walls
For single leaf walls and columns the effective thickness is the actual thickness.

For cavity walls and columns the effective thickness should be taken as two-thirds the sum of the actual thicknesses of the two leaves or the actual thickness of the thicker leaf, whichever is the greater.

28.4.2 Walls stiffened by piers or intersecting walls
Where a wall, which may be one leaf of a cavity wall, is stiffened by piers, the effective thickness, t_{ef} of the wall, or leaf of a cavity wall, is:

$$t_{ef} = t \times K$$

where

t is the actual thickness of the wall or leaf

t_{ef} is the effective thickness of the wall or leaf

K is the appropriate stiffness coefficient taken from table 5.

For a wall stiffened by intersecting walls, the appropriate stiffness coefficient may be determined from table 5 on the assumption that the intersecting walls are equivalent to piers of width equal to the thickness of the intersecting wall and of thickness equal to 3 times the thickness of the stiffened wall.

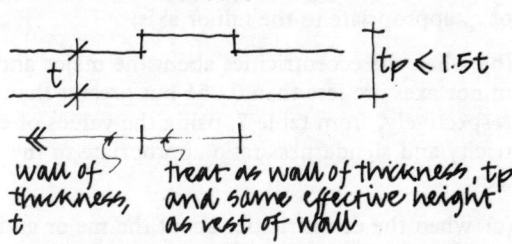

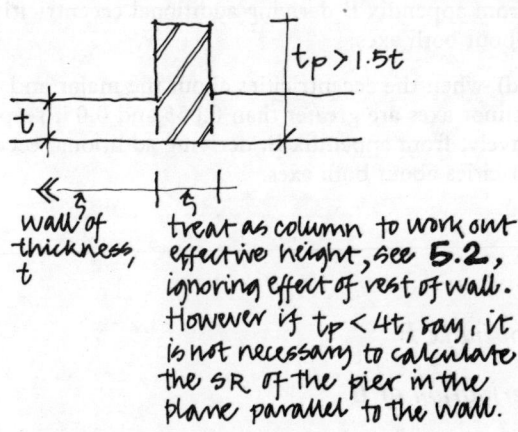

Fig. 5.18 Effective height of pier in a wall depends on its proportions: (a) $t_p \leqslant 1.5t$ and (b) $t_p > 1.5t$.

39

32.2.1 Design vertical load resistance of walls.

The design vertical load resistance of a wall per unit length is given by:

$$\frac{\beta t f_k}{\gamma_m}$$

where

β is a capacity reduction factor allowing for the effects of slenderness and eccentricity and is obtained from table 7;

f_k is the characteristic strength of the masonry obtained from clause 23;

γ_m is the partial safety factor for the material obtained from clause 27;

t is the thickness of the wall.

32.2.2 Design vertical load resistance of columns

The design vertical load resistance of a rectangular column is given by:

$$\frac{\beta b t f_k}{\gamma_m}$$

where

b is the width of the column;

t is the thickness of the column;

all other symbols are as given in 32.2.1. The value of β should be chosen as follows:

(a) when the eccentricities about the major and minor axes at the top of the column are less than $0.05b$ and $0.05t$ respectively, from the second column of table 7, basing the slenderness ratio on the value of t_{ef} appropriate to the minor axis;

(b) when the eccentricities about the major and minor axes are less than $0.05b$ but greater than $0.05t$ respectively, from table 7, using the values of eccentricity and slenderness ratio appropriate to the minor axis;

(c) when the eccentricities about the major and minor axes are greater than $0.05b$ but less than $0.05t$ respectively, from table 7, using the value of eccentricity appropriate to the major axis and the value of slenderness ratio appropriate to the minor axis or from appendix B deriving additional eccentricities about both axes;

(d) when the eccentricities about the major and minor axes are greater than $0.05b$ and $0.05t$ respectively, from appendix B, deriving additional eccentricities about both axes.

. . . . . .

Appendix B

Derivation of β

B.1 Assumptions for eccentricity and slenderness

The eccentricity is assumed to vary from the value e_x at the top of the wall, calculated in accordance with

40

clause 31, to zero at the bottom of the wall, subject to an additional eccentricity being considered to cover slenderness effects. No slenderness effect need be considered for walls or columns where the slenderness ratio is less than or equal to 6. The additional eccentricity may be assumed to vary linearly from zero at top and bottom of the wall, to a value e_a over the central fifth of the wall height where e_a is given by:

$$e_a = t \left[\frac{1}{2400} (h_{ef}/t_{ef})^2 - 0.015 \right] \tag{1}$$

where

t is the thickness of the wall (or depth of column);

t_{ef} is the effective thickness of the wall or column;

h_{ef} is the effective height of the wall or column.

The total design eccentricity, e_t, in the mid-height region of a slender wall is therefore given by:

$$e_t = 0.6e_x + e_a \tag{2}$$

where e_x is the eccentricity calculated at the top of the wall.

It should be noted that e_t can be less than e_x and plainly in such cases e_x, the eccentricity at the top of the wall, should govern the design, and should be taken as the design eccentricity.

B.2 Assumptions for design of wall made from solid units

For design eccentricities, e_m, of 0 to $0.05t$, calculated from B.1, equation (2), the design vertical load capacity of a member is given by;

$$\beta t \, (f_k/\gamma_m)$$

where

e_m is the larger of e_x and e_t

and $\beta = 1$

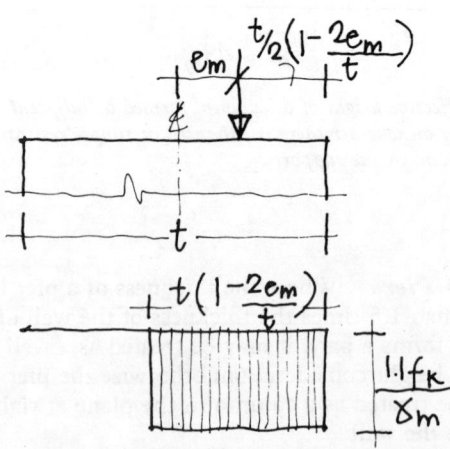

Fig. 5.19 *Stress block under ultimate conditions (after Fig. 10 of BS 5628: Part 1).*

For design eccentricities, e_m, of greater than $0.05t$, the eccentric load should be assumed to be resisted by a rectangular stress block with a constant stress of $1.1f_k/\gamma_m$ (see figure 10). It follows that the design vertical load capacity of the member is:

$$1.1 \left(1 - \frac{2e_m}{t}\right) \left(t \cdot \frac{f_k}{\gamma_m}\right) \qquad (3)$$

where

e_m is the larger of e_x and e_t, but not less than $0.05t$;

f_k is the characteristic strength of masonry as defined in clause 23;

γ_m is the partial safety factor as defined in clause 27.

Comparing the expressions for capacity given in 32.2 with equation (3), it will be seen that

$$\beta = 1.1 \, (1 - (2e_m/t)) \qquad (4)$$

The values of β in table 7 have been calculated from equation (4).

· · · · · ·

5.4 Examples

Example 5.1 Single-leaf wall

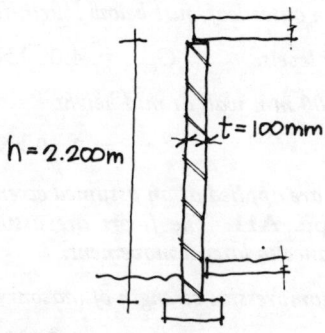

Fig. 5.20 Single-leaf wall.

A long external wall of a garage (Fig. 5.20) has a vertical load on it from the roof and a horizontal load from wind as follows:

Characteristic loads:
Vertical load from roof:

Dead load, G_k = 1.20 kN/m
Live load, Q_k = 1.60 kN/m

Vertical load from self-weight of wall, at bottom:

Dead load, G_k = 4.40 kN/m

Horizontal wind load:

Wind load, W_k = 0.25 kN/m^2

There is assumed to be no wind uplift on the roof. The roof loads are applied with an eccentricity at the top of the wall = 20 mm. The roof construction provides simple resistance to lateral movement.

Characteristic compressive strength of masonry,

$$f_k \quad = 6.00 \text{ N/mm}^2$$

Characteristic flexural strength of masonry (failure parallel to bed joints), ignoring effect of vertical load,

$$f_{ka} \quad = 0.40 \text{ N/mm}^2$$

Partial safety factor for material strength, from Table 5.1

$$\gamma_m \quad = 2.5$$

Check.

Vertical loading

Slenderness ratio (SR) = $\dfrac{h_{ef}}{t_{ef}} = \dfrac{1 \times 2.200}{0.100} = 22$ OK

Select Case (a) of Clause 22 of the Code: $1.4\,G_k + 1.6\,Q_k$:

At top of wall:

Design vertical load = $1.4 \times 1.20 + 1.6 \times 1.60 = 4.24$ kN/m
Design bending moment = $4.24 \times 0.020 = 0.084$ kN-m/m

$\therefore$ Resultant eccentricity, $e_x = \dfrac{0.084}{4.24} = 0.020$ m = $0.2\,t$

Hence, β, from Table 7 of the Code (Table 5.4) = 0.43
$\therefore$ Design vertical load resistance of wall

$$= \frac{\beta \cdot t \cdot f_k}{\gamma_m} = \frac{0.43 \times 100 \times 1{,}000 \times 6 \times 10^{-3}}{2.5} = 103 \text{ kN/m}$$

Because of self-weight, the design vertical load is slightly higher at the mid-height of the wall. The effect of this is to decrease the resultant eccentricity at mid-height from that assumed in the above calculation. Therefore the design vertical load resistance of the wall is at least that calculated above. By inspection, no further calculation is necessary. See Chapter 12 for more details.

Horizontal loading   Limiting dimensions of panel from Clause 36.3 of the Code (**6.2**),

$= 40 \times 0.100 \qquad = 4.000$ m

Select Case (b) of Clause 22 of the Code: $0.9\,G_k + 1.4\,W_k$:

At middle of wall:

Elastic design moment of resistance of wall, ignoring effect of vertical load,

$$= \frac{f_{ka} \cdot Z}{\gamma_m}$$

$$= \frac{0.4 \times 1{,}000 \times (100)^2 \times 10^{-6}}{2.5 \times 6}$$

$$= \frac{0.4}{2.5} \times 1.666 \qquad = 0.267 \text{ kN-m/m}$$

Elastic design moment

$$= \frac{W \cdot L}{8} \text{ from Table A4.1 Case (ii)}$$

$$= \frac{(1.4 \times 0.25 \times 2.2)\,2.2}{8} \qquad = 0.21 \text{ kN-m/m OK}$$

Example 5.2 Wall with piers

This example is the same as Example 5.1 *except that the horizontal wind force has now increased to 0.28 kN/m^2 and piers have been added, as shown* (Fig. 5.21).

Vertical loading   The wall will be checked for vertical load by ignoring the area of piers. This calculation has already been done for Example 5.1 and does not need to be repeated. However, the slenderness ratio of the wall with piers (SR), from Clause 28.4.2 of the Code (**5.3**), is

$$\text{SR} \quad = \frac{1 \times 2.200}{1.4 \times 0.100} = 16 \quad \text{OK}$$

41

SECTION

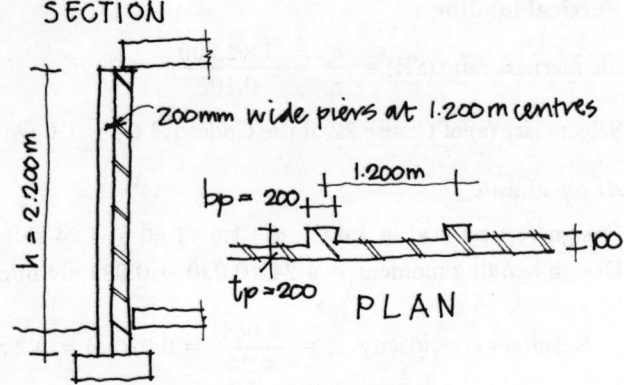

Fig. 5.21 *Section and plan on wall with piers.*

Horizontal loading   Wall spans vertically
Limiting dimensions of wall panel from Clause 36.3 (**6.2**) of
the Code are:

height $\leqslant 40t_{ef}$ $= 40 \times 0.140$ $= 5.60$ m
cf. actual height $= 2.20$ m   OK

Select Case (b) of Clause 22 of the Code: $0.9G_k + 1.4W_k$

From App. **A3**,

$$b = \frac{h_{ef}}{3} = \frac{2.200}{3} = 0.733 \text{ m}$$

for $\dfrac{b}{b_p} = 3.66$ and $\dfrac{t_p}{t} = 2,$

$y_2 = 0.357t_p$ $= 71$ mm
$y_1 = 200 - 71$ $= 129$ mm
$k = 1.751$

$$\therefore I_x = 1.751 \times \frac{(200)^4}{12} = 233 \times 10^6 \text{ mm}^4$$

Minimum section modulus,

$$Z_{min} = \frac{(233 \times 10^6)}{129} = 1.80 \times 10^6 \text{ mm}^3$$

$\therefore$ At middle of wall:

Elastic design moment of resistance of wall, per 1.200 m
length,

$$= \frac{0.4}{2.5} \times 1.80 = 0.288 \text{ kN-m}$$

Elastic design moment on wall, per 1.200 m length,

$$= \frac{(1.4 \times 0.28 \times 2.2) \times 2.2 \times 1.2}{8} = 0.284 \text{ kN-m}   \text{OK}$$

Example 5.3   Cavity wall

*A long external cavity wall, in one of the intermediate storeys of
a five-storey block, has a load-bearing inner leaf and an external facing leaf each of concrete block (Fig. 5.22).*

The averaged loads on the wall are as follows:

Characteristic loads

Vertical loads on inner leaf just below fourth-floor level

from upper levels, $G_k = 7.00$ kN/m
$Q_k = 5.50$ kN/m

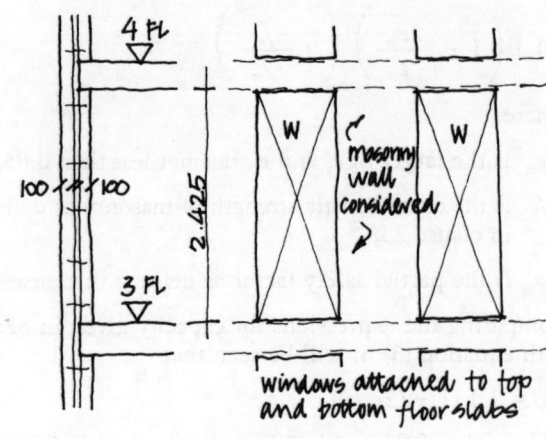

Fig. 5.22 *Section and elevation on storey-height masonry panel.*

from fourth-floor slab, $G_k = 12.00$ kN/m
$Q_k = 5.00$ kN/m

*The load from the levels above fourth-floor is assumed
to be axial (Clause 31 of the Code).*

Vertical loads on outer leaf, just below fourth-floor level,

from upper levels, $G_k = 4.00$ kN/m

Self-weight of 100 mm wall at mid-height

$= 1.50$ kN/m

*The floor loads are applied at an assumed eccentricity $= t/6 =
17$ mm, see App.* **A11**. *The floors are assumed to provide
enhanced resistance to lateral movement.*

Characteristic compressive strength of masonry,

$f_k = 3.20$ N/mm^2

*Characteristic flexural strength of masonry (failure parallel to
bed joints),*

$f_{ka} = 0.25$ N/mm^2

Partial safety factor for material strength,

$\gamma_m = 3.1$

Check.

Vertical loading

Effective thickness of cavity wall,

$t_{ef} = \frac{2}{3}(0.100 + 0.100)$
$= 0.133$ m

Slenderness ratio $= \dfrac{h_{ef}}{t_{ef}} = \dfrac{0.75 \times 2.425}{0.133}$

$= 14$ OK

Select Case (a) of Clause 22 of the Code: $1.4G_k + 1.6Q_k$

It is only necessary to check the inner leaf of wall.

At top of wall:

Design vertical load $= 1.4 \times (7.00 + 12.00)$
$+ 1.6 \times (5.50 + 5.00)$
$= 43.40$ kN-m/m

Design bending moment $= (1.4 \times 12.00 + 1.6$
$\times 5.00) \times 0.017$
$= 0.42$ kN-m/m

∴ Resultant eccentricity, $e_x = \dfrac{0.42}{43.40} = 0.010$ m

$= 0.1\,t$

Hence, β, from Table 7 of the Code (Table 5.4)

$= 0.83$

∴ Design vertical load resistance of wall

$$= \frac{0.83 \times 100 \times 1{,}000 \times 3.20 \times 10^{-3}}{3.1}$$

$= 85.67$ kN/m OK

Horizontal loading

See Example 6.3.

Example 5.4 Rectangular column

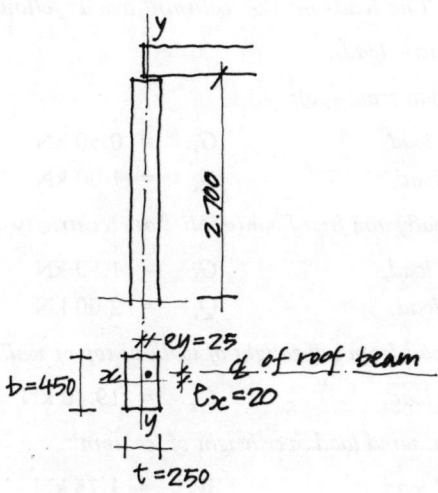

Fig. 5.23 Section and plan of column.

A solid masonry column takes load, from a single roof beam, which acts at a point 20 mm and 25 mm respectively from the major and minor axes of bending, as shown (Fig. 5.23). The loads on it are as follows:

Characteristic loads:

Vertical load from roof:

Dead load,	G_k	$= 25$ kN
Live load,	Q_k	$= 60$ kN

Vertical load from self-weight of column, at bottom:

Dead load,	G_k	$= 6$ kN

The roof beam, the only connection at the top of the column, provides enhanced resistance to lateral movement about the minor axis, y-y. The masonry has the same strengths and safety factors used in Example 5.1. Check column for vertical loading.

Vertical loading

Design slenderness ratio (SR) is greater of:

that about major axis, *x-x*,

and that about minor axis, *y-y*.

$$\text{SR}\ (x\text{-}x) = \frac{h_{ef}}{t_{ef}} = \frac{2 \times 2.700}{0.450}$$

$= 12$

$$\text{SR}\ (y\text{-}y) = \frac{1 \times 2.700}{0.250}$$

$= 10.8$

Design slenderness ratio $= 12$ OK

Select Case (a) of Clause 22 of the Code: $1.4G_k + 1.6Q_k$

At top of column:

Design vertical load $= 1.4 \times 25 + 1.6 \times 60 = 131$ kN

Resultant eccentricities are 20 mm $= 0.044b$ and 25 mm $= 0.100t$ about the major and minor axes respectively. This corresponds to Case (b) of Clause 32.2.2 of the Code (**5.3**). From Table 7 of the Code (Table 5.4), in this case using SR about major axis because it is greater than SR about minor axis, capacity reduction factor,

$$\beta = 0.87$$

∴ Design vertical load resistance of column

$$= \frac{0.86 \times 0.87 \times 250 \times 450 \times 6 \times 10^{-3}}{2.5}$$

$= 202$ kN OK

using area factor of 0.86 from Clause 23.1.1 of the Code (**5.3**).

For discussion on bi-axial bending of columns, see **5.2.1e**.

Example 5.5 Hollow-section column

A hollow-section masonry column takes load, from a single roof beam which acts at a point from the axes of bending as shown (Fig. 5.24). The loads on it are as follows:

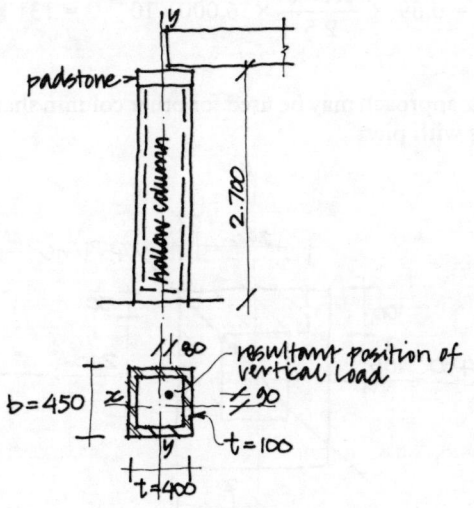

Fig. 5.24 Section and plan of column.

Characteristic loads:

Vertical load from roof:

Dead load,	G_k	$= 30$ kN
Live load,	Q_k	$= 50$ kN

Vertical load from self-weight of column, at bottom:

Dead load,	G_k	$= 6$ kN

The roof beam, the only connection at the top of the column, provides simple resistance to lateral movement about the y-y axis. The masonry has the same strengths and safety factors used in Example 5.1. Check column for vertical loading.

Select Case (a) of Clause 22 of the Code: $1.4G_k + 1.6Q_k$

Vertical loading

At top of column:

Design vertical load $= 1.4 \times 30 + 1.6 \times 50 = 122$ kN

Resultant eccentricities are 90 mm $= 0.2b$ and 80 mm $= 0.2t$ about the major and minor axes respectively. The use of Table 7 of the Code (Table 5.4) is not valid for a hollow section except if the vertical load happens to be axial. Use Appendix B of the Code (**5.3**) which shows that, at these eccentricities and for the slenderness ratio calculated, conditions are more critical at the top of the column than in the middle. Hence find the area of hollow section at the top of the column in equilibrium with the load; that is, the area of the section for which the point of application of the load would be the centre of gravity.

This may be calculated. However, the mathematics could be simplified in this case by assuming, conservatively, that the eccentricities are 105 mm and 80 mm about the major and minor axes respectively, so that there are equal lengths in compression in the *x*- and *y*-directions (Fig. 5.25). The area of section in uniform compression, in equilibrium with the load, would be

$$= 2 \times (280 \times 100) \qquad = 56,000 \text{ mm}^2$$

Hence vertical load resistance of column, including area factor of 0.89 from Clause 23.1.1 of the Code (**5.3**)

$$= 0.89 \times \frac{1.1 f_k}{\gamma_m} \times \text{(area in use)}$$

$$= 0.89 \times \frac{1.1 \times 6}{2.5} \times 56,000 \times 10^{-3} \quad = 131 \text{ kN}$$

OK

A similar approach may be used for other column shapes and for walls with piers.

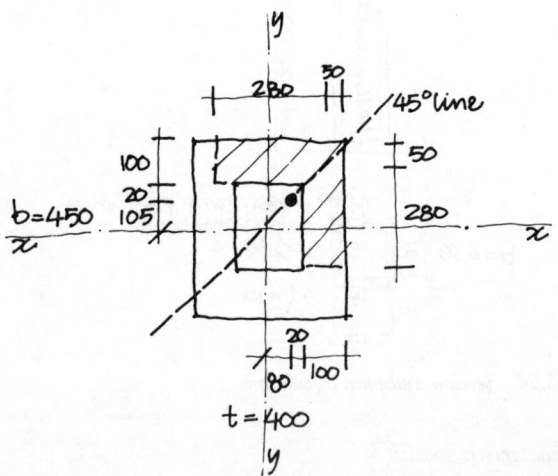

Fig. 5.25 Plan on column with hatched area showing assumed area in compression.

Example 5.6 Thick 'column' with no flexural strength

*An old house with a 200-mm-thick solid wall has a 'column' in the wall formed by a door and window opening (**5.2.1**c). The mortar is a weak non-hydraulic lime mortar with no tensile strength and acts chiefly as a bedding compound. The 'column' has a clear height of 2.380 m (Fig. 5.26) and requires*

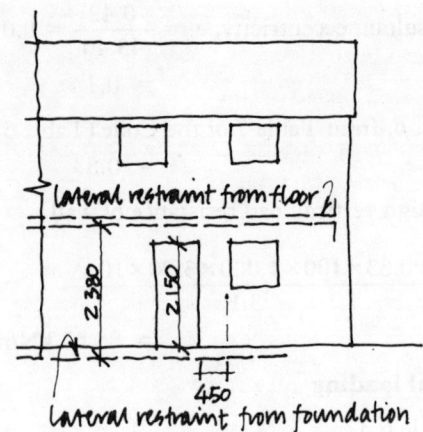

Fig. 5.26 Elevation on the old house.

checking. The loads on the 'column' are as follows:

Characteristic loads:

Vertical load from roof:

Dead load,	G_k	$= 0.50$ kN
Live load,	Q_k	$= 1.00$ kN

Vertical load from first floor (applied at eccentricity of 30 mm):

Dead load,	G_k	$= 1.50$ kN
Live load,	Q_k	$= 2.00$ kN

Vertical load from self-weight of wall, at top of wall:

Dead load,	G_k	$= 19.40$ kN

Horizontal wind load over height of 'column':

Wind load,	W_k	$= 1.75$ kN

Characteristic compressive strength of masonry:

$$f_k = 2.70 \text{ N/mm}^2$$

Partial safety factor for material strength:

$$\gamma_m = 3.5$$

Check.

Vertical and horizontal loading The masonry is a 'column' according to the Code because $b < 4t$. Assuming that enhanced resistance to lateral movement is provided at the lateral supports then (Clause 28.3.1.3 of the Code (**5.3**)),

$$h_{ef} = 0.75 \times 2.380 + 0.25 \times 2.150 \qquad = 2.323 \text{ m}$$

$$\therefore \text{ Slenderness ratio} = \frac{h_{ef}}{t_{ef}} = \frac{2.323}{0.200} = 12$$

OK

Because this is a 'column' the eccentricities about both major and minor axes need consideration. However, in this case there is no eccentricity of load about the major axis and, in general, Table 7 of the Code (Table 5.4) may be used to obtain the capacity reduction factor, β, using the eccentricity and slenderness ratio appropriate to the minor axis (Clause 32.2.2 of the Code (**5.3**)).

As the wall has no flexural strength, lateral loads must be resisted by a combination of the axial load and the compressive strength of the wall. It is not clear which of the three cases in Clause 22 of the Code is worst and hence, in principle, all three cases must be checked. Case (a) may be calculated as in Example 5.1.

For Case (b) the relevant combination is $0.9G_k + 1.4W_k$

Design vertical load

$$= 0.9\,(1.50 + 0.50 + 19.40) \qquad = 19.26 \text{ kN}$$

Bending moment in 'column' due to wind (Case (v) in Table A4.1) at top, is equal to $\dfrac{W.L}{8}$ and, at middle is equal to

$\dfrac{W.L}{14}$. Design moment due to wind,

$$\textit{at top of column} \qquad = \frac{1.4 \times 1.75 \times 2.380}{8}$$
$$= 0.73 \text{ kN-m}$$

and *near middle of column* $= \dfrac{1.4 \times 1.75 \times 2.380}{14}$
$$= 0.42 \text{ kN-m}$$

Hence eccentricity at top, e_x

$$= \frac{0.9 \times 1.50 \times 0.030 + 0.73}{19.26} = 0.040 \text{ m} = 40 \text{ mm}$$
$$= 0.20\,t$$

The use of Table 7 of the Code (Table 5.4) is not valid in this case because it does not take into account the eccentricity due to the wind load. For the calculation of β, the design eccentricity to be used, e_m, is either that at the top, e_x, or that $0.4h$ below the top, e_t, whichever is the larger (Appendix B of the Code (**5.3**)).

Eccentricity near middle due to wind moment, e_b

$$= \frac{0.42 \times 10^3}{19.26} = 22 \text{ mm} \qquad = 0.11\,t$$

Eccentricity near middle due to slenderness, e_a

$$= \left(\frac{(12)^2}{2400} - 0.015 \right) t \qquad = 0.045t$$

$\therefore$ Eccentricity in middle, e_t
$$= 0.6e_x + e_a + e_b$$
$$= 0.6 \times 0.20t + 0.045t + 0.11t \qquad = 0.275t$$
$$\therefore e_t > e_x \qquad \therefore e_m = e_t \qquad = 0.275t$$

Hence capacity reduction factor,

$$\beta = 1.1 \left(1 - \frac{2e_m}{t} \right)$$
$$= 1.1\,(1 - 0.55) \qquad = 0.495$$

and the design vertical load resistance of the column is

$$= \frac{\beta.b.t.f_k}{\gamma_m}$$
$$= \frac{0.495 \times 450 \times 200 \times (2.70 \times 0.83) \times 10^{-3}}{3.5}$$
$$= 28.52 \text{ kN}$$

with a reduction factor of

$0.70 + 1.5 \times 0.09 = 0.83$ (Clause 23.1.1 of the Code (**5.3**)).

cf. design vertical load $= 19.26$ kN
OK

This calculation should be repeated for Case (c) of Clause 22 of the Code.

CHAPTER 6

Masonry wall panels

Introduction

Masonry is very often used as cladding to a building. The building may have a steel or concrete frame or may have masonry piers which are integral with the cladding. This chapter is concerned with the design of such masonry wall panels. The particular features of this kind of masonry cladding are that it takes small vertical loads and that, in modern practice, it relies on its flexural strength to resist the horizontal loads on it. In some cases there may be axial forces to increase the effective flexural strength of the wall. Excerpts from the Code of Practice concerned with the design of wall panels are included in this chapter but the excerpts in Chapter 5 may be relevant too. Other information about masonry wall panels is given in App. **A** and **B**. Appendix A7 treats wall panels with openings and Appendix A14 reinforced panels. This chapter uses methods of design which do not necessarily assume that the wall panel is either pinned or fixed at its edges but allow the designer to assume any intermediate degree of fixity depending on the exact connection details. As elsewhere, this chapter is set out in the form of a series of instructions, as if for someone needing to assess a masonry building and undertake calculations. The format helps to separate the main points and provides a logical sequence in which the calculations for a wall panel may be done. With large walls it is often necessary to specially introduce extra lateral supports, usually extra vertical supports such as columns. In some cases it may be necessary to try several values of the wall thickness or spacing of lateral supports in a calculation. However, the calculations are easy to repeat and will quickly give satisfactory dimensions. Examples are given at the end of the chapter in order to clarify the main points.

6.1 Procedure for checking wall panels under mainly horizontal loads

(a) By inspection of the drawings decide which of the walls in the building is most critical, and at what level, when it is under a direct horizontal force, usually a wind force.

(b) Check the horizontal loading which would cause the maximum tensile or compressive stresses to occur.

(c) Limiting dimensions. Check that the limiting dimensions of the wall are not exceeded (Clause 36.3 of the Code (**6.2**)); see Fig. 6.1.

(d) Load combinations. Select Case (a), (b) or (c) from Clause 22 of the Code (**5.3**), whichever gives the worst combination of loads. For horizontal loads on a building usually only Cases (b) and (c) will be relevant. In low buildings tensile stresses are more likely to be critical than compressive stresses so that almost invariably Case (b) with the combina-

tion $0.9G_k + 1.4W_k$ will give the worst case. For panels whose removal would not affect the stability of the remaining structure, the Code allows Case (b) with the combination $0.9G_k + 1.2W_k$ to be taken instead.

(e) Work out the characteristic vertical load on the wall, or each leaf of the wall in the case of a cavity wall, as well as the horizontal load on the wall, keeping dead, live and wind loads separate. If the wind forces are low, the minimum horizontal load (Clause 20.1 of the Code (**5.3**)) may be the design horizontal force; see **5.1.2**d.

(f) Design moment of resistance. Decide how the wall will span between its supports – i.e. with a horizontal or vertical one-way span or a two-way span (Fig. B5.1). Work out the characteristic flexural strength of the wall in the direction in which the wall spans. For walls which span vertically it will first be necessary to calculate the design vertical load at the point under consideration. The design moment of resistance of the wall may then be calculated and this should be greater than the design moment. See also **5.1.2**e.

The design moment of resistance of a wall (Clause 36.4.3 of the Code (**6.2**)) is equal to

$$f_{kx} . Z/\gamma_m$$

where f_{kx} is the characteristic flexural strength of the wall (Table 2.3) or, for axially loaded walls which span vertically, it is equal to

$$f_{ka} . Z/\gamma_m$$

where f_{ka} is the effective flexural strength of the wall which is equal to $f_{kx} + \gamma_m . g_d$ or $f_{kx} + \gamma_m . g_A$ – i.e. including the effect of the design vertical dead load in the wall. This will be known as the 'elastic' design moment of resistance. The design moment of resistance of a freestanding wall due to the vertical load in it (Clause 36.5.3 of the Code (**7.2**)) is equal to

$$(n_w/2) . (t - n_w . \gamma_m/f_k)$$

where n_w is the design vertical load per unit length of wall. However, this design moment of resistance can also be used

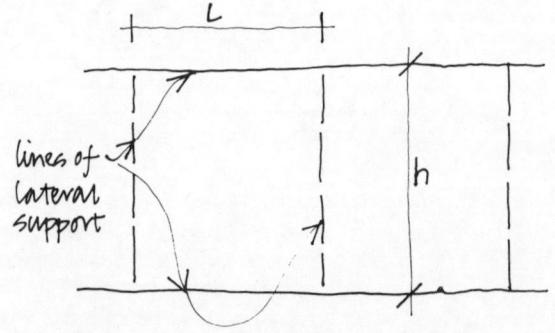

Fig. 6.1 Elevation on wall panel.

for other types of wall that span vertically. This will be known as a 'plastic' design moment of resistance.

For thick wall or walls in which the axial load is high relative to the bending moment, it may be that the resultant eccentricity due to lateral loading does not exceed $0.35t$ anywhere in the height of the wall; see **5.1.1**g and h. In this case it will not be necessary to rely on the flexural strength of the wall, if any, and the calculation proceeds as set in **5.1.1**.

(g) *Design moment.* The design moment on the wall should be calculated for the load combination selected. The design moment may be calculated using Tables A4.1 and A4.2, for one-way spans, or Table A6.1 for two-way spans. The design moment is of the general form:

$W_k . \gamma_f . L^2 \times$ (bending moment coefficient) per unit height for a wall spanning horizontally or in two directions, and

$W_k . \gamma_f . h^2 \times$ (bending moment coefficient) per unit width for a wall spanning vertically (Clause 36.4.2 of the Code (**6.2**)).

Continuity may be assumed at the lateral supports at the top and bottom of each storey-height of wall if the design loads do not cause any cracks to develop there, or any flexural stresses to develop there that exceed the design flexural strength of the wall, f_{kx}/γ_m or f_{ka}/γ_m. There may be continuity across a horizontal dpc, even if it has no tensile strength because of the effect of axial load in preventing tensile flexural stresses developing there. Otherwise it may be assumed there is only a simple support, or some intermediate value between these two extremes; see **6.1**g below.

One-way spans For walls which span in one direction Table A4.1 would normally be used to obtain a bending moment coefficient. The elastic bending moment patterns shown in Table A4.1 assume that the supports to the wall provide either: (i) geometrical fixity – i.e. the wall is not able to rotate at this edge – or (ii) simple support – i.e. the wall is free to rotate at this edge. Over the intermediate support the wall is assumed to be fully continuous.

Other assumptions may be made: the supports may provide moment restraint because of continuity or because of the joint details or, for horizontal joints, because of the vertical load across the joint, even if the wall is cracked. In such cases it is suggested that use be made of the notion of a 'plastic' design moment of resistance, in which the support is assumed to provide a constant moment of resistance against rotation.

Table A4.2 shows the elastic-plastic bending moments which would develop in the cases when i, the ratio of the plastic design moment of resistance at the support to the elastic design moment of resistance in the span, equals a half and when it equals one. In most cases i would be taken to have a value between zero and one, although it is possible for i to exceed one, at the bottom support of a high cantilevered wall or where the wall is only reinforced at the support for example. With only a rough estimate of i, one of the elastic-plastic bending patterns in Table A4.2 will usually still give a sufficiently accurate bending moment coefficient. For the particular case of a wall with vertical load in it, the 'plastic' design moment of resistance may be calculated as shown in **6.1**f. The assumption that a plastic moment can develop at joints is approximate and not always valid. See Heyman (1982, p.30) for further discussion. However, this method will often give more accurate results than the elastic assumptions. In some cases the possibility that cracks

could develop at a 'hinge' may not be acceptable. Examples of walls designed using 'elastic' and 'plastic' design moments of resistance are given in this chapter and Chapter 12. Typical joint details are given in App. **B4** and **B5**. See also Clause 36.8 of the Code.

The elastic-plastic bending moments may very easily be calculated for any other value of the ratio, i, not given in Table A4.2. In general, however, this will not be necessary. For example, with i having some value between a half and one, a safe value for the design bending moments may be obtained by assuming i to be equal to a half for the span moment calculation and then equal to one for the support moment calculation. However, this is over-conservative. By the lower bound theory of plasticity any one bending moment pattern which satisfies equilibrium and yield will give a safe result; the given bending moment patterns in Table A4.2 satisfy equilibrium and the condition of yield is satisfied if the given design bending moments are everywhere less than the calculated design bending moments of resistance.

In effect, the theory assumes that, ultimately, bending moment in the wall is distributed along its length in accordance with the strength of the wall and thus is distributed in an optimum way. Where the wall develops a 'plastic' hinge, it is assumed that the wall can rotate at this point but still develop a 'plastic' design moment of resistance.

Two-way spans For walls which span in two directions it will first be necessary to calculate the design vertical load at the point under consideration and hence the orthogonal ratio, μ, before using Table A6.1 (Table 9 of the Code). Note that Table A6.1 assumes that the supports to the wall provide either: (i) full continuity – i.e. the wall is restrained by a moment which at its maximum is equal to the flexural strength of the wall although a small amount of rotation may occur along an edge – or (ii) simple support – i.e. the wall is free to rotate along an edge. For supports which are intermediate between these two extremes it is suggested that the bending moment coefficients be obtained by interpolation or by use of the formulae in **A7**. This method is similar to that for one-way spanning walls; normally the fixity, i, varies between zero (simple support) and one (full continuity) for each edge and could be different for all four edge supports to the wall panel (Example 6.4).

(h) *Cavity walls.* For cavity walls, the vertical load and the design moment of resistance are calculated separately for each leaf. In general the design horizontal load can be assumed to be shared between the leaves in proportion to their design moment of resistance, as long as the wall ties are strong enough (see below). For each leaf of the wall the design moment of resistance must then be greater than the design moment. However the Code (Clause 36.4.5) allows an alternative and more convenient assumption, which is suitable in the majority of cases. This is that the design lateral strength of a cavity wall, that is the maximum design load (e.g. wind load) per unit area on the cavity wall, may be taken as the sum of the design lateral strengths of each individual leaf of the cavity wall (Example 12.6, p. 87). The orthogonal ratio defined in the Code, μ, is the ratio of the flexural strength of masonry when it spans vertically (including the effect of the design vertical load in the wall) to that when it spans horizontally – i.e. $= f_{ka}/f_{kb}$. It should not be taken as more than one. The orthogonal ratio can be obtained with the help of Table 2.3.

Cavity wall ties In order that the design horizontal load is properly shared between the two leaves of the cavity wall, the ties must be strong enough in tension and compression to transfer that calculated proportion of the total horizontal load from the outer to the inner leaf, or vice versa. Vertical twist ties are stronger than butterfly or double-triangle ties but less flexible; see **6.1**k. For a cavity wall with leaves of equal thickness having butterfly ties at the maximum spacings of 900 mm c/c horizontally and 450 mm c/c vertically (i.e. 2.5 ties per square metre) the maximum horizontal wind load with a 75 mm cavity, before overstressing the ties, would be

$$2.5 \times 0.5 \times 2/3 = 0.83 \text{ kN/m}^2; \text{ see } \mathbf{6.1}k.$$

For a leaf tied to a rigid backing, for example over lines of support, the maximum design horizontal wind load would be half this figure, if the same spacing and type of tie were to be used.

(i) Walls with piers. For walls with piers – i.e. walls behaving as T-sections – the position of the neutral axis and the moment of inertia may be quickly calculated using Tables A3.1 and A3.2. In general, T-section walls will need to be checked at their base and at about mid-height both with positive wind pressure and wind suction. Hence, in general, there are four conditions to consider for each load combination selected. For sections with a wide 'flange' and a deep 'web', the shear stresses along the line connecting the 'flange' to the 'web' may need checking (**8.3**f).

(j) Shear strength. It is sometimes necessary to check that the design shear strength of the masonry, f_v/γ_{mv} (Clause 25 of the Code (**5.3**)), is greater than the design shear stress, v_h, calculated for the load case from Clause 22 of the Code (**5.3**) which gives the highest value of design shear stress, normally Case (b). The design shear stress, v_h, is the horizontal design load divided by the horizontal area of wall taking the load (Clause 33 of the Code). The partial safety factor for material strength, γ_{mv}, equals 2.5 for any of the mortars listed in Table 2.1 (Clause 27.4 of the Code). Note that the design shear strength is likely to be low at flexible damp-proof courses (dpc).

(k) Connections. The connection of wall panels to the lateral support elements is extremely important. If possible, more connections should be provided than indicated by a static calculation of the number of ties, or other connections, needed for sufficient strength against wind forces. More connections provide greater fixity at the support which, in turn, reduces the bending moments in the middle of the wall panel. If wall ties are used in tension or shear, either in cavity walls or at connection points, their characteristic strengths may be found by referring to Table B1.2. The design strength of a tie is the characteristic strength divided by the partial safety factor, γ_m, which equals 3.0 for a wall tie (Clause 27.5 of the Code). If wall ties are used in compression, the same characteristic values as those given for tension in Table B1.2 may be used except with butterfly or double-triangle ties for which the characteristic compression strength is 0.5 kN and 1.25 kN respectively for up to a 75 mm cavity or 0.35 kN and 0.65 kN respectively for up to a 100 mm cavity with mortar designations (i), (ii) or (iii). See Table B1.1 and App. B1, B4 and B5.

(l) Openings in wall panels. Wall panels are considerably weakened by very large door or window openings. In such cases posts/beams spanning between the supports may be needed to provide support to some edges of the opening and restore the strength of the wall panel. However, if the openings are only small or of medium size, it will usually be

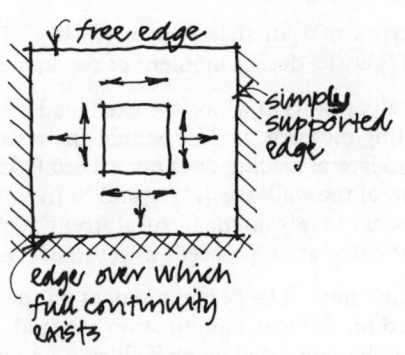

Fig. 6.2 *Wall panel with opening.*

possible to rely on the strength of the remaining masonry to carry the horizontal loads (Fig. 6.2). Appendix **A7** gives methods of calculations for walls with openings, with or without bed joint reinforcement in the wall, and includes examples; see also examples in this chapter.

6.2 Extracts from BS 5628:Part 1

36.3 Limiting dimensions In a laterally-loaded panel or freestanding wall built of masonry set in mortar designations (i) to (iv) and designed in accordance with clause 36, the dimensions should be limited as follows:

(a) *Panel supported on three edges*

(1) two or more sides continuous:
height × length equal to $1500t_{ef}^2$ or less

(2) all other cases:
height × length equal to $1350t_{ef}^2$ or less

(b) *Panel supported on four edges*

(1) three or more sides continuous:
height × length equal to $2250t_{ef}^2$ or less

(2) all other cases:
height × length equal to $2025t_{ef}^2$ or less

(c) *Panel simply supported at top and bottom*
Height equal to $40t_{ef}$ or less

(d) *Freestanding wall*
Height equal to $12t_{ef}$ or less

In cases (a) and (b) no dimension should exceed 50 times the effective thickness t_{ef}.

. . . . . .

36.4.2 Calculation of design moments in panels
Masonry walls are not isotropic and there is an orthogonal strength ratio, μ (see 3.16), depending on the brick or block and mortar used, as may be found from the characteristic flexural strengths given in clause 24.

The calculation of the design moment of a panel has to take into account the masonry properties referred to above and may be taken as either

$\alpha W_k \gamma_f L^2$ per unit height, when the plane of failure (see table 3) is perpendicular to the bed joints; or

$\mu \alpha W_k \gamma_f L^2$ per unit length, when the plane of failure (see table 3) is parallel to the bed joints where

- α is the bending moment coefficient taken from table 9;
- γ_f is the partial safety factor for loads (clause 22);
- μ is the orthogonal ratio;
- L is the length of the panel between supports;
- W_k is the characteristic wind load per unit area.

When a vertical load acts so as to increase the flexural strength in the parallel direction, the orthogonal strength ratio, μ, may be modified by using a flexural strength in the parallel direction of:

$$f_{kx} + \gamma_m g_d$$

where

- f_{kx} is the flexural strength in the parallel direction, taken from table 3;
- γ_m is the appropriate partial safety factor for materials (clause 27);
- g_d is the design vertical dead load per unit area.

The bending moment coefficient, $\mu\alpha$, at a damp-proof course may be taken as for an edge over which full continuity exists when there is sufficient vertical load on the damp-proof course to ensure that its flexural strength (see 24.1) is not exceeded.

Table 9 gives values of bending moment coefficients, α, for various values of μ, the orthogonal ratio derived from table 3, modified as necessary for vertical load.

For walls spanning vertically, the design moment per unit length of wall at mid-height of the panel may be taken as:

$$W_k \gamma_f h^2/8$$

unless the end conditions justify treating the panel as partially fixed. Piers should be treated in the same way, and the proportion of load being carried by the pier should be assessed from normal structural principles.

36.4.3 Calculation of design moment of resistance of panels
The design moment of resistance of a masonry wall is given by:

$$\frac{f_{kx}}{\gamma_m} Z$$

where

- f_{kx} is the characteristic flexural strength appropriate to the plane of bending (clause 24);
- γ_m is the partial safety factor for materials (clause 27);
- Z is the section modulus.

In assessing the section modulus of a wall including piers, the outstanding length of flange from the face of the pier should be taken as:

(a) 4 × thickness of wall forming the flange when the flange is unrestrained, or

(b) 6 × thickness of wall forming the flange when the flange is continuous,

but in no case more than half the clear distance between piers.

6.3 Examples

Example 6.1 Masonry wall panel supported on three sides

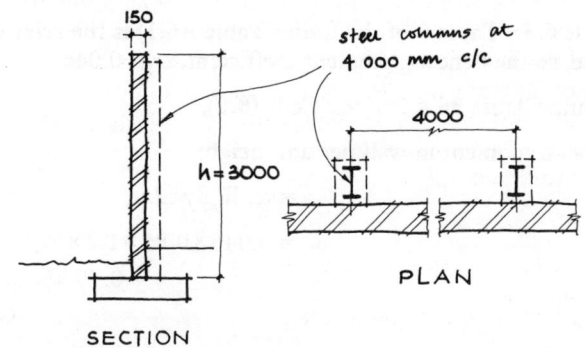

Fig. 6.3 Section and plan on wall.

A boundary wall 150 mm thick and 3.000 m high is connected to steel columns at 4.000 m centres which provide lateral stability (Fig. 6.3). The wall is of concrete block with a compressive strength of 10.5 N/mm²; it has a vertical load on it and a horizontal load from wind as follows:

Characteristic loads:

Vertical load from self-weight of wall, at bottom,

    Dead load, G_k = 4.40 kN/m

Horizontal load:

    Wind load, W_k = 0.85 kN/m²

Characteristic flexural strength of masonry, f_{kx}:

    f_{kb}, *for failure perpendicular to bed joints,*

        = 0.75 N/mm²

and f_{ka}, for failure parallel to bed joints,

        = 0.25 N/mm²

from Table 3 of the Code (Table 2.3). Mortar designation (iii) is used. Factor of safety for material strength, γ_m
        = 3.1

Check masonry wall panel for horizontal loading and shear.

Horizontal loading Limiting dimensions of panel from Clause 36.3 of the Code (**6.2**) given by:

height or length ≤ 50 × 0.150	= 7.500 m
height × length ≤ 1500 × (0.150)²	= 33.7 m²
cf. actual height × length	= 13.5 m²

        OK

Select Case (b) of Clause 22 of the Code (**5.3**): $0.9G_k + 1.2W_k$

Wall spans in two directions; worst moment is near free edge at the top of the wall.

Using Clause 36.4.3 of the Code (**6.2**),

Elastic design moment of resistance of wall per unit height, at top of wall,

$$= \frac{f_{kb} \cdot Z}{\gamma_m}$$

$$= \frac{0.75 \times 10^3 \times (150)^2 \times 10^{-6}}{3.1 \times 6}$$

$$= 0.90 \text{ kN-m/m}$$

Orthogonal strength ratio, $\mu = \dfrac{0.25}{0.75} = 0.33$

and $\dfrac{h}{L}$ for masonry panel $= \dfrac{3}{4} = 0.75$

Case C in Table 9 of the Code (Table A6.1) is the relevant case, so the bending moment coefficient, $\alpha = 0.046$

Using Clause 36.4.2 of the Code (**6.2**),

Design moment in wall per unit height

$$= \alpha . W_k . \gamma_f . L^2$$
$$= 0.046 \times 0.85 \times 1.2 \times (4)^2$$
$$= 0.75 \text{ kN-m/m}$$
$$\text{OK}$$

Shear

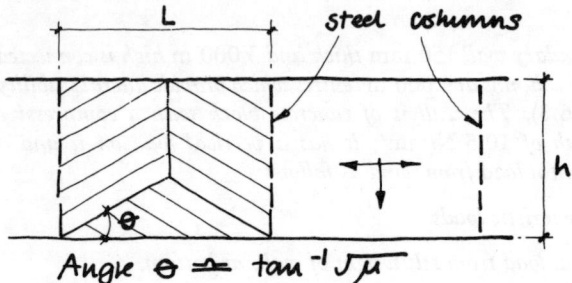

$$\text{Angle } \theta = \tan^{-1} \sqrt{\mu}$$

Fig. 6.4 Elevation on wall under horizontal load, showing areas of wall panel supported by sides and bottom support.

Use same load case then (Fig. 6.4),

Design vertical compressive stress at bottom of wall, g_d

$$= \dfrac{0.9 \times 4.40 \times 10^3}{1,000 \times 150}$$
$$= 0.026 \text{ N/mm}^2$$

$\therefore$ Design shear strength of wall for designation (iii) mortar

(Clause 25 of the Code (**5.3**)), $\dfrac{f_v}{\gamma_{mv}}$,

at bottom of the wall $= \dfrac{0.15 + 0.6 \times 0.026}{2.5}$

$$= 0.066 \text{ N/mm}^2$$

and at side wall $= \dfrac{0.15}{2.5} = 0.060 \text{N/mm}^2$

Assume area of wall panel under horizontal load supported by column, on each side $= h . L/2$ and by foundation $= L^2/4$ then,

Design shear stress, v_h

at bottom of the wall $= \dfrac{W_k . \gamma_f . L}{4t}$

$$= \dfrac{0.85 \times 1.2 \times 4}{4 \times 150}$$
$$= 0.007 \text{ N/mm}^2$$

at the side of the wall $= \dfrac{W_k . \gamma_f . L}{2t} = 0.014 \text{ N/mm}^2$

$$\text{OK}$$

Example 6.2 Masonry wall panel supported on four sides

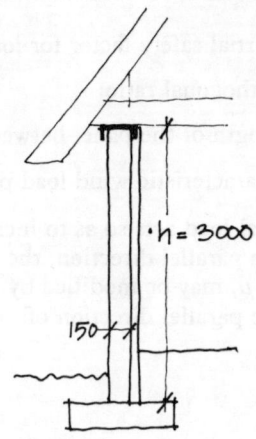

Fig. 6.5 Section of wall supporting roof.

This example is the same as Example 6.1 *except that there is now extra dead load on the wall from a steep roof (Fig. 6.5). The roof construction provides lateral support to the top edge of the wall panel.*

Characteristic loads:

Vertical load from roof:

Dead load, $G_k = 5.60 \text{ kN/m}$

Check masonry wall panel for horizontal loading and shear.

Vertical loading

Limiting dimensions of wall panel from Clause 36.3 of the Code (**6.2**) given by:

height or length $\leqslant 50 \times 0.150$ $= 7.500 \text{ m}$
height $\times$ length $\leqslant 2,025 \times (0.150)^2$ $= 45.6 \text{ m}^2$
cf. actual height $\times$ length $= 13.5 \text{ m}^2$

$$\text{OK}$$

Select Case (b) of Clause 22 of the Code: $0.9G_k + 1.4W_k$

Wall spans in two directions; worst moment is near middle of wall.

Design vertical compressive stress near middle of wall, g_d

$$= \dfrac{0.9 \times (2.20 + 5.60) \times 10^3}{1,000 \times 150} = 0.046 \text{ N/mm}^2$$

$\therefore$ Effective characteristic flexural strength of wall, f_{ka}

$$= f_{kx} + \gamma_m . g_d = 0.25 + 3.1 \times 0.046 = 0.39 \text{ N/mm}^2$$

from Clause 36.4.2 of the Code (**6.2**).

Using Clause 36.4.3 of the Code,

Elastic design moment of resistance of wall per unit height, near middle of wall, as for Example 6.1,

$$= \dfrac{f_{kx} . Z}{\gamma_m}$$

$$= \dfrac{0.75 \times 10^3 \times (150)^2 \times 10^{-6}}{3.1 \times 6}$$

$$= 0.90 \text{ kN-m/m}$$

Orthogonal strength ratio, $\mu = \dfrac{0.39}{0.75} = 0.52$

$$\frac{h}{L} = \frac{3}{4} = 0.75$$

Assuming Case G in Table 9 of the Code (Table A6.1) to be the relevant case, the bending moment coefficient, $\alpha = 0.030$

Using Clause 36.4.2 of the Code (**6.2**),

Design moment in wall per unit height

$$= \alpha \cdot W_k \cdot \gamma_f \cdot L^2$$
$$= 0.030 \times 0.85 \times 1.4 \times (4)^2$$
$$= 0.58 \text{ kN-m/m}$$
$$\text{OK}$$

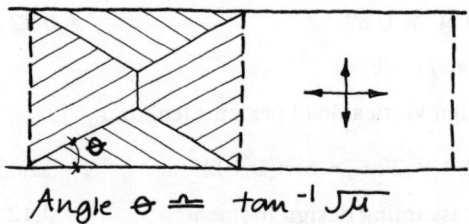

Angle $\theta \triangleq \tan^{-1}\sqrt{\mu}$

Fig. 6.6 Elevation on wall under horizontal load, showing areas of wall panel supported by sides and top and bottom supports.

The calculation for shear is similar to that already done in Example 6.1

Example 6.3 Cavity wall

The wall considered in Example 5.3 *is now checked under horizontal load. The horizontal wind load* = 0.65 kN/m²

Horizontal loading Limiting dimensions of panel from Clause 36.3 of the Code (**6.2**)

$$= 40 \times 0.100 = 4.000 \text{ m} \quad \text{OK}$$

Characteristic vertical load at mid-height,

on inner leaf, $G_k = 19.00 + 1.50 = 20.50$ kN/m
on outer leaf, $G_k = 4.00 + 1.50 = 5.50$ kN/m

Select Case (b) of Clause 22 of the Code: $0.9G_k + 1.4W_k$

at mid-height:

Design compressive stress in wall, g_d

for inner leaf $= \dfrac{0.9 \times 20.50 \times 10^3}{1,000 \times 100} = 0.184$ N/mm²

for outer leaf $= \dfrac{0.9 \times 5.50 \times 10^3}{1,000 \times 100} = 0.049$ N/mm²

Effective characteristic flexural strength, $f_{ka} = f_{kx} + \gamma_m \cdot g_d$

for inner leaf $= 0.25 + 3.1 \times 0.184 = 0.82$ N/mm²

for outer leaf $= 0.25 + 3.1 \times 0.049 = 0.40$ N/mm²

from Clause 36.4.2 of the Code (**6.2**).

∴ Elastic design moment of resistance $= \dfrac{f_{ka} \cdot Z}{\gamma_m}$

for inner leaf $= \dfrac{0.82 \times 1,000 \times (100)^2 \times 10^{-6}}{3.1 \times 6}$

$$= \frac{0.82}{3.1} \times 1.66 = 0.44 \text{ kN-m/m}$$

for outer leaf $= \dfrac{0.40}{3.1} \times 1.66 = 0.21$ kN-m/m

Hence elastic design moment of resistance for both leaves $= 0.65$ kN-m/m

Assuming that the window frames carry the wind load on the glass and that there is fixity at the top and bottom of the cavity wall then from Case (iv) in Table A4.1,

Elastic design moment $= \dfrac{W \cdot L}{12}$

$$= \frac{1.4 \times (1.000 \times 2.425 \times 0.65) \times 2.425}{12}$$

$$= 0.45 \text{ kN-m/m}$$
$$\text{OK}$$

The elastic design moment of resistance has been calculated at mid-height but is not significantly different at the top or bottom of the wall where the design moments are assumed to occur. If ties are spaced at 900 mm c/c horizontally and 450 mm c/c vertically – i.e. 2.5 ties per square metre – then the characteristic compressive resistance required in each tie would be

$$= \frac{1.4 \times 0.65 \times 3}{2.5 \times 2} = 0.55 \text{ kN; see } \textbf{6.1}\text{h and k.}$$

Double-triangle ties are often used with concrete blocks and would be adequate (**6.1**k). Otherwise vertical twist ties may be used.

Example 6.4 Cavity wall with partial restraint on one side

An assembly hall with an octagonal plan shape has clay brick cavity walls supporting a pyramid roof made of laminated timber with tie rods connected to its four springing points (Fig. 6.7). The roof is connected to all the wall thus stabilising the building and providing lateral support at the top of each wall. A mortar of designation (iii) is used. The brick weighs 22.5 kN/m³ and has a water absorption of 9%. The factor of safety for material strength, γ_m = 2.5.

The horizontal wind load is 0.60 kN/m². Check wall panel without openings under horizontal load, assuming the vertical support on one side of the panel provides partial moment restraint and vertical load in the walls from the roof is small. There is a bitumen dpc at ground level.

Horizontal loading Limiting dimensions of wall panel from Clause 36.3 of the Code given by:

height or length $\leqslant 50 \times 0.136 = 6.800$ m
height $\times$ length $\leqslant 2,025 \times (0.136)^2 = 37.45$ m²

$$\text{OK}$$

Select Case (b) of Clause 22 of the Code: $0.9G_k + 1.4W_k$

Wall spans in two directions (Fig. 6.8):

Design vertical compressive stress near middle of wall, g_d

$$= 0.9 \times 22.5 \times 3 \times 10^{-3} = 0.06 \text{ N/mm}^2$$

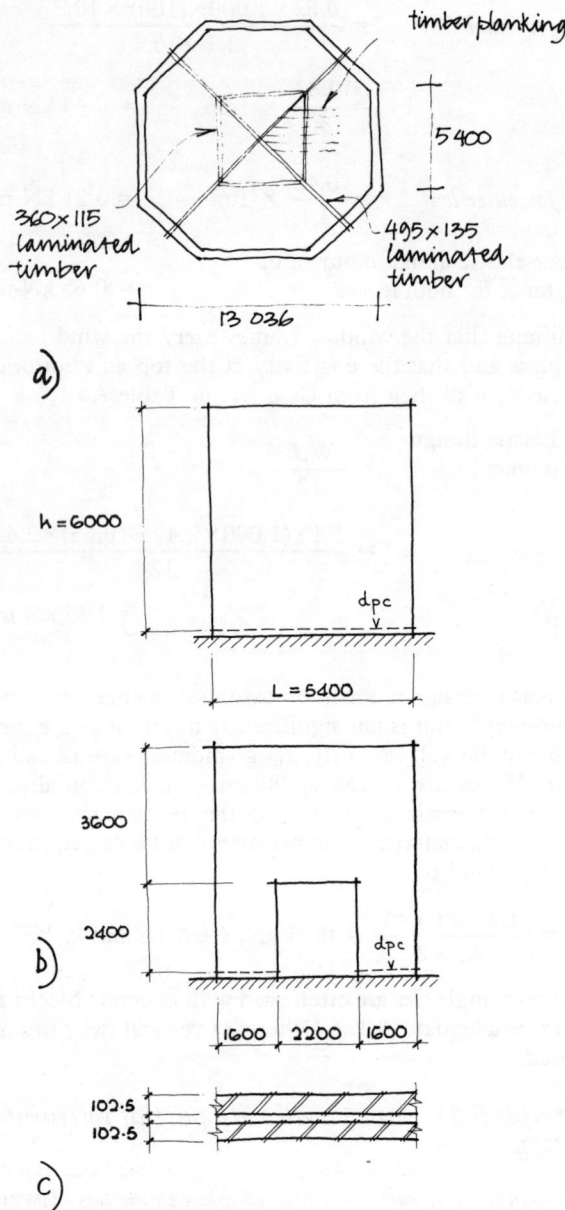

a)

b)

c)

Fig. 6.7 (a) Plan on assembly hall, (b) elevation on a side with and without openings in it and (c) horizontal section through wall.

∴ Effective flexural strength of wall, f_{ka}

$$= 0.4 + 2.5 \times 0.06 \qquad = 0.55 \text{ N/mm}^2$$

Elastic design moment of resistance of cavity wall per unit height

$$= 2 \times \frac{1.1 \times 1{,}000 \times (102.5)^2 \times 10^{-6}}{2.5 \times 6} \quad = 1.54 \text{ kN-m/m}$$

Orthogonal strength ratio, $\mu = \dfrac{0.55}{1.10} \qquad = 0.50$

$$\frac{h}{L} = \frac{6.000}{5.400} \qquad = 1.11$$

Assuming panel is continuous about the bottom edge then Case H in Table 9 of the Code (Table A6.1) – see Fig. 6.8 – gives a bending moment coefficient, $\alpha = 0.034$ with the design moment in wall per unit height (horizontal direction), $m = \alpha \cdot W_k \cdot \gamma_f \cdot L^2$

$$= 0.034 \times 0.60 \times 1.4 \times (5.400)^2 \qquad = 0.83 \text{ kN-m/m}$$

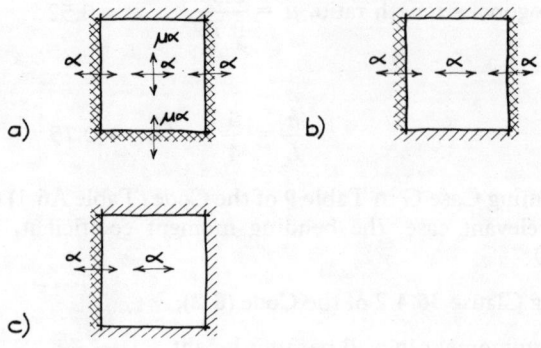

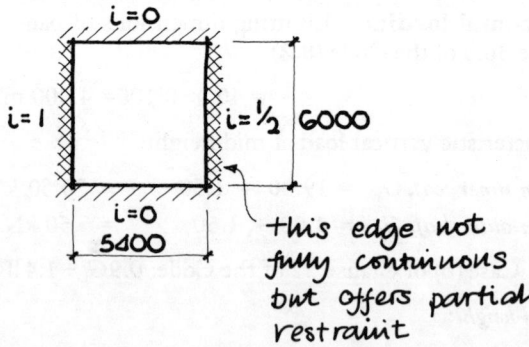

Fig. 6.8 (a) Case H, (b) Case G and (c) Case F from Table A6.1.

and the design moment in wall per unit length (vertical direction)

$$= \mu \cdot m$$
$$= 0.50 \times 0.83 \qquad\qquad = 0.42 \text{ kN-m/m}$$

at bottom of wall:

Design vertical load per unit length, g_d

$$= 0.9 \times 22.5 \times 6 \times 0.1025 \qquad = 12.45 \text{ kN/m}$$

and assuming design moment $= 0.42 \text{ kN-m/m}$

then resultant eccentricity of load at bottom of wall

$$= \frac{0.42 \times 10^3}{12.45} = 34 \text{ mm} = 0.33t > t/6$$

However, as the dpc provides no resistance to bending and the resultant load is outside the middle third of the wall, it may be assumed that under the design loads the panel is cracked and therefore not continuous about the bottom edge.

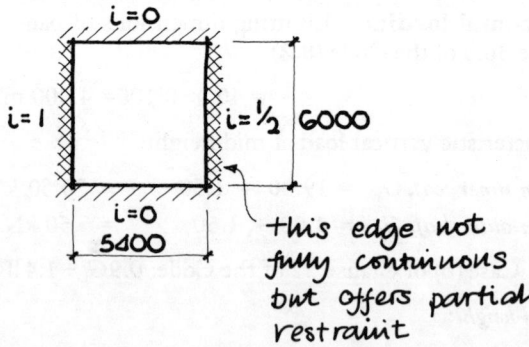

Fig. 6.9 Wall panel with partial restraint on one edge.

Assuming the right-hand vertical support to provide restraint equal to about half that provided by a continuous edge (Fig. 6.9), then the bending moment coefficient for this case is obtained by interpolating between Cases F and G of Table 9 of the Code (Table A6.1); see Fig. 6.8.

hence $\alpha = \dfrac{0.047 + 0.038}{2} \qquad = 0.043$

∴ Design bending moment in wall

$$= 0.043 \times 0.60 \times 1.4 \times (5.400)^2 \qquad = 1.06 \text{ kN-m/m}$$

OK

Example 6.5 Cavity wall with opening

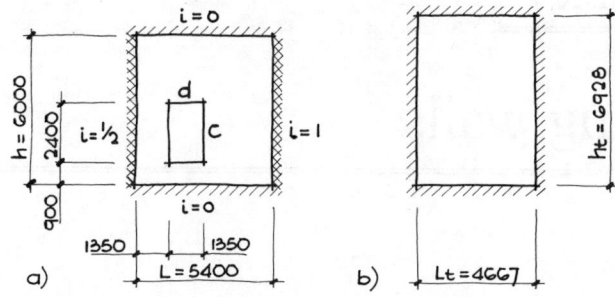

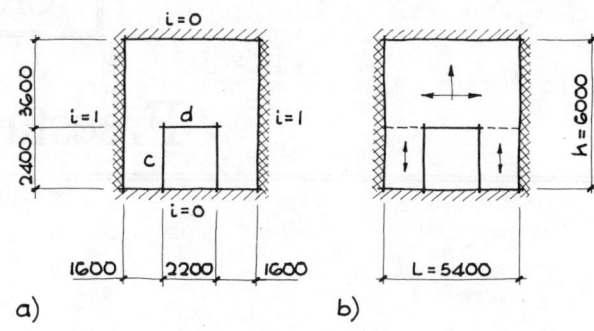

Fig. 6.10 (a) Wall panel with opening and (b) equivalent simply supported panel.

Fig. 6.11 (a) Wall panel with opening and (b) divided into parts for calculation.

The wall panel adjacent to that treated in Example 6.4 is similar except that the wall has an opening in it (Fig. 6.10). Check wall panel under horizontal load.

Horizontal loading Select Case (b) of Clause 22 of the Code: $0.9G_k + 1.4W_k$

Wall spans in two directions. Use method given in App. **A7.2** since $c < 0.6h$ and $d < 0.35L$. In addition none of the edge distances is less than a third of the adjacent opening dimensions and therefore these edges are considered to be self-supporting (Fig. A7.9).

Dimensions of equivalent simply supported slab

$$L_t = \frac{2 \times 5.400}{\sqrt{(1+1-0.4)} + \sqrt{(1+0.5-0.4)}} = 4.667 \text{ m}$$

$$h_t = \frac{2 \times 6.000}{\sqrt{(1+0-0.25)} + \sqrt{(1+0-0.25)}} = 6.928 \text{ m}$$

$$\therefore \frac{h_t}{L_t} = \frac{6.928}{4.667} = 1.48$$

Orthogonal strength ratio, $\mu = 0.50$, as before; hence from Case E of Table 9 of the Code (Table A6.1), $\alpha = 0.074$

hence,

Design bending moment $= 0.074 \times 0.60 \times 1.4 \times (4.667)^2$
$$= 1.35 \text{ kN-m/m}$$

Elastic design moment of resistance of cavity wall,
$$= 1.54 \text{ kN-m/m, as before.}$$

OK

Example 6.6 Cavity wall with medium-sized openings

The wall panel adjacent to that treated in Example 6.4 is similar except that there is an opening in the wall and both vertical supports provide full continuity at the edge (Fig. 6.11). Check wall panel under a wind load of 0.75 kN/m² with $\gamma_m = 2.8$.

Horizontal loading Select Case (b) of Clause 22 of the Code: $0.9G_k + 1.4W_k$

Equivalent load on top part of wall panel (Fig. 6.11) is, from Appendix A7,

$$W_k\left(1 + \frac{2.400}{3.600}\right) = 1.67 \ W_k \text{ kN/m}^2$$

where W_k is the characteristic wind load.

For Case C in Table 9 of the Code (Table A6.1) orthogonal strength ratio, $\mu = 0.50$, as before and

$$\frac{h}{L} = \frac{3.600}{5.400} = 0.67 \text{ so that } \alpha = 0.041$$

Design bending moment

$$= 0.041 \times 1.4 \times 1.67 \ W_k \times (5.400)^2 = 2.8 \ W_k \text{ kN-m/m}$$

Elastic design moment of resistance of cavity wall

$$= 1.37 \text{ kN-m/m}$$

$$\therefore \max W_k = 0.49 \text{ kN/m}^2 < 0.75 \text{ kN/m}^2$$

Hence greater flexural strength is required; brickwork reinforcement may be added in alternate bed joints above the lintel (Fig. 6.12) or a light support beam acting both as a lintel and as a lateral support should be provided. See Examples 10.2, 10.3 and A14.1.

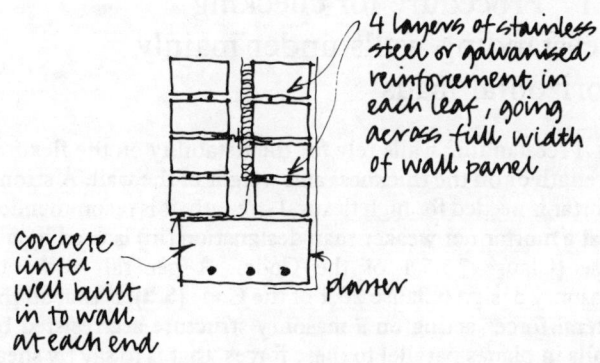

Fig. 6.12 Steel reinforcement in bed joints of masonry to increase flexural strength.

CHAPTER 7

Freestanding walls

Introduction

Most walls receive lateral support either from vertical elements, such as columns or walls spaced at intervals, or from horizontal elements, such as floors or roofs at the top and bottom of the wall. Freestanding walls, which may include retaining walls, have only one lateral support, that at the bottom, and must cantilever up from that level. In modern practice freestanding walls will generally rely on the moment of resistance of the wall at the bottom to resist overturning. However, the wall may also be designed as a mass wall which is sufficiently thick and heavy that it is able to resist rotation about one edge without the need for any significant tensile strength. In both cases it is advantageous if the section modulus of the wall, in plan, can be increased by staggering the wall or using piers, in effect forming Z- or T-section walls in plan. As freestanding walls are generally in exposed positions durable brick or block units and a strong mortar are necessary. In general, therefore, no dampproof course is necessary and hence there need be no plane of weakness at the bottom of the wall so that high flexural strength can be obtained here. Excerpts from the Code of Practice are included in this chapter but excerpts in Chapters 5 and 6 and Appendix A13 are relevant too. As elsewhere, this chapter is set out in the form of a series of instructions, as if for someone needing to assess a masonry building and undertake calculations. This format helps to separate the main points and provides a logical sequence in which the calculations for freestanding walls may be done. Examples are given at the end of the chapter.

7.1 Procedure for checking freestanding walls under mainly horizontal loads

(a) Freestanding walls rely for their stability on the flexural strength or on the thickness and weight of the wall. A strong mortar is needed for high flexural strength; it is recommended that a mortar not weaker than designation (iii) is used in any case (Clause 36.5.1 of the Code). A general axiom of masonry design (Clause 20.1 of the Code (**5.3**)) is that all the lateral forces acting on a masonry structure are resisted by walls in planes parallel to these forces, that is to say by shear walls. Freestanding walls will usually need piers, which may be considered to act like shear walls. However, for a low freestanding wall this may be unnecessary.

(b) Consider the horizontal loading which would cause the maximum tensile stresses.

(c) Limiting dimensions. Check that the limiting dimensions of the wall or wall panel are not exceeded (see **6.2** and Appendix A13). The wall panel is that portion of the wall between vertical or horizontal lateral supports (Fig. 7.1).

(d) Slenderness ratio. For walls with piers check the slenderness ratio of the pier alone or the slenderness ratio of the wall and pier acting together; see **5.1.1**c.

(e) Load combinations. Select Case (a), (b) or (c) from Clause 22 of the Code, whichever gives the worst combination of loads. For lateral load on a wall usually only Cases (b) and (c) will be relevant for freestanding wind walls and $0.9G_k + 1.4E_n$ for retaining walls. Almost invariably Case (b) with the combination $0.9G_k + 1.2W_k$ will give the worst case for freestanding wind walls; it is not necessary to consider the combination $0.9G_k + 1.4W_k$.

(f) Vertical loads. Work out the characteristic vertical loads on the wall, each leaf of the wall in the case of a cavity wall, keeping the dead, live and wind loads separate.

(g) Design moment of resistance. Work out the characteristic flexural strength of the wall at the level to be considered, usually the base. It will first be necessary to calculate the design vertical load at this level. The design moment of resistance may now be calculated and this should be greater than the design moment. The 'elastic' design moment of resistance of a freestanding wall (Clause 36.5.3 of the Code (**7.2**)) is equal to

$$(f_{kx}/\gamma_m + g_d)Z$$

and the 'plastic' design moment of resistance of a freestanding wall due to the vertical load in it is equal to

$$(n_w/2) . (t - n_w . \gamma_m/f_k).$$

The latter formula should be used for walls with little or no tensile strength; see **A13** for reinforced walls.

(h) For walls having piers, or walls that are staggered on plan — i.e. walls that behave as T-, I-, L- or Z-sections — the position of the neutral axis and the moment of inertia may quickly be calculated using Tables A3.1 and A3.2. For sections with a wide 'flange' and a deep 'web' the shear stresses along the lines connecting the 'flange' to the 'web' may need checking (**8.3**f). See Clause 25 of the Code.

(i) Design moment. The design moment on the wall should be calculated for the load combination selected. The design

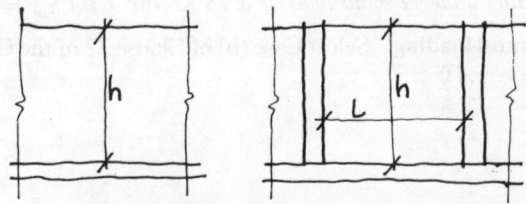

Fig. 7.1 Elevation on freestanding walls with and without piers showing relevant dimensions for use in Clause 36.3 of BS 5628: Part 1.

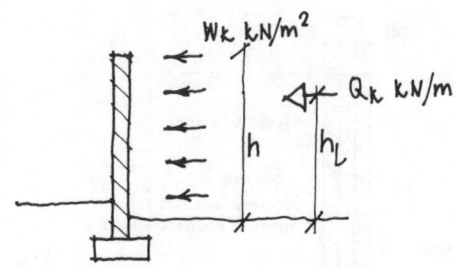

Fig. 7.2 *Forces acting on freestanding wall which cause bending moment.*

moment is of the general form:

$$W_k \cdot \gamma_f \cdot h^2/2 + Q_k \cdot \gamma_f \cdot h_L$$

(Clause 36.5.2 of the Code); see Fig. 7.2.

(j) For walls with piers check the ability of the wall panel to span between the piers; see **6.1**.

(k) Shear strength. Shear should be checked at the bottom of the wall; see **6.1**j and examples in Chapter 6.

7.2 Extracts from BS 5628:Part 1

36.5.3 Calculation of design moment of resistance of freestanding walls The design moment of resistance across the bed joints is given by

$$\left(\frac{f_{kx}}{\gamma_m} + g_d \right) \times Z$$

where

f_{kx} is the characteristic flexural strength at the critical section, which may be the damp-proof course (clause 24);

γ_m is the partial safety factor for materials (clause 27);

Z is the section modulus, which may take into account any variation on the plan arrangement, e.g. chevron, curved or zig-zag walls (in the case of walls with piers, see 36.4.3);

g_d is the design vertical dead load per unit area.

In cases where the flexural strength of the masonry cannot be relied upon (see 24.1) a freestanding wall can only be used when there is sufficient vertical load acting. The design moment of resistance per unit length may then be assessed by assuming that the vertical load is resisted by a rectangular stress block when the moment in a wall of uniform thickness is:

$$\frac{n_w}{2} \left[t - \frac{n_w \gamma_m}{f_k} \right]$$

where

t is the thickness of the wall;

n_w is the design vertical load per unit length of wall;

f_k is the characteristic compressive strength of masonry (clause 23).

. . . . . .

7.3 Examples

Example 7.1 Freestanding wall

A freestanding wall is built with masonry having the same strengths and factors of safety as Example 5.2 (Fig. 7.3).

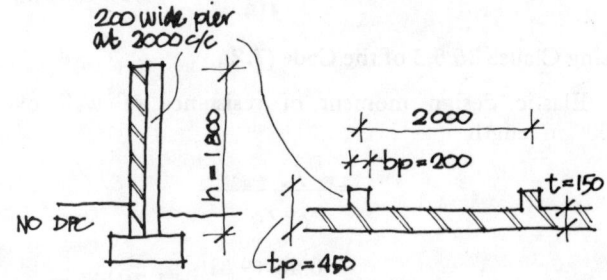

Fig. 7.3 *Section and plan on wall.*

Characteristic loads:

Vertical load from self-weight of wall, at bottom, excluding weight of piers

    Dead load, G_k = 4.00 kN/m

Horizontal load

    Wind load, W_k = 0.4 kN/m²

Check horizontal loading.

Horizontal loading Effective thickness of wall, t_{ef}, from Clause 28.4.2 of the Code (**5.3**),

                = 1.4 × 150   = 210 mm

Limiting dimensions of wall panel from Clause 36.3 of the Code (**6.2**) are:

    height $12t_{ef}$ = 12 × 0.210   = 2.520 m

    cf. actual height   = 1.800 m

                                         OK

Select Case (b) of Clause 22 of the Code: $0.9G_k + 1.2W_k$

At bottom of wall,

    Design compressive stress in wall, g_d

$$= \frac{0.9 \times 4.00 \times 10^3}{1,000 \times 150}$$

                           = 0.024 N/mm²

$\therefore$ Effective f_{kx} *(parallel to bed joints)* = f_{ka}

                 = 0.40 + 2.5 × 0.024

                 = 0.46 N/mm²

from Clause 36.4.2 of the Code (**6.2**).

Effective height of wall, h_{ef} = 2.0 × 1.800 = 3.600 m

From App. **A3** $b = \dfrac{h_{ef}}{3}$ = 1.200 m

    for $\dfrac{b}{b_p} = 6, \dfrac{t_p}{t} = 3, y_2 = 0.292t_p$ = 131 mm

               $y_1 = 450 - 131$ = 319 mm

                    $k = 2.019$

$$\therefore I_x = \frac{2.019 \times 200 \times (450)^3}{12}$$

$$= 3{,}066 \times 10^6 \text{ mm}^4$$

Minimum section modulus, Z_{min}

$$= \frac{3{,}066 \times 10^6}{319} = 9.61 \times 10^6 \text{ mm}^3$$

Using Clause 36.5.3 of the Code (**7.2**),

Elastic design moment of resistance of wall over 2.000 m length

$$= \frac{f_{ka} \cdot Z_{min}}{\gamma_m}$$

$$= \frac{0.46 \times 9.61}{2.5} = 1.76 \text{ kN-m}$$

Using Clause 36.5.2 of the Code,

Design moment in wall per 2.000 m length, at bottom,

$$= W_k \cdot \gamma_f \cdot \frac{h^2}{2}$$

$$= \frac{0.40 \times 1.2 \times (1.800)^2}{2} \times 2.000$$

$$= 1.56 \text{ kN-m}$$
$$\text{OK}$$

The 150-mm-thick wall will span horizontally between the piers; its strength does not need to be checked. This calculation is conservative in that, to a degree depending on the proportions, the wall between the piers also carries horizontal load to ground.

Example 7.2 Grouted cavity retaining wall

A grouted brickwork cavity wall is to be built to retain earth (Fig. 7.4); take $\gamma_{mm} = 2.0$ *and* $f_k = 11.4$ N/mm² *(Table 2.2). Check.*

A retaining wall of this height requires reinforcement and a structural mesh is provided in the cavity. See Appendix A13 for information on vertically reinforced walls.

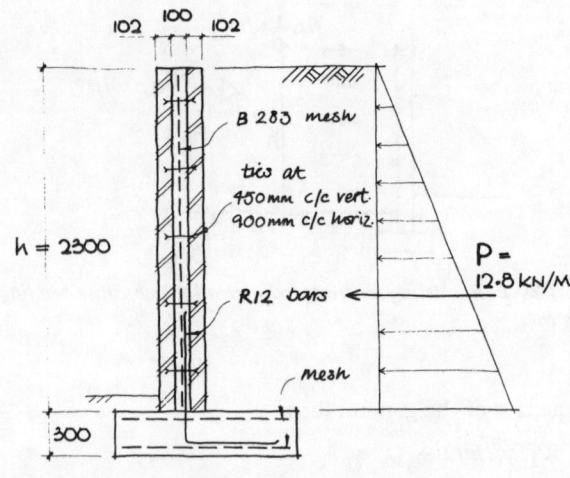

Fig. 7.4 *Section on wall with earth pressure diagram.*

Span to effective depth ratio $= (2\,300 + 152/2)/152 = 15.6$

cf. allowable $= 18$ (Table A13.1)

OK

Design vertical load $< 0.1 f_k . A_m$

OK

Min area of secondary steel

$= (0.05/100) \times 152 \times 10^3 = 76$ mm²/m

Using B283 mesh, lever arm given by (**A13.2.1**)

$$z = \left(1 - \frac{0.5 \times 283 \times 485 \times 2.0}{10^3 \times 152 \times 11.4 \times 1.15}\right) d = 0.93\,d$$

$$\therefore M_d = 283 \times 485 \times 0.93 \times 152 \times 10^{-6}/1.15$$
$$= 16.8 \text{ kN-m/m}$$

In this case it is not necessary to check bending strength in compression.

Selecting the Case (b) load combination,

Design moment $= 1.4 \times 12.8 \times 2.30/3 = 13.8$ kN-m/m

OK

Design shear stress $= 18/152 = 0.12$ N/mm²

cf. allowable shear stress, ignoring enhancements,

$f_v/\gamma_{mv} = 0.35/2.0 = 0.17$ N/mm²

OK

CHAPTER 8

Lateral stability

Introduction

Lateral stability is perhaps the single most important topic in the design of masonry buildings. In this chapter the subject is split up into two parts, that concerned with ensuring there is lateral support to the individual elements of masonry and that concerned with the overall stability of the building against horizontal forces. In both cases attention focuses on the connections which are often between completely dissimilar materials but are still required to be strong and reasonably rigid too. Many building failures have originated at the connection points. For the overall stability of a masonry building some walls must act as shear walls and these walls will be more effective if they also carry axial vertical load. In every case the building should be looked at as a whole and ways devised of taking the horizontal forces to ground. The first part of this chapter consists of a general discussion on lateral stability; further discussion is given in Chapter 11, especially that concerned with lateral stability after accidental or even deliberate damage has occurred to the building. As elsewhere, this chapter is set out as a series of instructions, as if for someone needing to assess a masonry building and undertake calculations. The format helps to separate the main points and provides a logical sequence in which the calculations for shear walls may be done. Examples are given at the end of the chapter.

8.1 Stability of individual masonry elements

(a) Overall stability and stability of individual elements. By inspection of the drawings, or by calculation if necessary, decide which walls are the critical ones to consider when horizontal forces are applied to the building. There are two main points to examine:

(i) The stability of individual masonry elements under direct horizontal forces, as explained in Chapters 5, 6 and 7.

(ii) The stability of the whole building under lateral forces which is assured by connecting all the different elements of the building together so as to form one or several stable units. In each unit the walls which in plan are parallel to the direction of the horizontal forces and which resist these forces are known as shear walls (Figs 8.1 and 8.6.)

(b) Stability of individual elements. An ideal structural arrangement for masonry buildings is that they be laid in plan as a cellular construction in which each wall spans horizontally (Table 2.3) and is stabilised by other walls bonded at right angles to it (Fig. 8.1). A wall is very resistant to forces in its own plane. If there are no return walls then piers may be specially built. Clearly the return walls, or piers, must be large enough to take the horizontal forces put

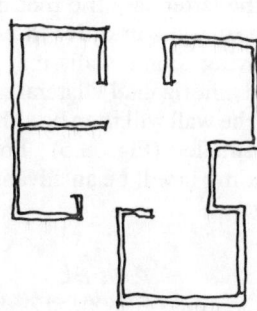

Fig. 8.1 Plan on building of cellular construction in which, generally, slenderness ratio of individual walls = L_{ef}/t_{ef}.

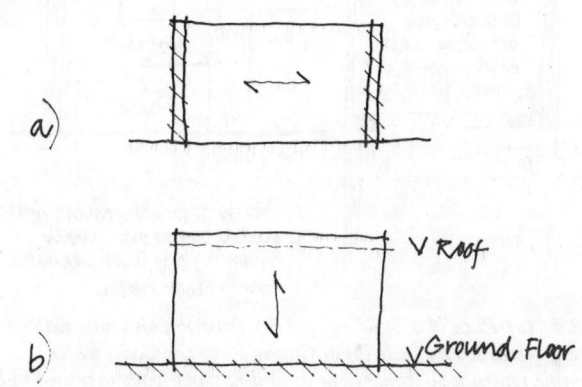

Fig. 8.2 Elevation on individual walls stabilised (a) by return walls or piers for which slenderness ratio = L_{ef}/t_{ef} and (b) by roof and floor slabs for which slenderness ratio = h_{ef}/t_{ef}.

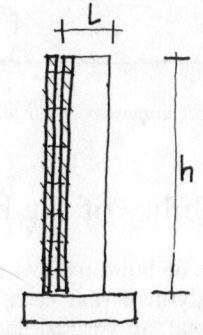

Fig. 8.3 Cavity wall having piers with h/L > 5.

on them (Fig. 8.2) and in low-rise buildings such return walls would need to be, as an order of magnitude, about 500 mm long. Return walls or piers which have the ratio, height divided by overall depth, greater than about five may be designed as walls with piers by the methods suggested in Chapters 5, 6 and 7 (Fig. 8.3). Walls or piers with a lower value of this ratio will usually be designed as shear walls by the methods suggested in **8.3**; see also App. **A12**.

Minimum horizontal load The design horizontal forces are given by Clause 22 of the Code, as stated in **5.1.2**c and **7.1**e. However, as noted in **5.1.2**d, a building should be able to resist a minimum design horizontal force equal to 1.5% of the characteristic dead load above any level (Clause 20.1 of the Code (**5.3**)). The importance of this provision is that it gives a basis for determining those minimum plan dimensions which ensure that a building, or part of a building, is stable. The provision may be relevant even where the design horizontal wind force is greater, overall, than the minimum horizontal force, for example because of the difficulty of providing adequate connections (Fig. 8.4). A wall may also be stabilised by connection to a floor or roof at its top and bottom (Fig. 8.2). In the latter case the roof must be properly connected to the shear walls in order that the roof itself is prevented from moving. Some walls may be stabilised by both vertical and horizontal lateral supports. The slenderness ratio of the wall will then be either L_{ef}/t_{ef} or h_{ef}/t_{ef} whichever is the smaller (Fig. 8.5). For a wall taking horizontal wind loading it will be an advantage to have support on all four edges.

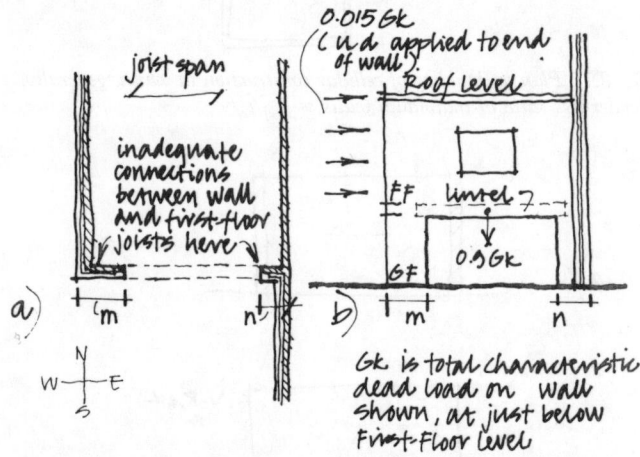

Fig. 8.4 (a) Plan of a building and (b) elevation on south wall showing design loads (Case (b) of Clause 22 of the Code) which determine return wall dimensions m and n; wind loads assumed to be taken by floor and roof constructions; connections between south wall and first floor assumed to have inadequate stiffness or strength.

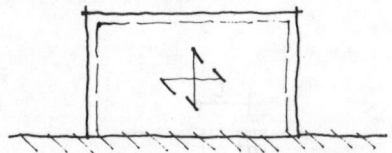

Fig. 8.5 Elevation of wall supported on all four edges.

8.2 Overall stability of the building

(a) Overall resistance of building to horizontal forces. In addition to the requirement that each individual wall be stabilised by horizontal or vertical lateral supports, the building acting as a whole must be able to resist the lateral forces on it. The horizontal forces on a building are resisted by walls in planes parallel to those forces, acting as shear walls (Clause 20 of the Code); see Fig. 8.6. Hence buildings such as those shown in Fig. 8.7 are inadmissible even if the horizontal wind forces are small and the walls are able to develop adequate flexural strength to resist them. For example, wind forces in a north–south direction that are resisted by a masonry building must eventually be carried to ground by walls going in this direction, and wind forces in an

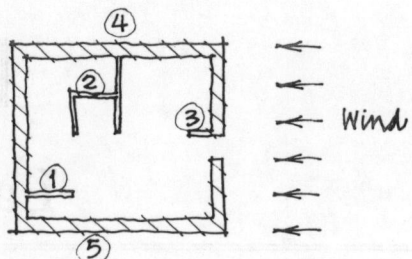

Fig. 8.6 Plan on building with wind forces taken by walls 1 to 5.

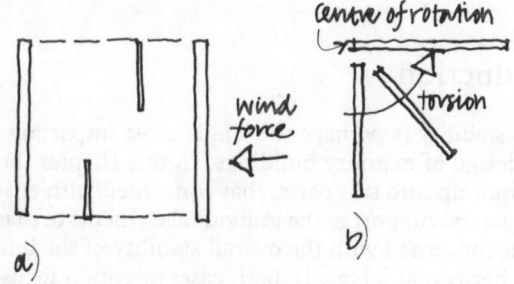

Fig. 8.7 Plans on buildings with wall arrangements unable to resist (a) lateral force or (b) torsion.

east–west direction by walls going in that direction (Clauses 20 and 22 of the Code).

(b) Structural model of overall behaviour of building under horizontal load. The structural model of the behaviour of the complete building under lateral load may be represented as either:

(i) A horizontal beam, representing the roof or floor system, on unyielding points of support, if the roof or floor is relatively flexible in a horizontal direction compared with the shear walls – model A or

(ii) An infinitely stiff beam, representing the roof or floor system, on flexible supports, if the roof or floor is relatively stiff in a horizontal direction compared with the shear walls – model B (Fig. 8.8) or some combination of the two.

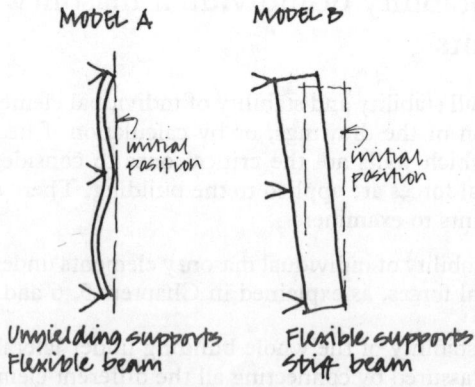

Fig. 8.8 Schematic plan showing model A with unyielding supports and flexible beam and model B with flexible supports and stiff beam.

Floors or roofs acting as horizontal diaphragms behave like deep beams. These elements or other horizontal diaphragms can be used to provide lateral support to walls. However, diaphragms that support masonry walls need to be stiff, otherwise cracks may develop in the walls as they move with the diaphragm. In practice, therefore, model B is the more important of the two cases. A reinforced concrete slab makes a stiff horizontal diaphragm and can also make a good connection with the walls. In other cases the horizontal deflection may be limited by specifying a maximum span to width ratio for the diaphragm – i.e. the maximum horizontal span of the diaphragm between shear walls divided by the overall width of the diaphragm in plan. For metal deck roofs made up of continuous sheets on steel joists, this ratio should be no more than about four, and about half this value for roofs with overlapping, discontinuous sheets. For plywood sheets nailed on all edges to timber joists and to blocking pieces between the joists, the ratio should not exceed four. For timber constructions using timber cross-bracing instead of sheet materials, the ratio should not exceed two or three. These are only approximate figures. An estimate of the deflection of a horizontal diaphragm may be made but this is not usually worthwhile because of the uncertainty concerning the point at which damage or visible cracking starts to occur in the wall.

(c) Torsion and the shear centre of a building. Under lateral forces a building may undergo a twisting movement as well as a lateral displacement. This could be caused by an uneven distribution of wind pressure with a symmetrical arrangement of shear walls, for example, or by an unsymmetrical arrangement of the shear walls with an even distribution of wind pressure, or by some combination of these two conditions (Fig. 8.9). If model A is the correct one then no forces due to torsion would develop on the shear walls. However, if model B is the correct one, it will be necessary to find the shear centre of the building, in order to calculate these forces due to torsion. The shear centre is that point through which the resultant of all the horizontal forces must act in order that there is no twisting, only lateral displacement, under the horizontal force.

Looking at Fig. 8.10 and taking moments about a reference

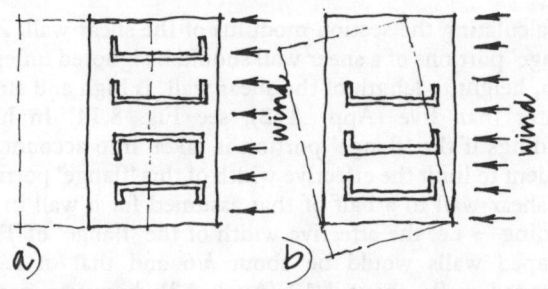

Fig. 8.9 Plans on building under an evenly distributed lateral force with (a) lateral displacement only and (b) with lateral displacement and twist caused by an unsymmetrical arrangement of shear walls.

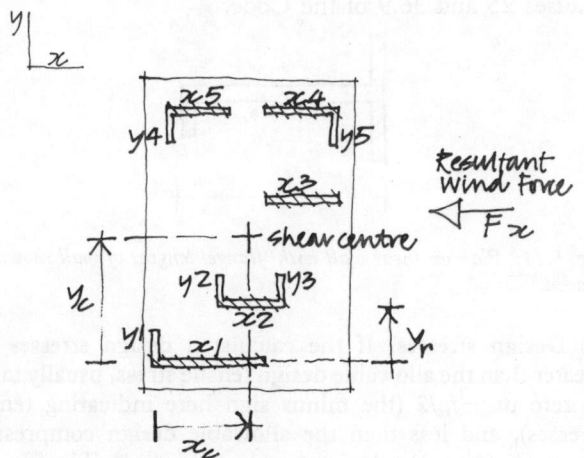

Fig. 8.10 Plan on building showing shear walls, of equal height, and position of shear centre.

point, for example the centre of shear wall x1, and assuming y_2, y_3, etc. to be the distances from this point to the centres of the straight walls x2, x3, etc. which line up with the x-direction:

$$y_c = (I_{x1} \cdot y_1 + I_{x2} \cdot y_2 + \ldots)/(I_{x1} + I_{x2} + \ldots)$$
$$= \Sigma(I_{xr} \cdot y_r)/\Sigma(I_{xr})$$

where y_c is the distance from the reference point to the shear centre in the y-direction, y_r is the distance from the reference point to the centre of any straight wall in the x-direction and I_{xr} is the moment of inertia of each straight length of wall about its x-x axis, ignoring the 'flanges' attached to it if necessary (Fig. 8.14). The procedure is repeated for shear walls y1 to y5 which are at right angles, in order to find x_c, the distance from the reference point to the shear centre in the x-direction. Thus the shear centre is defined by the dimensions y_c and x_c from a reference point which in this example is the intersection of the centre lines of walls x1 and y1.

(d) Forces in each wall due to horizontal loads. The resultant of all wind forces acts at the centre of the building unit, for an evenly distributed wind force. In general, however, the shear centre of the building does not coincide with the centre point of the building. Hence with a wind force at right angles to the long side – i.e. in the x-direction – there

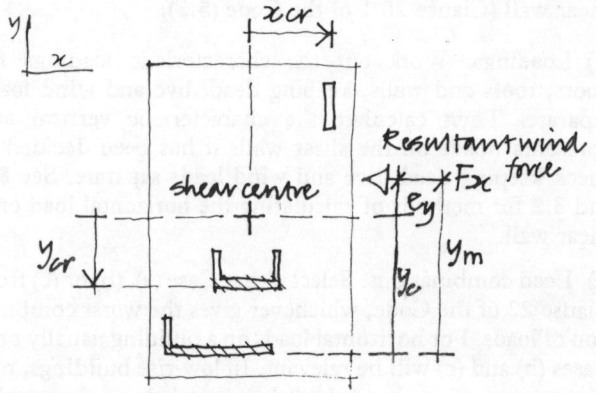

Fig. 8.11 Plan on building showing shear centre and position of resultant wind force on long side of building.

is a twisting moment on the building (Fig. 8.11) equal to

$$F_x \cdot e_y = F_x(y_m - y_c)$$

and the load resisted by each wall in the x-direction, f_{xr}, is as follows:

$$f_{xr} = F_x \cdot I_{xr}/(I_{x1} + I_{x2} + \ldots)$$
$$+ F_x \cdot e_y \cdot I_{xr} \cdot y_{cr}/(I_{x1} \cdot y_{c1}{}^2 + I_{x2} \cdot y_{c2}{}^2 + \ldots$$
$$+ I_{y1} \cdot x_{c1}{}^2 + I_{y2} \cdot x_{c2}{}^2 + \ldots)$$
$$= F_x \cdot I_{xr}/\Sigma I_{xr} + F_x \cdot e_y \cdot I_{xr} \cdot y_{cr}/\Sigma(I_{xr} \cdot y_{cr}{}^2 + I_{yr} \cdot x_{cr}{}^2)$$

where I_{xr} and I_{yr} are the moments of inertia about the major axis of bending for those walls in the x- and y-directions respectively (**A1.2**), x_{cr} and y_{cr} are the distances from the shear centre to each wall in the x- and y-directions respectively and F_x is the total design wind load on the building unit in the x-direction. The term $\Sigma(I_{xr} \cdot y_{cr}{}^2 + I_{yr} \cdot x_{cr}{}^2)$ is known as the torsional rigidity. The load developed in each wall is made up of that due to pure lateral displacement and that due to twist (Fig. 8.9). The term in the formula due to twist is the only one there is for walls running in the y-direction when the wind blows in the x-direction.

Similarly if the wind blows in the y-direction, the load resisted by each wall in the y-direction, f_{yr}, is as follows:

$$f_{yr} = F_y . I_{yr}/\Sigma I_{yr} + F_y . e_x . I_{yr} . x_{cr}/\Sigma(I_{xr} . y_{cr}^2 + I_{yr} . x_{cr}^2)$$

where F_y is the total design wind load on the building unit in the y-direction and e_x is the distance between the resultant of this wind load and the shear centre.

(e) The calculation of the horizontal forces on each shear wall in **8.2**d assumes that all the shear walls are solid and of equal height; these assumptions are adequate for most practical calculations. For other information on stiffness deflections and the treatment of shear walls with openings see App. **A12**.

8.3 Procedure for checking shear walls under mainly horizontal loads

(a) By inspection of the drawings, or calculation if necessary, decide which part of any of the walls in the building is most critical and at what floor level.

(b) Horizontal loading on individual shear walls. Consider the horizontal loading which would cause the maximum tensile or compressive stresses.

(c) Slenderness ratio. Check the slenderness ratio of the shear wall (Clause 28.1 of the Code (**5.3**)).

(d) Loadings. Work out the characteristic loadings for floors, roofs and walls, keeping dead, live and wind loads separate. Then calculate the characteristic vertical and horizontal loads on the shear walls it has been decided to check keeping dead, live and wind loads separate. See **8.1** and **8.2** for methods of calculating the horizontal load on a shear wall.

(e) Load combinations. Select either Case (a), (b) or (c) from Clause 22 of the Code, whichever gives the worst combination of loads. For horizontal loads on a building usually only Cases (b) and (c) will be relevant. In low-rise buildings, tensile stresses are more likely to be critical than compressive stresses so that almost invariably Case (b) with the combination $0.9G_k + 1.4W_k$ will give the worst case. For tall buildings Case (c) and Case (b) with the combination $1.4G_k + 1.4W_k$ would be examined too.

(f) Design forces. For the load combination selected, calculate the design axial force, the design moment and hence the resultant eccentricity of the load at the critical level, usually the bottom of the wall. This eccentricity is that about the major axis, e (Fig. 8.12). If this eccentricity is less than $L/6$, the maximum design compressive stress can be calculated using the standard elastic bending formula in which the maximum design compressive stress is given by (Fig. 8.13):

$$P/A + M/Z$$
$$= P/A(1 + 6e/L)$$

where P and M are the design axial force and bending moment respectively. If this eccentricity is greater than $L/6$, the wall may still be acceptable but then either tensile stresses would develop or the wall would crack, if the design loads were to be applied. The design compressive stresses must be checked using the formula appropriate for the cracked section; the maximum design compressive stress is given by (Figs 8.12 and 8.13):

$$2P/[3t(L/2 - e)]; \text{ see App. } \textbf{A1.5}.$$

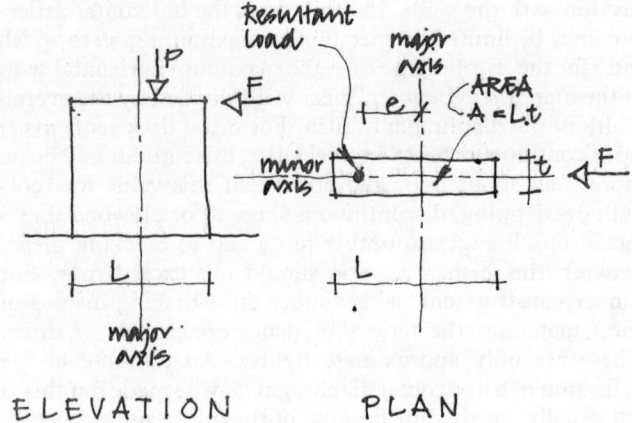

ELEVATION   PLAN

Fig. 8.12   Elevation and plan of typical shear wall showing position of resultant vertical load in plan, at bottom of wall.

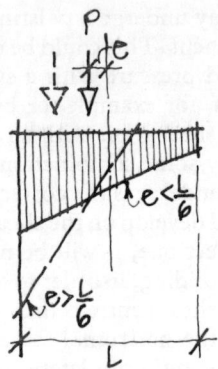

Fig. 8.13   Distribution of vertical stress in shear wall with $e < L/6$ and with $e > L/6$.

In calculating the section modulus of the shear wall, Z, the 'flange' portions of a shear wall should be ignored unless the ratio, height to length of the shear wall, is high and anyway greater than five (App. **A12**); see Fig. 8.14. In higher buildings if the 'flange' portion is taken into account, it is prudent to limit the effective width of this 'flange' portion of the shear wall to a half of that assumed for a wall in local bending – i.e. the effective width of the 'flange' of T- and I-shaped walls would be about $h/6$ and that of L- and Z-shaped walls about $h/12$ (App. **A3**); however, see also Hendry (1990, p. 116). Shear stresses along the line connecting the 'flange' to the 'web' should be checked if the 'flange' portion exceeds about 40% of the length of the 'web'. In low-rise building the 'flange' portion is best ignored. See Clauses 25 and 36.9 of the Code.

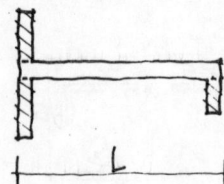

Fig. 8.14   Plan on shear wall with 'flange' lengths of wall shown shaded.

(g) Design stresses. If the calculated design stresses are greater than the allowable design tensile stress, usually taken as zero or $-f_{ka}/2$ (the minus sign here indicating tensile stresses), and less than the allowable design compressive stress, $(\beta . f_k)/\gamma_m$, the design may be acceptable. The factor β is taken from Table 5.4 where the slenderness ratio used is based on the effective thickness and height of the wall. In practice it is often nearly equal to one. It is not satisfactory to have a very eccentric load which may cause excessive cracking and this should be considered too. A check on whether cracking is likely in service may be made by recalculating the eccentricity using characteristic loads instead of design loads. The design is usually satisfactory if this eccentricity is something less than $L/6$. Because shear walls may also function as load-bearing walls, they may have eccentricities about the minor as well as the major axis of bending. Shear walls of short length may be treated as in bi-axial bending in a similar manner to columns in bi-axial bending (**5.2.1e**). However, for longer shear walls the bending stresses due to horizontal forces can be considered as being caused by extra axial loads at either end of the wall so that there will only be an eccentricity about the minor axis (**12.7**).

(h) **Shear strength.** The design shear strength of the masonry, f_v/γ_{mv}, should be calculated and this should be greater than the design shear stress, v_h, calculated for the load case from Clause 22 of the Code which gives the highest value of design shear stress; see **5.1.2h**. Note that shear is more likely to be critical than bending in shear walls which have 'flanges' taking part in the bending action.

(i) **Connections.** In practice most problems arise with connections, which are essential in order to ensure the stability of the building. The connections must be designed for the most critical of the load combinations given in Clause 22 of the Code. For example, a roof connection to a shear wall may have to take a large horizontal design wind load as well as a design vertical uplift load due to wind. In general the connections will need to be designed. For example, a steel strap taking a design vertical uplift load of $1.4W_k - 0.9G_k$ must have a design strength greater than this; the design strength of a steel strap is $f_y . A/\gamma_m$ where f_y is the characteristic strength of the steel which is equal to about 250 N/mm^2 for mild steel, A is the area of the steel and γ_m is the partial safety factor for strength of the steel which is equal to 1.15 (Appendix C of the Code). More information about connections is given in App. **B4** and **B5**.

8.4 Examples

Example 8.1 Shear walls in two-storey building

A two-storey masonry building built with concrete slabs at first-floor and roof level is open at ground-floor level. The shear walls at ground-floor level take all the horizontal wind loads. Calculate the horizontal loads on the shear walls just below first-floor slab level, when wind blows on the long side of the building. The shear walls have overall lengths of x1, x2, ... *and* y1, y2, ... *in the* x *and* y *directions respectively.*

Horizontal loading

Characteristic loads:

Wind load, $W_k = 0.85$ kN/m^2

Shear centre of building (Fig. 8.15):

Model B is the relevant one in this case (**8.2**). To find y co-ordinates of shear centre, take moments about centre line of south wall:

Moment of inertia of 4 m length of wall in x-direction,

$$I_{x1} = \frac{0.150 \times (4)^3}{12} = 0.800 \text{ m}^4$$

and Moment of inertia of 3 m length of wall in x-

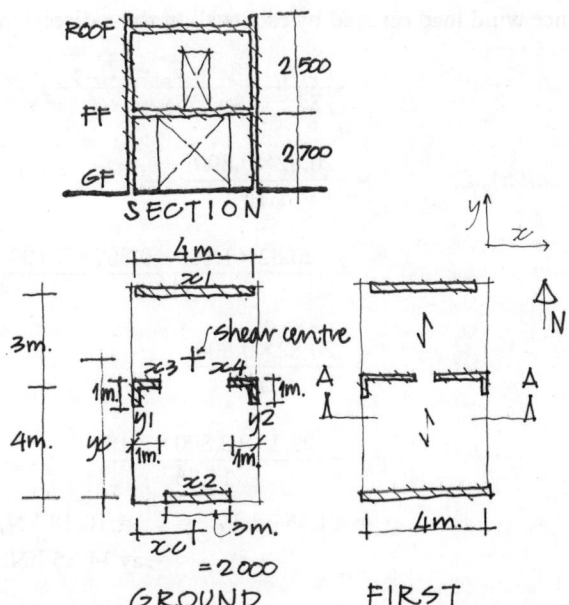

Fig. 8.15 Section, ground and first-floor plans of building, showing crosswall construction.

direction,

$$I_{x2} = \frac{0.150 \times (3)^3}{12} = 0.337 \text{ m}^4$$

and Moment of inertia of 1 m length of wall in x-direction,

$$I_{x3} = I_{x4} = \frac{0.150 \times (1)^3}{12} = 0.012 \text{ m}^4$$

Hence $y_c = \dfrac{(0.800 \times 7) + (2 \times 0.012 \times 4)}{0.800 + (2 \times 0.012) + 0.337}$

$$= 4.900 \text{ m}$$

Moment of inertia of 1 m length of wall in y-direction

$$I_{y1} = I_{y2} = \frac{0.150 \times (1)^3}{12} = 0.012 \text{ m}^4$$

by symmetry

$$x_c = \frac{4.000}{2} = 2.000 \text{ m}$$

Consider Case (b) of Clause 22 of the Code (**5.3**): $0.9G_k + 1.4W_k$

Total design wind load on top storey of building, F_x

$$= 1.4 \times 7.000 \times 2.500 \times 0.85$$
$$= 20.82 \text{ kN}$$

This acts at 3.500 m from centre line of south wall and, hence, has an eccentricity, e_y, with respect to the shear centre of 1.400 m.

$$\Sigma I_{xr} = 0.800 + (2 \times 0.012) + 0.337$$
$$= 1.16 \text{ m}^4$$

Torsional rigidity, $\Sigma(I_{xr} . y_{cr}^2 + I_{yr} . x_{cr}^2)$

$$= 0.800 \times (2.100)^2 + 2 \times 0.012 \times (0.900)^2$$
$$+ 0.337 \times (4.900)^2 + 2 \times 0.012 (2.000)^2 = 11.73 \text{ m}^6$$

61

Hence wind load resisted by each wall in the *x*-direction, f_{xr}

$$= \frac{F_x \cdot I_{xr}}{\Sigma I_{xr}} + \frac{F_x \cdot e_y \cdot I_{xr} \cdot y_{cr}}{\Sigma(I_{xr} \cdot y_{cr}^2 + I_{yr} \cdot x_{cr}^2)}$$

for wall x1, f_{xr}

$$= \frac{20.82 \times 0.800}{1.16}$$

$$- \frac{20.82 \times 1.400 \times 0.800 \times 2.100}{11.73}$$

$$= \frac{20.82 \times 0.800}{1.16}$$

$$- \frac{29.15 \times 0.800 \times 2.100}{11.73}$$

$$= 14.35 - 4.17 \qquad = 10.18 \text{ kN},$$
$$\text{say } 14.35 \text{ kN}$$

for walls x3 *and* x4, $f_{xr} = \dfrac{20.82 \times 0.012}{1.16}$

$$+ \frac{29.15 \times 0.012 \times 0.900}{11.73}$$

$$= 0.21 + 0.02 \qquad = 0.23 \text{ kN}$$

for wall x2, f_{xr}

$$= \frac{20.82}{1.16} \times 0.337$$

$$+ \frac{29.15 \times 0.337 \times 4.900}{11.73}$$

$$= 6.05 + 4.10 \qquad = 10.15 \text{ kN}$$

and wind load resisted by each wall in the *y*-direction, f_{yr}

$$= 0 + F_x \cdot e_y \cdot \frac{I_{yr} \cdot x_{cr}}{\Sigma(I_{xr} \cdot y_{cr}^2 + I_{yr} \cdot x_{cr}^2)}$$

for walls y1 *and* y2, $f_{yr} = \dfrac{29.15 \times 0.012 \times 2.000}{11.73}$

$$= 0.06 \text{ kN for}$$
$$\text{each of walls y1 and y2}$$

Shear should also be checked.

Example 8.2 Return walls in crosswall construction

A load-bearing masonry building of crosswall construction contains four shop units on the ground floor and accommodation on the first floor (Fig. 8.16).

The first floor is of timber construction and spans between the crosswalls. It is necessary to establish suitable minimum dimensions L_1 and L_2 for the wall returns on the south face of the building (Fig. 8.16). There is a continuous reinforced concrete lintel at first-floor level on the south wall. The total dead load of the building is 1,800 kN and the maximum wind load on the building is 0.75 kN/m². The dead load is 35 kN on the two outer return walls and double this on the inner ones.

Horizontal loading The plan of the building may be simplified to that shown in Fig. 8.17.

This building can accept large horizontal forces from the north–south direction; however, a horizontal force in the east–west direction could cause excessive movement and instability on the south face of the building.

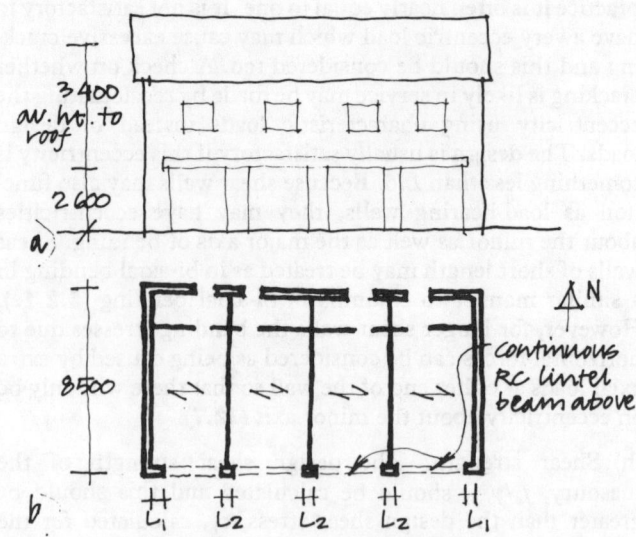

Fig. 8.16 (a) Elevation on south face and (b) ground-floor plan of building showing direction of span of first-floor timber joists and internal leaves of return walls having lengths L_1 and L_2.

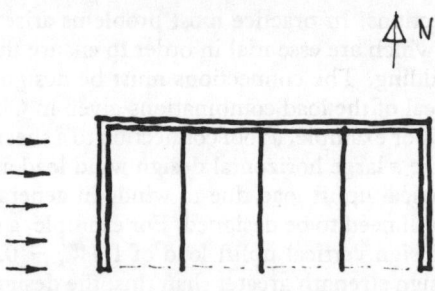

Fig. 8.17 Simplified ground floor plan of the building with a horizontal force from the east–west direction.

A floor structure with in-plane stiffness and rigid connections to all the wall, such as could be provided by an *in situ* concrete floor slab, may be satisfactory for a building of only moderate depth assuming the small twisting movement which would occur is acceptable (model B in **8.2**). However, with timber joists and boarding, the floor structure has insufficient in-plane stiffness and insufficiently good connections with the walls for this model to be satisfactory; the return walls on the south face should therefore be designed to resist half the horizontal forces on the building from the east–west direction (model A in **8.2**).

The design horizontal force on the south wall, F, acting at first-floor level, is resisted by the inner leaf of each wall return in proportion to its moment of inertia which, in turn, is proportional to the cube of the return wall length, L_1 or L_2 (Fig. 8.18). The walls all have equal thickness.

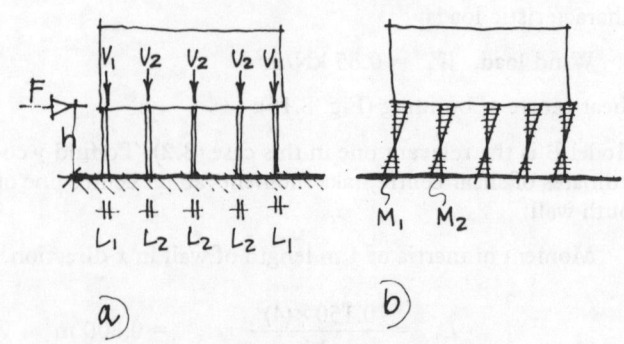

Fig. 8.18 Diagrams (a) of load and (b) of bending moment in return walls.

Hence if the design horizontal force taken by an outer and inner return wall is F_1 and F_2 respectively, then

$$\frac{F_1}{F_2} = \left(\frac{L_1}{L_2}\right)^3 = \frac{0.5F_1 \cdot h}{0.5F_2 \cdot h} = \frac{M_1}{M_2}$$

where M_1 and M_2 are the moments at the bottom of the outer and inner return walls respectively (Fig. 8.18).

The eccentricity, e, of the vertical load at the bottom of a return wall is given by $e = M/V$, where V is the design vertical load in each return wall (Fig. 8.18). If the eccentricities are limited to a sixth of the length of a return wall then

$$e_1 = \frac{M_1}{V_1} = \frac{L_1}{6}$$

and $e_2 = \dfrac{M_2}{V_2} = \dfrac{L_2}{6}$

so that $\dfrac{L_1}{L_2} = \dfrac{M_1}{M_2} \cdot \dfrac{V_2}{V_1} = 2\left(\dfrac{L_1}{L_2}\right)^3$

and therefore $L_2 = \sqrt{2} \cdot L_1$

Hence $F_1 = \dfrac{F \cdot L_1{}^3}{2L_1{}^3 + 3L_2{}^3} = \dfrac{F}{2 + 6\sqrt{2}} = 0.095\,F$

and $F_2 = \dfrac{F \cdot L_2{}^3}{2L_1{}^3 + 3L_2{}^3} = \dfrac{2\sqrt{2}F}{2 + 6\sqrt{2}} = 0.270\,F$

Consider Case (b) of Clause 22 of the Code: $0.9G_k + 1.4W_k$: the total design wind load on the building, from the

east–west direction, is equal to 53 kN, being greater than $0.015G_k$, equal to 27 kN.

Hence F $= 1.4 \times (3.4 + 1.3) \times \dfrac{8.5}{2} \times 0.75$ $= 21$ kN

and $F_2 = 0.270 \times 21$ $= 5.67$ kN

$M_2 = 5.67 \times \dfrac{2.600}{2}$ $= 7.38$ kN-m

$V_2 = 0.9 \times 70$ $= 63$ kN

$\therefore \quad \dfrac{L_2}{6} = \dfrac{7.38}{63}$ i.e. $L_2 = 702$ mm, say $= 700$ mm

and $L_1 = 497$ mm, say $= 500$ mm

Shear Check shear stress in return wall of length, L_2 with same load case.

Design shear strength for mortar designation (iii), $\dfrac{f_v}{\gamma_{mv}}$

$= \dfrac{0.15 + 0.6 \times 0.9}{2.5} = 0.27$ N/mm^2

Design shear stress, v_h

$= \dfrac{5.67 \times 10^3}{700 \times 100}$ $= 0.08$ N/mm^2

OK

The design vertical load resistance of the return walls should also be checked.

CHAPTER 9

Concentrated loads

Introduction

An important aspect of the detailed design of masonry is that of concentrated loads on walls. In these cases heavy loads, usually from beams or lintels, bear on to a small area of wall and cause high compressive stresses to occur there. This chapter is concerned with the design of those areas under the action of a concentrated load. Higher stresses are allowed on local areas than would be generally acceptable in the wall, if the concentrated load is able to disperse itself rapidly at lower levels. The increased stresses allowed depend on the details adopted; allowable stresses are given in App. **A8**. Concentrations of load occur quite commonly in general practice and are particularly likely to occur after a building has been damaged. The ability of a wall to accept a local concentrated load may be vital for the stability of the whole building. There is discussion on accidental damage in Chapter 11. As elsewhere, this chapter is set out in the form of a series of instructions, as if for someone needing to assess a masonry building and undertake calculations. This format helps to separate the main points and provides a logical sequence in which calculations for concentrated loads may be done. Examples are provided at the end of the chapter.

9.1 Procedure for checking walls under concentrated loads

(a) Places to check. A wall with a concentrated load should be checked at two places (Clause 34 of the Code):

(i) immediately under the bearing,
(ii) 0.4h below the bearing, where h is the clear height of the wall.

(b) Consider the vertical loading which would cause the maximum compressive stress.

(c) Work out the characteristic vertical loads:

(i) at the bearing on the wall, assuming the load to be uniformly distributed over the bearing area;
(ii) 0.4h below the bearing, assuming the concentrated load to be uniformly distributed at any level over an area contained within lines extending downwards from the edge of the loaded area at 45° to the horizontal. Add in any other loadings, for example uniformly distributed (ud) loadings which are present at this level too (Fig. 9.1).

Keep dead, live and wind loads separate.

(d) Load combinations. Select either Case (a), (b) or (c) from Clause 22 of the Code, whichever gives the worst combination of loads. For the local area of wall immediately below a concentrated load the most frequent case would be Case (a) with the combination $1.4G_k + 1.6Q_k$. For the area of wall

0.4h below the bearing the same case is likely to be the critical one but see **5.1.1**f.

(e) Design loads. For the local area of wall immediately under a concentrated load, calculate the local design load for the load combination selected; this must be less than the local design bearing resistance. For the area of wall 0.4h below the bearing, the procedure to follow is that outlined in Chapter 5; the eccentricity, e_x, of the resultant of the concentrated design load, P, and the ud design load $w(x + 0.8h)$ is calculated first; the calculation then proceeds as if the wall was subject to a ud load of $P/(x + 0.8h) + w$ kN/m at an eccentricity, e_x, at the top of the wall (Fig. 9.1).

(f) Local design bearing resistance. The local design bearing resistance depends on the bearing detail adopted. The Code distinguishes three bearing types (Clause 34 of the Code). A detail which meets any of the conditions shown in Fig. A8.1 is a bearing type 1 and the local design bearing strength = $1.25 f_k/\gamma_m$; for those shown in Fig. A8.2 it is a bearing type 2 and the local design bearing strength = $1.5 f_k/\gamma_m$.

Where a beam bears on to the end of a wall as shown in Fig. A8.1 Case (b), and the local design stress exceeds $1.25f_k/\gamma_m$, it is possible to use a spreader beam underneath instead. This case, shown in Fig. A8.3, is a bearing type 3 (Clause 34 of the Code) and the maximum allowable stress under the spreader beam may go up to $2f_k/\gamma_m$. The stress under the spreader beam may be calculated by assuming the stress distribution to be triangular or similar to that for a beam on an elastic foundation (Fig. A8.6).

For bearing details which do not fit into any of these categories, for example because the allowable bearing area is exceeded, the local design strength can be taken as equal to $1.1f_k/\gamma_m$ ($= f_k/\gamma_m$ if the load has an eccentricity of 0.05t or less); see Appendix B of the Code (**5.3**).

It is suggested that in any critical case, the proposed details are also checked against the empirical rules devised by Page and Hendry (1988); see **A8.3**. These rules tend to be more conservative, for the common cases, than the Code provisions but are based on experimental work with masonry.

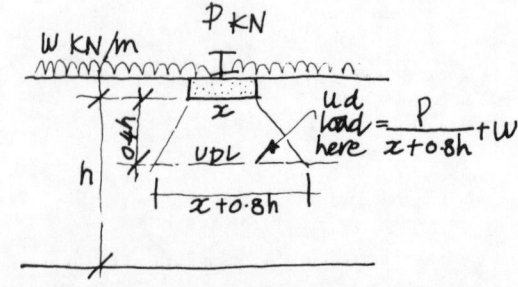

Fig. 9.1 Elevation on wall of uniform thickness showing dispersion of concentrated load below bearing pad.

9.2 Examples

Example 9.1 Bearing type 1

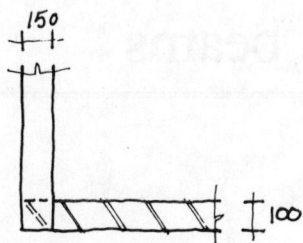

Fig. 9.2 Plan on bearing.

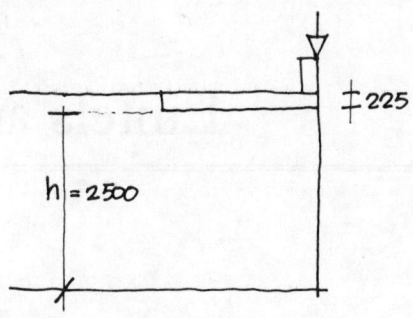

Fig. 9.3 Elevation on bearing.

A concrete beam 150 mm wide bears on to a masonry wall 100 mm wide as shown (Fig. 9.2). The loads from the beam are as follows:

Characteristic loads:

Dead load, G_k = 20 kN

Live load, Q_k = 10 kN

The masonry has the same strengths and safety factors used in Example 5.1. Check local design strength.

Select Case (a) of Clause 22 of the Code: $1.4G_k + 1.6Q_k$

Hence local design load
$$= 1.4 \times 20 + 1.6 \times 10 = 44 \text{ kN}$$

From Fig. A8.1 Case (b), it can be seen that this is a bearing type 1 according to Clause 34 of the Code.

Hence local design bearing resistance

$$= \frac{1.25 f_k}{\gamma_m} \times \text{bearing area}$$

$$= \frac{1.25 \times 6 \times 150 \times 100 \times 10^{-3}}{2.5}$$

$$= 45 \text{ kN}$$
OK

Example 9.2 Bearing type 3

This is similar to Example 9.1 except that the applied loads are doubled. Check local design strength, using spreader beam as shown (Fig. 9.3).

Select Case (a) of Clause 22 of the Code: $1.4G_k + 1.6Q_k$

Local design load
$$= 1.4 \times 40 + 1.6 \times 20 = 88 \text{ kN}$$

The wall beneath the beam requires a spreader beam and hence from App. **A8**, it can be seen that this is a bearing type 3, according to Clause 34 of the Code.

Take E_b = 30 kN/mm²

E_w = 15 kN/mm² (higher than actual E value of wall)

$$I_b = \frac{100 \times (225)^3}{12} = 95 \times 10^6 \text{ mm}^4$$

$$k = \frac{15}{2.500 \times 10^3} = 0.006 \text{ kN/mm}^2 \text{ per mm of deflection.}$$

$$\gamma = \left(\frac{100 \times 0.006}{4 \times 30 \times 95 \times 10^6} \right)^{\frac{1}{4}} = 0.269 \times 10^{-2} \text{ 1/mm}$$

∴ Maximum design stress

$$= k \times \frac{88 \times 10^3}{2 \times 1.954 \times 10^{-8} \times 30 \times 95 \times 10^6}$$

$$= k \times 790$$

$$= 0.006 \times 790 = 4.74 \text{ N/mm}^2$$

Taking an increase in f_k because the wall is assumed to be in brick and of the same thickness as the standard brick width (Clause 23.1.2 of the Code)

Allowable design stress
$$= \frac{2(1.15 f_k)}{\gamma_m}$$

$$= \frac{2 \times 1.15 \times 6}{2.5} = 5.52 \text{ N/mm}^2$$
OK

Example 9.3 Additional check

Check the previous two examples against the Page and Hendry rules, assuming the overall height of the wall is 2.725 m (Fig. 9.3). For Example 9.1, the strength enhancement factor, with a_1 = o, is

$$= \frac{0.55}{(150/1\,512)^{0.33}} = 1.18$$

The factor only rises to 1.25, as assumed previously, for a wall 3.30 m high. For Example 9.2, ignoring the beam width and assuming a 30° load dispersal line through the spreader beam, the strength enhancement factor is greater of

$$= \frac{0.55}{(390/1\,527)^{0.33}} = 0.86 \text{ or } 1$$

Hence local design bearing resistance

$$= \frac{1 \times 1.15 \times 6 \times 390 \times 100 \times 10^{-3}}{2.5} = 107 \text{ kN}$$

This answer is less conservative than that previously calculated.

CHAPTER 10

Lintels and composite beams

10.1 Composite action

Composite action refers to the ability of different materials to combine and behave as one composite system which is very much stronger than the individual strengths of the two materials added together. Mostly composite action can only be reliably achieved under special conditions. The composite action discussed here is that between a beam and the masonry immediately above, that is resting on the beam; the two elements can combine to form an arch system. In practice a composite beam would work in bending, as a beam, as well as in tension, as a tie to the arch system (Fig. 10.1).

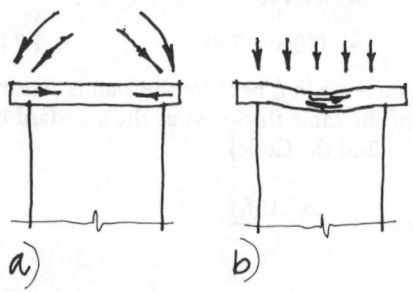

Fig. 10.1 Lintel supporting masonry which acts both (a) in tension and (b) in bending.

Composite beams are made of a number of materials which all behave in different ways. Wood (1952) showed that with brick masonry on reinforced concrete beams, composite action could be considered to occur if the height of the wall was at least $0.6L$, where L is the span of the beam and the depth of the beam is between about $L/20$ and $L/15$. Wood recommended that the beam be designed for an equivalent bending moment of $W.L/100$ for a wall with no openings in it or openings only near the middle of the wall and $W.L/50$ for a wall with an opening near supports, where W is the total design load on the beam, consisting of the self-weight of the wall, W_1, and any other load, W_2, which must be applied at least $0.6L$ above the top of the beam (Fig. 10.2). Any loads applied below this level must be taken by the lintel acting as a beam − i.e. the bending moment would be $W_3.L/8$ for a simply supported lintel where W_3 is that part of the load not taken by composite action. If props are not used to support the beam while the masonry is being laid then bending moment on the beam during construction may need consideration. Wood showed this equivalent bending moment to be about $W_1L/25$, where W_1 is the self-weight of an unhardened wall built to a height between about $0.6L$ and $0.75L$. The unhardened brickwork above a height of $0.75L$ could be considered as being supported by full composite action (Fig. 10.2). Wood simplified this matter by assuming all the loads in unpropped construction to be supported by full composite action but used only 70% of the allowable steel

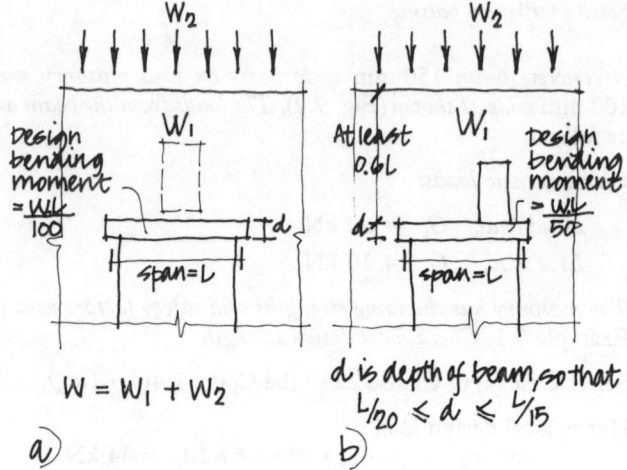

Fig. 10.2 Equivalent bending moments in concrete lintels acting compositely with brickwork (a) with no openings, or openings near middle, and (b) with openings near support.

stress that was used for propped construction. Composite action is made use of in lintels and in ground beams which support brick walls (Table 10.1). When other materials are used it may be assumed that the effects are broadly similar although the design figures given may vary to some extent.

Table 10.1 Beam depth and reinforcement for composite concrete/brickwork beam with loading of 50 kN/m, assuming bending moment = W.L/50 (after Table 1 of BRE Digest 242)

Span	Beam width (mm)		
(m)	280	330	380
2	200 440	200 480	200 520
3	200 670	200 720	200 780
4	250 890	250 960	200 1035
5	300 1110	300 1200	250 1300
6	350 1330	300 1450	300 1550

Beam depth (mm) [upper-left value in each cell]
Steel area (mm²) [lower-right value in each cell]

In particular if a steel beam is used it may not be possible for sufficient friction to develop between the steel and the masonry, unless special details are adopted (Fig. 10.3). For another method of composite design, see Davies and Ahmed (1980). In many cases test data are available from the manufacturers of lintels. If in any particular case the conditions for composite action are not met, for example because a significant part of the load is applied less than $0.6L$ above the lintel, then that part of the load on the lintel must be taken in bending.

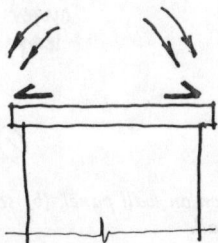

Fig. 10.3 Frictional force of the masonry on the supporting beam, where there is composite action.

Note that arching in masonry may occur without composite action if there is horizontal support from the adjacent parts of the wall. For example, a traditional method of lintel design assumes the lintel only supports a triangular load from the wall itself; arch action is assumed to occur (Fig. 10.4). However, any applied loads within the 60 degree triangle should also be taken as being supported by the lintel. The method is conservative for a total load above the lintel of less than about 20 tonnes, assuming that there is a height of masonry above the lintel at least equal to $0.6L$ and that the horizontal support can be mobilised. Another slightly different method is given in BS 5977: Part 1 (1981) for lintels up to 4.5 m span (single-storey buildings) in which horizontal support from adjacent parts of the wall allows arch action. In any case where the conditions for composite or arch action are not met, the full value of the load above should be assumed to be supported by the lintel.

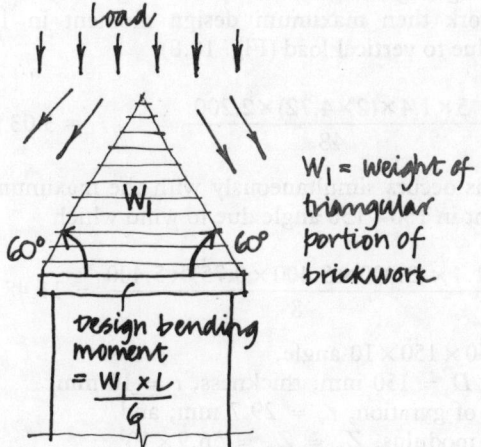

Fig. 10.4 Arch action in masonry above lintel either because of buttressing provided by surrounding masonry or because of composite action with the lintel; assumed triangular load on beam is shown.

10.2 Procedure for checking lintels under vertical and horizontal loads

The following sections cover the design of lintels with composite or non-composite action.

10.2.1 Vertical loads

(a) Consider the vertical loading which would cause the maximum compressive stress in the masonry and the maximum bending stress in the lintel. The masonry will have already been checked over its height between lateral supports (Ch. 5) and for concentrated load under the lintel bearing (Ch. 9). However, in a composite beam the masonry immediately above the end of the lintel also has a concentration of forces on it and the compressive stress may need to be considered (Clause 35 of the Code); see Fig. 10.1.

(b) Work out the characteristic vertical loads.

(c) Load combinations. Select either Case (a), (b) or (c) from Clause 22 of the Code (**5.3**), whichever gives the worst combination of loads. Normally this will be Case (a) with the combination $1.4G_k + 1.6Q_k$.

(d) Design loads. Calculate the design bending moment on the lintel, using the equivalent bending moment in the case of a composite beam. This must be less than the design moment of resistance of the beam. Check shear in a similar manner. If steel lintels are used it may be necessary in a small number of cases to check the bearing and web buckling of the lintel. For a composite beam ensure that the compressive stress in the masonry immediately above the end of the lintel does not exceed f_k/γ_m; an indication of whether this stress is likely to be exceeded will already have been provided by the calculation for concentrated load under the lintel bearing. Normally this would only be an important consideration for an isolated wall; in most other cases the lintel may simply be extended in length to overcome any difficulties (Fig. 10.5).

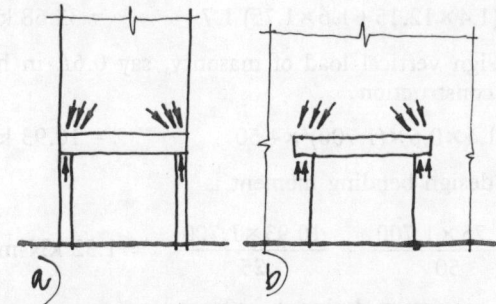

Fig. 10.5 (a) Isolated and (b) continuous wall showing concentration of forces immediately above the end of the lintel.

10.2.2 Horizontal loads

(a) In some cases beams are required to support not only vertical loads but horizontal ones too. A common case is where there is a large opening in a masonry wall so that a secondary support beam is necessary; see **6.1**l and App. **A7**.

(b) Consider the horizontal loading which would cause the maximum bending stresses in the supporting beam.

(c) Work out the characteristic horizontal loads.

(d) Load combinations. Select either Case (a), (b) or (c) from Clause 22 of the Code (**5.3**), whichever gives the worst combination of loads. Normally this will be Case (b) with the combination $0.9G_k + 1.4W_k$ or $0.9G_k + 1.2W_k$ for wall panels whose removal would not affect the stability of the building. Sometimes Case (c) may be more critical.

(e) Design loads. Calculate the design bending moment on the beam. This must be less than the design moment of resistance of the beam. With beams made of thin steel sheet or similar material it may be necessary to check shear, bearing and web buckling too.

10.3 Examples

Example 10.1 Composite concrete lintel beam

A 100-mm-deep concrete lintel supports a cavity wall and a roof load, applied to both leaves of the cavity wall, over a clear distance of 1.600 mm. The characteristic loads are as shown in Fig. 10.6. Calculate the design bending moment and shear on the lintel, assuming the lintel is not propped during construction.

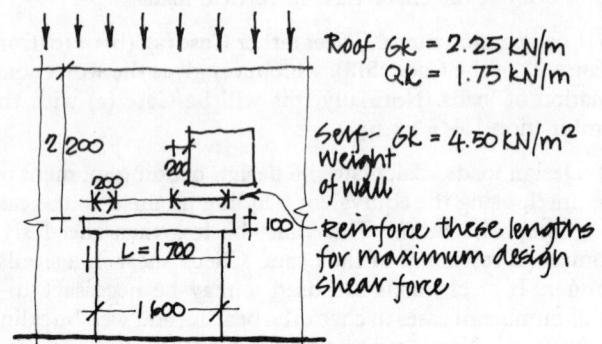

Fig. 10.6 Elevation on wall with opening above lintel.

Because of the depth of the lintel and wall above, composite action may be assumed to occur. The span is the clear distance plus the depth of the lintel = 1.700 m

Consider Clause (a) of Clause 22 of the Code: $1.4G_k + 1.6Q_k$

Total design vertical load on lintel, ignoring opening,

$$= (1.4 \times 12.15 + 1.6 \times 1.75) \, 1.7 \qquad = 33.68 \text{ kN}$$

and design vertical load of masonry, say $0.6L$ in height, during construction

$$= 1.4 \times 0.6 \times (1.700)^2 \times 4.50 \qquad = 10.93 \text{ kN}$$

Hence design bending moment is

$$\frac{22.75 \times 1.700}{50} + \frac{10.93 \times 1.700}{25} = 1.52 \text{ kN-m}$$

and the maximum design shear force

$$= \frac{33.68}{2} \qquad = 16.84 \text{ kN}$$

Note that the reinforcement for this shear force should extend to the left of the opening (Fig. 10.6) and would generally be in the form of vertical steel link bars. Normally for walls which do not have openings near a support, shear reinforcement is unnecessary.

Example 10.2 Lintel taking horizontal and vertical loads

For the panel considered in Example 6.6 *provide a suitable lateral support.*

Steel angles will be used to provide a lateral support against wind and also act as a lintel over the opening for vertical load (Fig. 10.7).

Because the height of the masonry above the opening is greater than 0.6 times the span of the opening and because the masonry each side of the opening is able to act as a buttress, it may be assumed that an arch forms to carry vertical load. Hence it may be conservatively assumed that the lintel only carries a triangular load (Fig. 10.4).

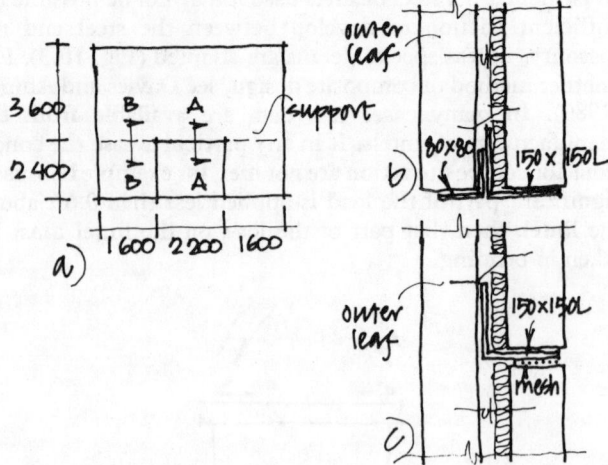

Fig. 10.7 (a) Elevation on wall panel, (b) section A-A and (c) section B-B through wall.

If weight of 102.5-mm-thick leaf of brickwork = 2.25 kN/m² then weight of each leaf of a 60° triangle of brickwork

$$= (2.200)^2 \times \frac{\sqrt{3}}{4} \times 2.25 \qquad = 4.72 \text{ kN}$$

Consider Case (b) of Clause 22 of the Code: $1.4G_k + 1.4W_k$

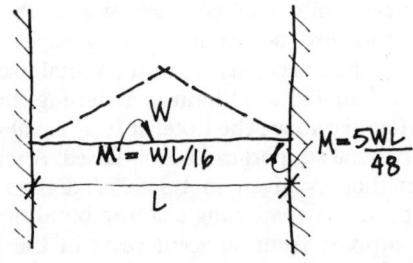

Fig. 10.8 Bending moments in fixed end beam with triangular load, W.

Assuming larger angle to carry the weight of both leaves of brickwork then maximum design moment in 150×150 angle due to vertical load (Fig. 10.8)

$$= \frac{5 \times 1.4 \times (2 \times 4.72) \times 2.200}{48} \qquad = 3.03 \text{ kN-m}$$

and this occurs simultaneously with the maximum design moment in 150×150 angle due to wind which

$$= \frac{1.4 \times (3.000 \times 5.400 \times 0.75) \times 5.400}{8} = 11.48 \text{ kN-m}$$

For $150 \times 150 \times 10$ angle,
depth, $D = 150$ mm; thickness, $t = 10$ mm;
radius of gyration, $r_v = 29.7$ mm; and
elastic modulus, $Z_x = Z_y = 56.9 \times 10^3$
or $= 154 \times 10^3$ mm³

For 150×150 angle, using BS 5950:Part 1, allowable compressive stress in each leg governed by two conditions:

(a) when spanning across opening, and (b) when spanning across panel; note that horizontal leg of angle is restrained by masonry over most of the panel length. The lintel can be classified as semi-compact, at all positions, the two angles being welded together. Therefore, considering central portion,

$$L/r_v = 2\,200/29.7 \qquad = 74$$

$$\therefore M_b = 0.8 p_y.Z = 0.8 \times 0.275 \times 56.9 = 12.52 \text{ kN-m}$$

and $p_y.Z = 0.275 \times 154.8 \qquad = 42.57$ kN-m

$$\frac{M_x}{M_b} + \frac{M_y}{P_y.Z} = \frac{11.48}{12.52} + \frac{3.03}{42.57} = 0.99 < 1 \quad \text{OK}$$

This calculation ignores the contribution made to the lateral strength of the wall by the masonry; see Example 10.3.

Example 10.3 Lintel acting compositely with masonry

Repeat Example 10.2 *taking account of the strength of the brickwork in bending about the vertical axis.*

Try use of $125 \times 75 \times 8$ angle, instead of $150 \times 150 \times 10$ angle, with same loads as before; considering bending due to vertical loads on angle over central portion, using BS 5950:Part 1,

$$\frac{M_y}{p_y.Z_y} = \frac{3.03}{0.275 \times 40.2} = 0.28, \text{ giving}$$

max. $M_x = (1 - 0.28) M_b = 0.72 \times 0.7 \times 0.275 \times 29.6$
$$= 4.10 \text{ kN-m}$$

with $L/r_v = 2\,200/16.3 = 135$ and

the lintel taken as semi-compact.

Using type A failure in **A7.1.2** with

$k_1 = k_3 = \dfrac{4.10}{1.37 \times 3.6} = 0.83; K_1 = K_3$
$\qquad\qquad\qquad\qquad = \sqrt{(1 + 1 + 0.83)};$

$\beta = 1.20/3.60; i_2 = 0; \mu = 0.5;$

$w_e = 1.33 \, w; a_e = 3.60\sqrt{(1.33)}$

$\therefore m = 1.25 \, w = 1.25 \, W_k.\gamma_f$

so max. $W_k = 1.37/1.75 = 0.78$ kN/m² \qquad OK

Compare stiffness of, say, 1 m depth of cavity wall with steelwork; for brickwork with $f_k = 7.1$ N/mm²

$$E_b.I_b = 0.70 \, f_k \times 89.7 \times 10^6 \times 2 = 0.89 \times 10^9 \text{ kN-mm}^2$$

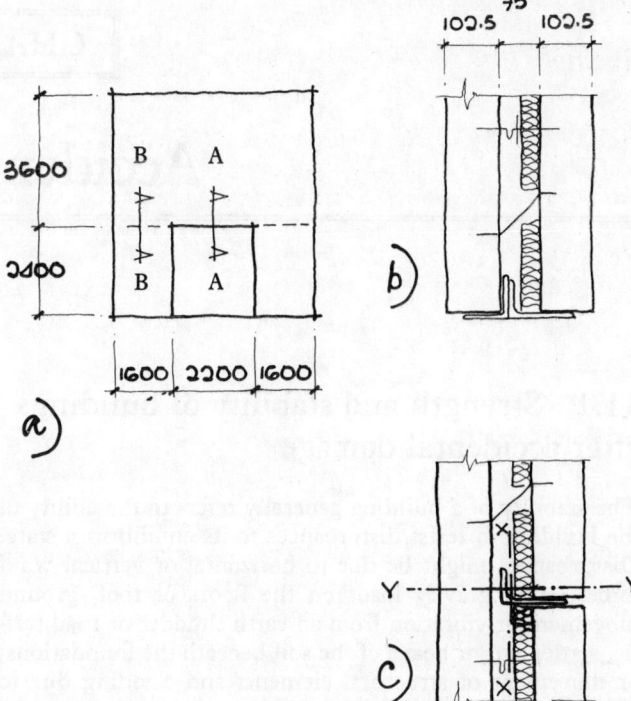

Fig. 10.9 (a) Elevation on wall panel, (b) section A–A and (c) section B–B through wall.

Composite inertia of the two steel angles over the central section $= 600$ cm⁴; for end rotation of beam under uniform bending,

equivalent uniform inertia, $I_e = 0.6 \times 247 + 0.4 \times 600$
$$\qquad\qquad\qquad\qquad\qquad = 388 \text{ cm}^4$$

$\therefore$ take $E_s.I_s = 205 \times 388 \times 10^4 = 0.80 \times 10^9$ kN-mm²

This last calculation indicates that the steel angles will make a significant contribution to stiffness at working loads.

CHAPTER 11

Accidental damage

11.1 Strength and stability of buildings after accidental damage

The stability of a building generally refers to the ability of the building to resist disturbances to its equilibrium state. Disturbances might be due to horizontal or vertical wind forces, extra gravity loads on the floors or roof, ground movement or vibration from an earth shudder or road traffic, settlement or heave of the soil beneath the foundations, or movement of structural elements and cladding due to changes in temperature and moisture content of the air. These are the kinds of disturbance that in moderate degree a building must be designed to contain without affecting its usefulness or safety. A building whose structure is designed in such a way that it is able to cope with such disturbances, as defined, is said to be stable. A building which has large changes in forces or deflections in it as a result of such disturbances could be said to be sensitive to them. If the building is so affected by these disturbances that it becomes unsafe then the building is said to be unstable.

The stability of masonry building depends on tying together the floors, roofs, walls and columns so that any disturbances can be resisted by the building, or a part of the building, acting as one unit; the disturbing forces must be taken to ground and are carried there, mainly, by forces acting in the planes of the walls. One of the most important aspects of stability is the lateral stability and this is discussed in Chapter 8. As a test of the lateral stability of a building Clause 20.1 of the Code (**5.3**) requires that it be able to resist a ud horizontal disturbing force equal to 1.5% of the characteristic dead load above any level, if this is more than the design horizontal wind forces that it would take anyway (**5.1.2**d).

However, in rare cases the disturbance to a building may no longer be of moderate degree but is such as to cause local damage. Typical cases would be damage to a building caused by vehicle collisions, explosions, large foundation movements or other unexpected occurrences. Clause 20.2 of the Code refers to these as accidental forces and specifically requires that there is a reasonable probability that such local damage will not cause total or progressive collapse or, in general, damage disproportionate to the original cause (Fig. 11.1).

A building which is designed in such a way that it is able to cope with such accidental damage is said to be 'robust'. If the building is to be robust in this sense then thought must be given to ways in which local damage may be caused in each particular building and alternative ways of supporting the loads after removal of a part of the structure must be found. In addition to this general requirement, Clause 37 of the Code has more specific requirements for buildings of five storeys or more; the Clause of the Code defines the

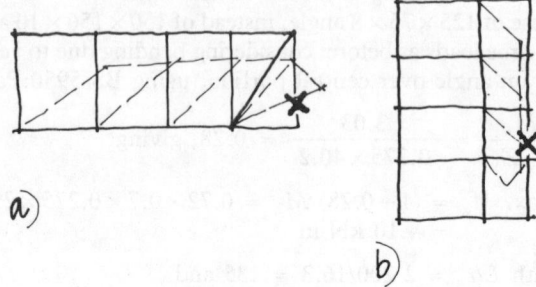

Fig. 11.1   Progressive collapse (a) of hall caused by failure of end column and (b) of masonry building, by failure of first-floor wall and subsequent debris loading on first-floor slab.

extent of the local damage that must be allowed for and requires that the building be designed so that this damage is unlikely to spread significantly. At design stage this means removing load-bearing elements, one at a time all round the building, and assessing the structure of the building after this has happened. A special load case is given in Clause 22 of the Code (**5.3**) for use after accidental damage has occurred and reduced partial safety factors for material strength, γ_m, are allowed. An alternative with some elements is to make them sufficiently strong so that they are unlikely to be damaged in an accident; Clause 37 of the Code assumes that any element that has been designed to resist a gas pressure in the room of 34 kN/m^2, at a reduced factor of safety from that normally required, is a so-called 'protected' element. Protected elements do not need to be considered as having been removed in the design appraisal.

This procedure is relevant for damage caused by explosions but damage to elements from vehicles is better prevented by bollards, earth banks or repositioning of the element. Both these methods of preventing general damage after an accident, what might be termed the individual design of each area, are covered by Option 1 in Clause 37 of the Code. Option 3 consists of providing general vertical and horizontal ties in all parts of the building to resist forces specified by Clause 37 of the Code. The ties are expected to contain any damage. Option 2 is a combination of these two approaches whereby general horizontal ties are incorporated but vertical ties or columns are inserted only as and when required; this approach is particularly suitable for masonry construction. Although the requirements of Clause 37 of the Code are only mandatory, under Paragraph A.3/4 of Schedule 1 of the Building Regulations 1991 (UK), to Category 2 buildings of five storeys or more and to public buildings which incorporate a clear span exceeding nine metres, the methods are clearly applicable to any other building at the discretion of the designer.

Three general ideas about the design of buildings after accidental damage has occurred, which are relevant to

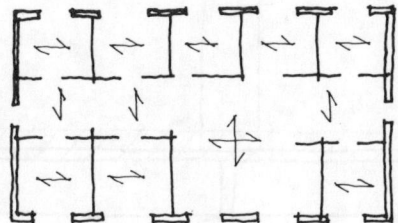

Fig. 11.2   Floor plan on building with part of floor spanning in two directions, after wall removed at lower level.

masonry construction, are mentioned here. Firstly, if a vertical load-bearing element has been removed after an accident, it is often possible for the remaining structure to cantilever or span over the damaged part even if it is in a slightly distorted state (Fig. 11.2). This is particularly true where damage is to internal supports and where the floors above incorporate horizontal ties, similar to that proposed by Clause 37 of the Code; often the floor is able to act as a catenary as well as in bending (Fig. 11.3). Secondly,

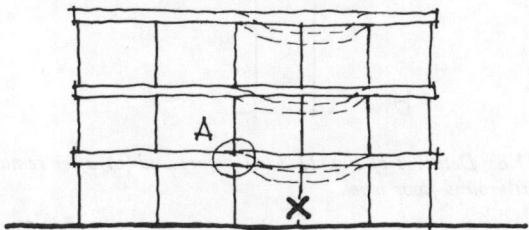

Fig. 11.3   Section on building with some floor slabs acting as catenaries, after removal of part of wall at ground-floor level.

although it is important that the various parts of a building are properly tied together and able to support each other, in some cases this may be disadvantageous because a damaged portion of the building may dislodge or bring down the other parts of the building to which it is tied. In such cases, the damage may be better contained by separating the parts of the building or ensuring that any ties between units are weak ones (Figs 11.4, 11.5 and 11.6). Note that if in Fig. 11.4 the distance between the vertical supports were reduced then the forces put on the remaining structure would be smaller and a general collapse would be less likely. Because of the relatively weak bond between unreinforced masonry units, a building of masonry construction has the major advantage that it is able to accept local damage without the forces causing damage being transferred to other parts of the building (Fig. 11.7). Thirdly, it is important that the connections between components which are designed to work after accidental damage has occurred maintain their strength even when the components are in a distorted or slightly damaged condition (Figs 11.6 and 11.8).

Further discussion on stability and accidental damage is provided by Sutherland (1978), Morton (1985) and Moore (1978). A study of load-bearing masonry buildings has identified three situations where special provisions against accidental damage may be necessary. These are an outside wall without return walls, or only one internal return wall; an internal wall without returns; when the removal of a wall, or part of a wall, causes high concentrated loads (**9.1**) to occur on the remaining walls; see Hendry (1990, p. 205). In practice most masonry buildings can be made 'robust' by there being sufficient horizontal ties in the floor structure, an easy matter for a reinforced concrete slab, and some additional ties or columns on the external corners of the buildings where there is less likely to be an alternative

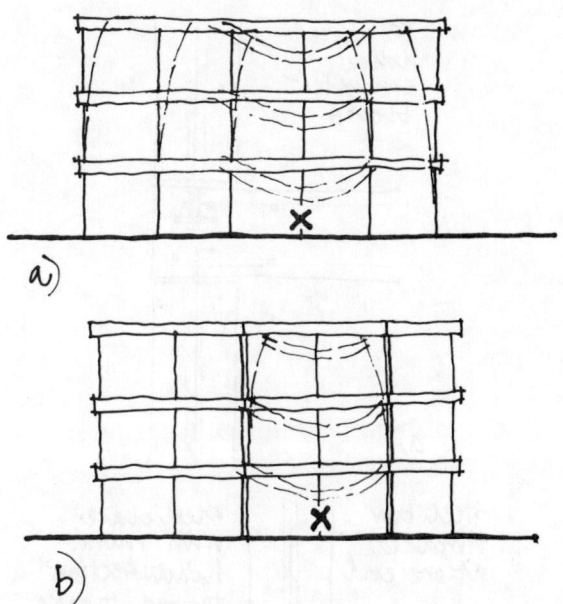

Fig. 11.4   (a) Section on building where removal of column causes general collapse; (b) damage can be limited by separating building into independent units.

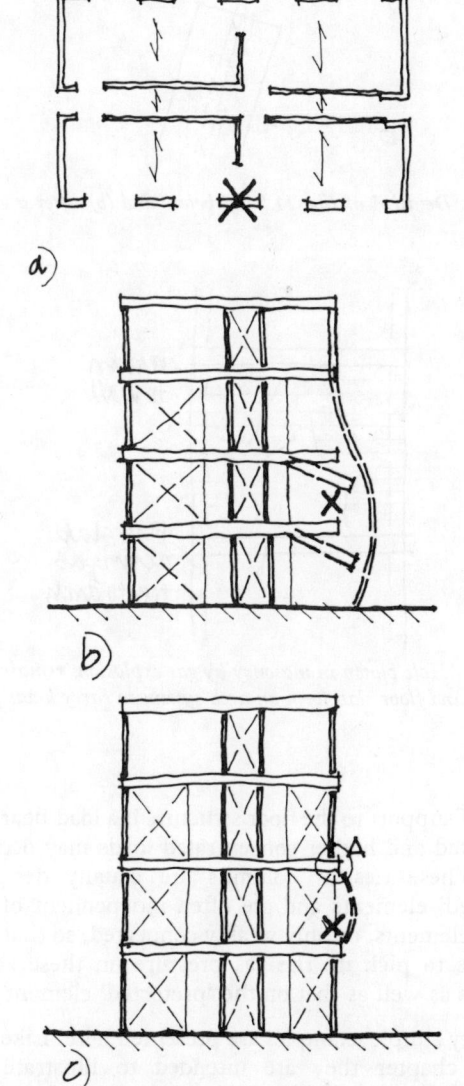

Fig. 11.5   (a) Plan on building with a wall removed at first-floor level; (b) section showing that damage is caused at all levels by continuously reinforced wall but (c) is restricted to first floor by introducing weak links in the wall.

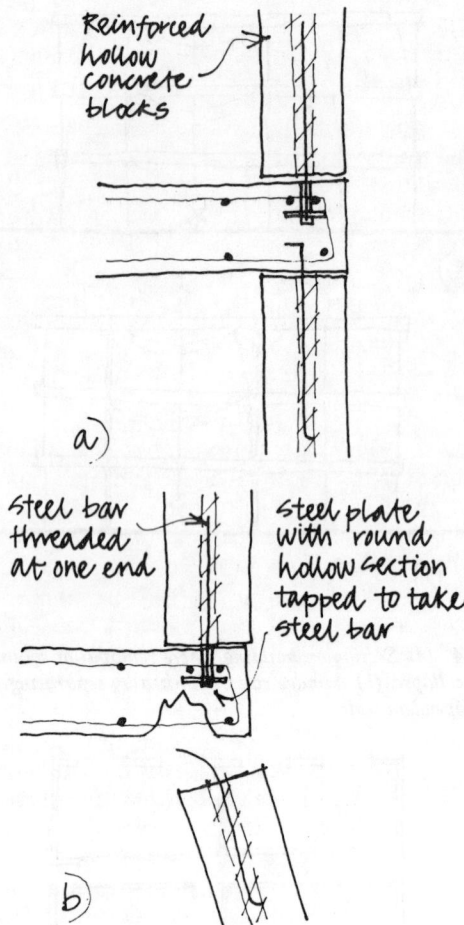

Fig. 11.6 Detail A of Fig. 11.5 (a) before and (b) after a gas explosion.

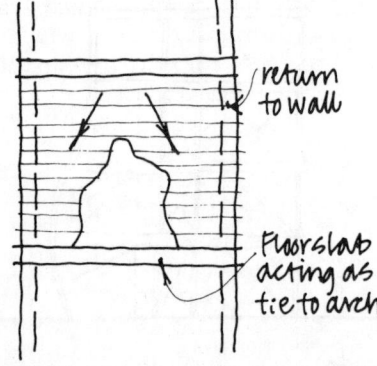

Fig. 11.7 Hole blown in masonry by gas explosion; remaining masonry and floor slab form an arch system to carry loads from above.

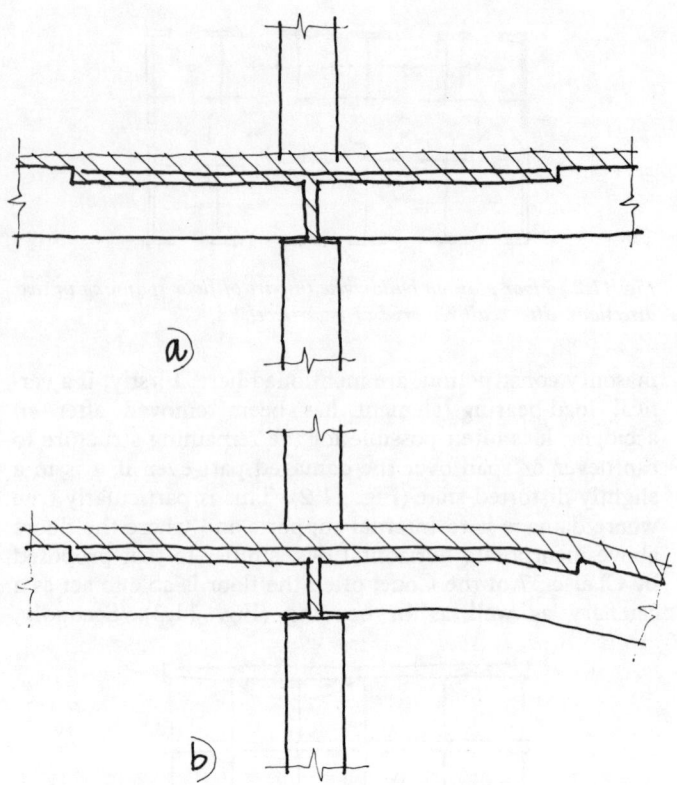

Fig. 11.8 Detail A of Fig. 11.3 (a) before and (b) after removal of wall at ground-floor level.

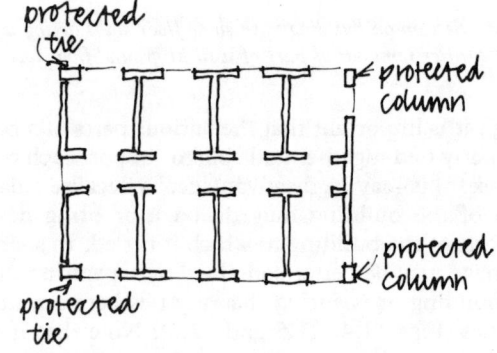

Fig. 11.9 Plan on load-bearing masonry building showing independent 'protected' elements designed for case of accidental damage.

11.2 Examples

Example 11.1 Double garage

A double garage is to be built (Fig. 11.10). It is required that precautions are taken in the event of a vehicle collision with the wall containing the entrances.

A collision with the central pier would cause collapse of the whole roof. Hence the two separate lintels should be built as one continuous lintel which is able to support the steel beam at a reduced factor of safety, if the central pier were to be removed. A collision at either end of the wall would cause only local damage. Note that although the garage could be built without a central pier, using a much deeper lintel, a collision at either end of the wall would then cause collapse of the whole roof – i.e. the extra, redundant pier is necessary for design against accidental damage. It is a general feature of design for accidental damage that the structure be redundant or, in other words, that an alternative means of supporting the loads is available.

means of support to the floor structure if a load-bearing wall is removed and higher concentrated loads may occur (Fig. 11.9). These ties or columns are usually designed as 'protected' elements and are often independent of the adjoining elements, or only weakly connected, so that they do not have to pick up the gas pressure on these adjoining elements as well as that on the 'protected' element itself.

Two very simple examples are presented here. Like the rest of this chapter they are intended to illustrate typical approaches to accidental design where damage is caused by a vehicle collision or a gas explosion. However, many other aspects of design for safety could be approached in a similar way, including damage caused by fire, corrosion or errors in execution.

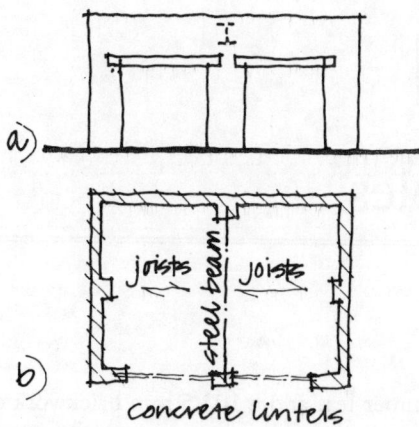

Fig. 11.10 (a) Elevation and (b) plan on double garage.

Example 11.2 Two-storey crosswall construction

The load-bearing masonry building of Example 8.2 (Fig. 8.16) *has four shop units on the ground floor and accommodation on the first floor. It is required that precautions are taken against progressive collapse and unacceptable damage as a result of a vehicle collision on the east wall or of a gas explosion in one of the kitchen areas at the north end of the shop units.*

One approach to this problem would be to accept the loss of the ground and first floor of one unit and prevent the spread of damage to other units, for example by not connecting the floors of adjacent units together. However, this approach, as well as allowing a relatively large amount of damage to occur, is unsuitable and would make it difficult for the building to have proper lateral stability under normal conditions (Example 8.2).

An alternative approach is to provide a frame structure with a concrete floor and infill masonry panels. This is an ideal arrangement in that the infill panels can allow gas explosions to be vented and so limit damage from blast while the frame, being tied together in the horizontal and vertical directions, is inherently robust and able to accept large amounts of damage without this causing further collapse. In addition the frame is strong and rigid and gives lateral stability to the

building under normal loads. However, this implies a substantial change in the normal method of construction and increased cost.

The approach adopted here is to use load-bearing masonry walls and timber floors, but to provide around the perimeter of the building, or part of it, an *in situ* concrete capping beam placed on the inner skin of masonry just below first-floor level (Fig. 11.11).

This beam could act as a lintel as well as carrying the masonry and floor loads at first-floor level in the event of a vehicle collision with the east wall at ground-floor level. Even damage by a large vehicle to the beam itself would not necessarily cause further collapse.

However, in general gas explosions will cause more extensive damage than vehicle collisions and need more thought at design stage. In this case provision should be made to vent the explosion through the openings in the north wall as well as through the shop front on the south wall, in order to prevent excessive damage to the east wall or any of the internal crosswalls. In the event of a gas explosion in the kitchen unit adjacent to the east wall the return walls 1, 7 and 8 can stabilise and help to prevent the collapse of walls 9 and 10 (Fig. 11.11). Wall 1 would need to be fully tied over its height to wall 10 by steel reinforcement in the bed joints to **remain** in position after an explosion but the loss of wall 7 **can** be accepted because wall 8, being on the other side of the crosswall, is likely to remain. If necessary a reinforced concrete blockwork column can be incorporated in the inner skin of masonry at the corner to support the perimeter concrete beam if there is a danger of the rest of wall 10 collapsing (Fig. 11.11). The very worst damage likely to be caused by such a gas explosion is the removal of walls 2, 3, 4, 6 and 7 depending on the amount of venting; in general, however, the load-bearing elements to be considered as removed at any one time can be obtained by following Option 1 in Clause 37 of the Code. There is likely to be damage to the timber floor immediately above the gas explosion but the connection details adopted (Fig. 11.12) should ensure that generally the walls remain stable after the gas explosion.

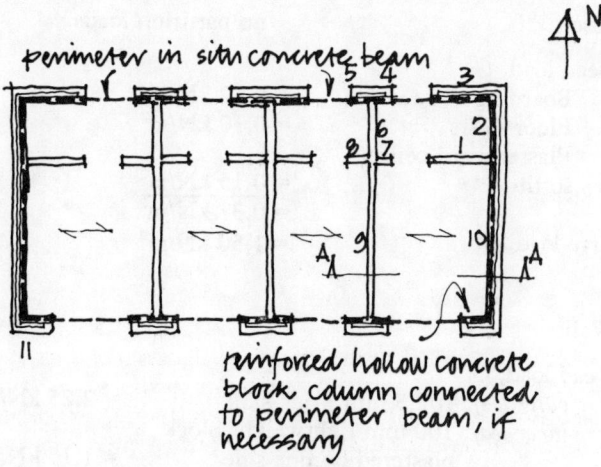

Fig. 11.11 Plan on ground floor of building.

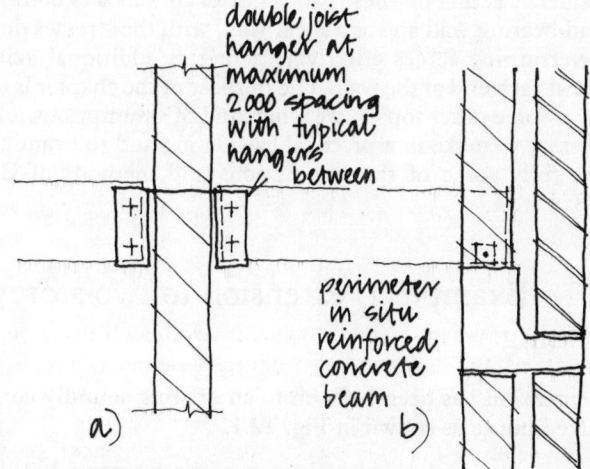

Fig. 11.12 Section A-A showing (a) internal crosswall and (b) perimeter beam and external wall.

CHAPTER 12

Design examples

Introduction

The whole of this chapter is given over to examples showing how the principles of masonry design are applied in two typical cases – that of the extension to an existing two-storey building (**12.1** to **12.6**) and that of a new nine-storey building (**12.7** to **12.11**).

Section **12.2** derives characteristic loads, including wind loads from CP3: Chapter V: Part 2, for the two-storey building. Section **12.3** considers the vertical and horizontal loading on a small area of wall between window openings. The wall spans vertically under horizontal loads although its 'elastic' flexural strength is found to be inadequate and a method of increasing the flexural strength is considered. Section **12.4** considers methods of calculating the connection forces, although relatively small in this case. Section **12.5** considers the effect of a wall that is bowed and out-of-plumb on its strength under vertical and horizontal load. Section **12.6** considers a small area of wall under high vertical load. The wall spans vertically under horizontal loads and relies on its 'plastic' flexural strength to do this. A larger but irregularly shaped wall which spans in two directions under horizontal loads is also considered.

Section **12.8** gives characteristic loads for the nine-storey building. Section **12.9** calculates the stresses due to the characteristic loads on an internal and external wall at first-floor level and **12.10** and **12.11** consider the various combinations of vertical and horizontal load it is necessary to consider as acting on these two walls. Each wall acts both as a load-bearing wall and as a shear wall, with the stresses due to overturning forces effectively acting as additional axial loads at each end of the wall. The purpose of the chapter is to look at some other topics, see what kind of assumptions it is necessary to make in a practical calculation and to examine more fully some of the assumptions and methods of BS 5628: Part 1.

12.1 Example 1: extension to two-storey building

An extension has been built on to an existing soundly constructed house as shown in Fig. 12.1.

It is necessary to decide which parts of the extension to this house need to be checked and show that these parts are built to the requirements of the Code of Practice BS 5628:Part 1.

After construction a part of the bottom wall on the east elevation where there are no windows is found to be out-of-plumb by 75 mm over the height of the ground floor and bowed within this height by 20 mm. Hence it is necessary to check whether the wall is still satisfactory.

The extension to the house has a cavity wall with a 100 mm blockwork inner leaf and a 102.5 mm brickwork outer leaf. The brickwork has a unit compressive strength of 27.5 N/mm² and a water absorption of 9%. The blockwork is hollow and insulated in the voids and has a unit compressive strength over the gross area of the block, including the voids, of 3.5 N/mm². A 1:1:6 cement:lime:sand mortar mix, mortar designation (iii), is used. Only the inner leaf of the cavity wall takes roof and floor loads. See Table 2.2 for the characteristic compressive strengths, f_k. A solid block of the same strength is used for the wall in Section **12.6**.

The house and extension are shown in Fig. 12.1. The Clauses referred to in this example are those in BS 5628:Part 1, extracts of which are given in previous chapters.

12.2 Characteristic loadings

Roof:

Dead load, G_k

Tiles	= 0.52 kN/m²
Battens and felt	= 0.10 kN/m²
Wood rafters	= 0.15 kN/m²
	= 0.77 kN/m² on slope
or	= 0.77/Cos θ kN/m² on plan
	= 0.77/Cos 30°
	= 0.89 kN/m² on plan

Live load, Q_k = 0.75 kN/m² on plan

Ceiling: None

Floor: no partition loads

Dead load, G_k

Boarding	= 0.12 kN/m²
Floor joists	= 0.10 kN/m²
Plasterboard ceiling soffit	= 0.15 kN/m²
	= 0.37 kN/m²

Live load, Q_k = 1.50 kN/m²

Wall:

Dead load, G_k
outer leaf: 102.5 mm brick skin = 2.25 kN/m²
inner leaf: 100 mm lightweight block
plastered on one side = 1.15 kN/m²

Wind load, W_k:
Typical values from CP3: Chapter V: Part 2: 1972
(as an alternative to BS6399:Part 2)
S_1 = 1.0 (usual value)
S_2 = 0.74 (building less than 10 m height on outskirts of town)
S_3 = 1.0 (usual value)

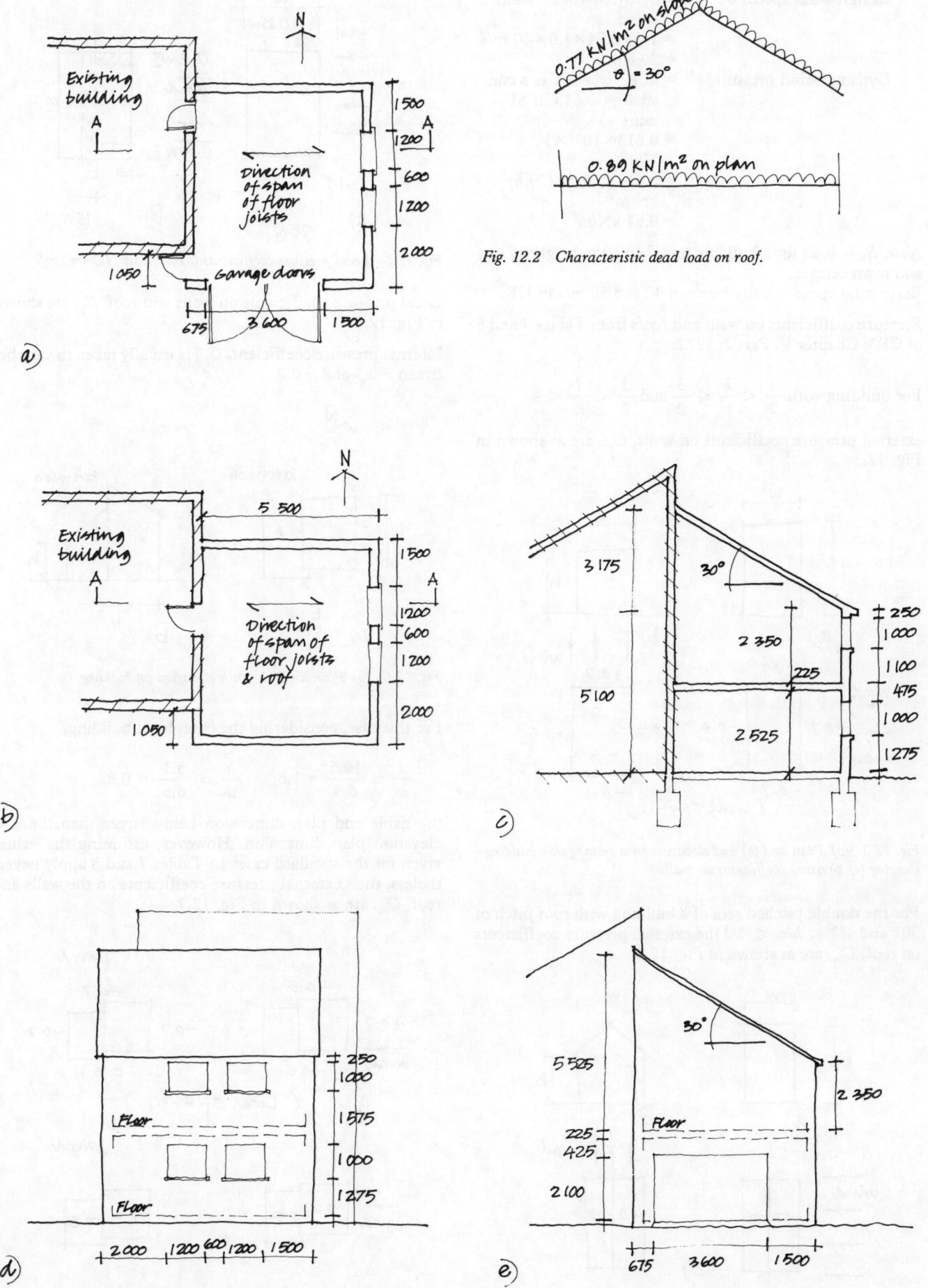

Fig. 12.2   Characteristic dead load on roof.

Fig. 12.1   (a) Ground-floor plan, (b) first-floor plan, (c) section A-A, (d) east elevation and (e) south elevation.

Design wind speed, V_s = $S_1 . S_2 . S_3 \times$ basic wind speed

= $1.0 \times 0.74 \times 1.0 \times 40$ m/s

= 29.6 m/s

Dynamic wind pressure, q = $k . V_s^2$ where k is a constant = 0.613 in SI units

= $0.613 \times 10^{-3} \times V_s^2$ kN/m^2

= $0.613 \times 10^{-3} \times (29.6)^2$ kN/m^2

= 0.54 kN/m^2

Note. $S_2 = 0.62$ for a building less than 10 m height in city and town centres.

Basic wind speed usually between 40 and 50 m/s in UK.

Pressure coefficients on walls and roofs from Tables 7 and 8 of CP3: Chapter V: Part 2: 1972.

For building with $\frac{1}{2} < \frac{h}{w} \leqslant \frac{3}{2}$ and $\frac{3}{2} < \frac{l}{w} < 4$,

external pressure coefficients on walls, C_{pe}, are as shown in Fig. 12.3.

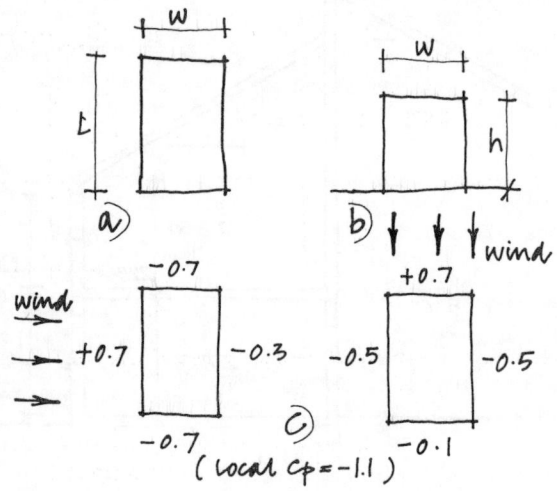

Fig. 12.3 (a) Plan and (b) end elevation on a rectangular building showing (c) pressure coefficients on walls.

For the double pitched roof of a building with roof pitch of 30° and 1/2 < h/w ≤ 3/2 the external pressure coefficients on roof, C_{pe}, are as shown in Fig. 12.4.

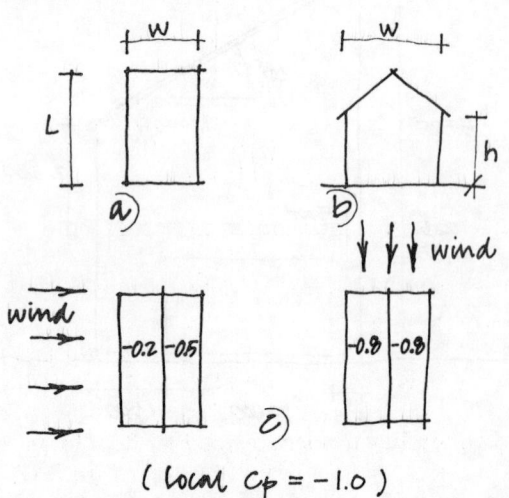

Fig. 12.4 (a) Plan and (b) end elevation on rectangular building with roof showing (c) pressure coefficients on roof.

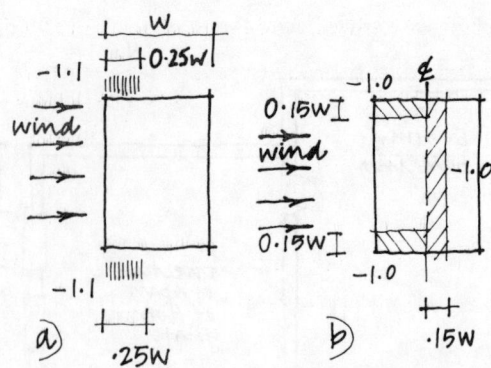

Fig. 12.5 Local pressure coefficients (a) on walls, (b) on roof.

Local pressure coefficients on walls and roof, C_p, are shown in Fig. 12.5.

Internal pressure coefficient, C_{pi}, is usually taken to vary between -0.3 and $+0.2$.

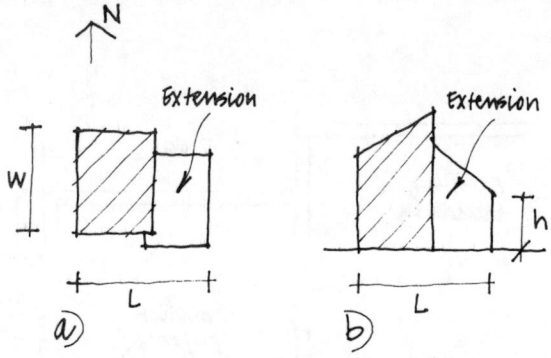

Fig. 12.6 (a) Plan and (b) south elevation on building.

For this case, considering the completed building,

$$\frac{l}{w} \simeq \frac{10.5}{6.5} = 1.6; \qquad \frac{h}{w} \simeq \frac{5.1}{6.5} = 0.8,$$

the gable end plan dimension being larger than the east elevation plan dimension. However, assuming the values given for the standard cases in Tables 7 and 8 apply nevertheless, then external pressure coefficients on the walls and roof, C_{pe}, are as shown in Fig. 12.7.

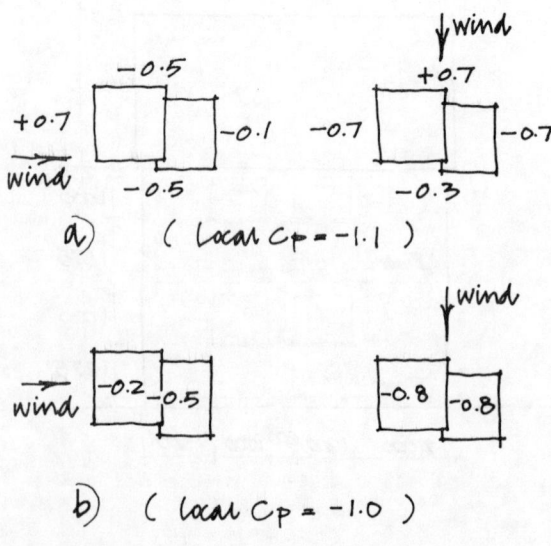

Fig. 12.7 Pressure coefficients (a) on walls, (b) on roof with wind from north and west.

Considering case where wind blows in north–south direction

General uplift pressure on roof $= (C_{pe} - C_{pi}) \times q$
$$= (-0.8 - 0.2) \times 0.54$$
$$= -0.54 \text{ kN/m}^2$$

General suction on wall $= (-0.7 - 0.2) \times 0.54$
$$= -0.9 \times 0.54$$
$$= -0.49 \text{ kN/m}^2$$

Considering case where wind blows in east–west direction,

General uplift pressure on roof $= (C_{pe} - C_{pi}) \times q$
$$= (-0.5 - 0.2) \times 0.54$$
$$= -0.38 \text{ kN/m}^2$$

General pressure on wall $= (+0.7 - (-0.3)) \times 0.54$
$$= 1.0 \times 0.54$$
$$= 0.54 \text{ kN/m}^2$$

12.3 East wall – vertical and horizontal loads

12.3.1 Vertical loading of wall on east elevation

The worst position for vertical loading – that is where the allowable compressive stress may be exceeded – is the 600-mm-long inner leaf of lightweight blockwork between the windows, which is load-bearing and, in addition, has very much less compressive strength than the brickwork outer leaf. Check this element between ground and first floor.

Slenderness ratio of wall:
The inner leaf has a thickness of 100 mm and a length of 600 mm so that $b > 4t_{ef}$ and this piece of masonry is classified as a 'wall' not a 'column', according to the Code.

Effective thickness, $t_{ef} = \frac{2}{3}(102.5 + 100)$
$$= 135 \text{ mm}$$

Effective height, $h_{ef} = 0.75 \times 2,525$
$$= 1,894 \text{ mm}$$

(enhanced resistance to lateral movement)

Slenderness ratio (SR) $= \dfrac{h_{ef}}{t_{ef}} = \dfrac{1,894}{135}$

$$= 14$$

cf. allowable SR $= 27$ OK

Characteristic vertical load on inner leaf just below first-floor level:

from roof, $G_k = 1.800 \times \dfrac{5.500}{2} \times 0.89$

$$= 4.95 \times 0.89$$
$$= 4.41 \text{ kN}$$

$$Q_k = 1.800 \times \dfrac{5.500}{2} \times 0.75$$

$$= 4.95 \times 0.75$$
$$= 3.71 \text{ kN}$$

from first floor, $G_k = 4.95 \times 0.37$
$$= 1.83 \text{ kN}$$
$$Q_k = 4.95 \times 1.50$$
$$= 7.42 \text{ kN}$$

from self-weight of inner leaf of wall (at head of window),

$G_k = 1.825 \times 1.200 \times 1.15$
$$+ 2.825 \times 0.600 \times 1.15$$
$$= 3.885 \times 1.15$$
$$= 4.47 \text{ kN}$$

Summary of characteristic vertical loads on inner leaf

Dead loads, $G_k = 4.41 + 1.83 + 4.47$
$$= 10.71 \text{ kN}$$

Live loads, $Q_k = 3.71 + 7.42$
$$= 11.13 \text{ kN}$$

Design vertical loads The combination of loads which gives the maximum compressive stress in the inner leaf of blockwork is that of Case (a) of Clause 22: $1.4G_k + 1.6Q_k$

For a wall loaded from one side only and from above, the combination to be considered is that of $1.4G_k + 1.6Q_k$, both for the axial load, P_1 and the eccentric load, P_2 (Fig. 12.8).

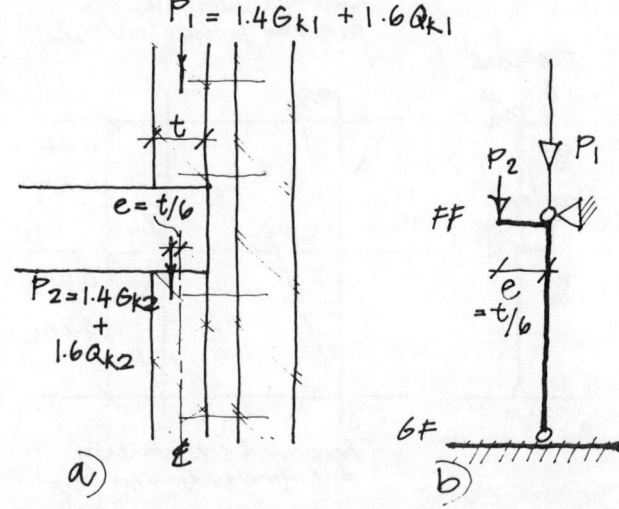

Fig. 12.8 (a) First-floor junction and (b) model assumed for inner leaf of ground-floor wall under vertical load.

Eccentricity of first-floor loads, $e = t/6 = 100/6 = 17$ mm $= 0.017$ m and the stress distribution will be similar to that shown in Fig. 12.9.

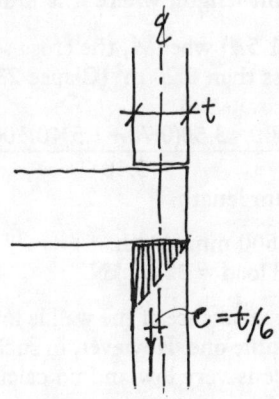

Fig. 12.9 Approximate stress distribution on inner leaf.

Hence, at position just below first floor for Case (a)

Design vertical load,
$P_1 + P_2 = 1.4 \times 10.71 + 1.6 \times 11.13$
$$= 15.00 + 17.81$$
$$= 32.81 \text{ kN}$$

Design bending moment due to first-floor loads
$$= (1.4 \times 1.83 + 1.6 \times 7.42)$$
$$\times 0.017$$
$$= 14.43 \times 0.017$$
$$= 0.25 \text{ kN-m}$$

Resultant eccentricity

at top of wall, $\quad e_x = \dfrac{0.25 \times 10^3}{32.81}$

$\qquad\qquad\qquad = 7.6$ mm

$\qquad\qquad\qquad = 0.08\,t$

Slenderness ratio $\quad = 14$

$\therefore \qquad\qquad \beta = 0.85$, from Table 7 of the Code (Table 5.4, Clause 32.2.1) which takes account of slenderness of wall and eccentricity of load at top of wall (Fig. 12.10).

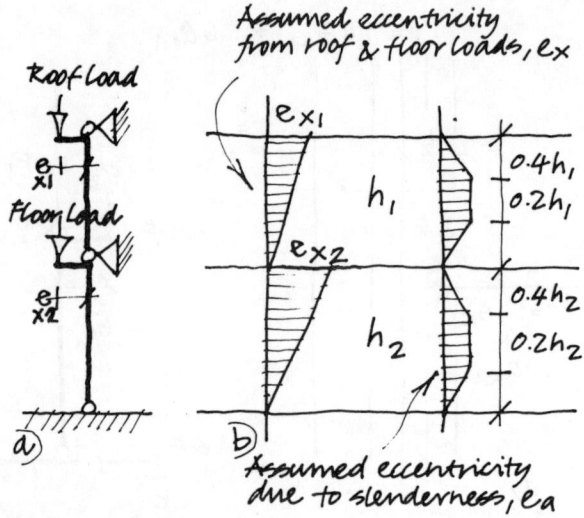

Fig. 12.10 (a) Model and (b) eccentricities used to calculate capacity reduction factor, β.

Design vertical load resistance of wall

$= \dfrac{\beta \cdot t \cdot f_k}{\gamma_m}$ per unit length, where f_k is multiplied by a factor $(0.70 + 1.5A)$ when A, the cross-sectional area of the wall, is less than 0.20 m^2 (Clause 23.1.1)

$= \dfrac{0.85 \times 100 \times 600 \times 3.50(0.70 + 1.5 \times 0.100 \times 0.600) \times 10^{-3}}{3.10}$ kN per 600 mm length.

$= 45.48$ kN per 600 mm length.

cf. design vertical load $= 32.81$ kN $\qquad$ OK

The design shear resistance of the wall is lowest at dpc level, if the dpc is a flexible one. However, in such cases as this the design shear force is very low and no calculation is usually necessary.

Remarks: the masonry requires no flexural strength for this calculation to be valid.

12.3.2 Horizontal loading of wall on east elevation

The worst position for horizontal loading, that is the position where the allowable tensile or compressive stress is most likely to be exceeded is on the 600 mm length of wall between the windows at first-floor level, which has the same wind pressure but less axial vertical load than the similar wall at ground-floor level. The wind load is resisted by both leaves of the wall.

Check on limiting dimensions:

Max. height allowed

$\qquad\qquad = 40 t_{ef}$ $\quad$ (Clause 36.3)

$\qquad\qquad = 40 \times 0.135 \qquad = 5.400$ m

Actual height $\qquad\qquad\qquad = 2.525$ m

$\qquad\qquad\qquad\qquad\qquad\qquad$ OK

Consider Case (b) of Clause 22: $0.9G_k + 1.4W_k$ which will give the maximum tensile stresses.

Characteristic vertical loads on wall element at mid-height (at window-sill level):

on inner leaf:

from roof, $G_k \qquad\qquad\qquad = 4.41$ kN

$\qquad\qquad Q_k \qquad\qquad\qquad = 3.71$ kN

from weight of inner leaf of wall,

$\quad G_k = 0.250 \times 1.200 \times 1.15$

$\qquad\quad + 1.250 \times 0.600 \times 1.15$

$\qquad = 0.34 + 0.86 \qquad = 1.20$ kN

from wind uplift on roof,

$$W_k = 1.800 \times \frac{5.500}{2} \times - 0.54 \text{ say}$$

$\qquad = 4.95 \times -0.54 \qquad = -2.67$ kN

on outer leaf:

from weight of outer leaf of wall,

$\quad G_k = 0.250 \times 1.200 \times 2.25$

$\qquad\quad + 1.250 \times 0.600 \times 2.25$

$\qquad = 0.67 + 1.68 \qquad = 2.35$ kN

Characteristic horizontal load on both leaves of wall at first-floor level:

$$W_k = 1.800 \times 2.450 \times 0.54$$

$\qquad\qquad\qquad = 2.38$ kN

Summary of characteristic vertical loads on cavity wall:

on inner leaf:

Dead loads, $G_k = 4.41 + 1.20 \qquad = 5.61$ kN

Live loads, $Q_k \qquad\qquad\qquad = 3.71$ kN

Wind loads, $W_k \qquad\qquad\qquad = -2.67$ kN

on outer leaf:

Dead loads, $G_k \qquad\qquad\qquad = 2.35$ kN

Design loads Considering Case (b) of Clause 22: $0.9G_k + 1.4W_k$

Design horizontal load $\quad = 1.4 \times 2.38 \qquad = 3.33$ kN

Design vertical load

on inner leaf: $\quad = 0.9 \times 5.61$

$\qquad\qquad\qquad - 1.4 \times 2.67 \qquad = 1.31$ kN

on outer leaf: $\quad = 0.9 \times 2.35 \qquad = 2.11$ kN

From Table 3 of the Code (Table 2.3), elastic flexural strength of wall,

f_{kx} *(parallel to bed joints)* $\quad = f_{ka}$

which *for inner leaf* $\qquad\qquad = 0.25$ N/mm^2

and *for outer leaf* $\qquad\qquad = 0.40$ N/mm^2

Design compressive stress in wall, g_A:

for inner leaf $\quad = \dfrac{1.31 \times 10^3}{600 \times 100} \qquad = 0.02$ N/mm^2

and

for outer leaf $\quad = \dfrac{2.11 \times 10^3}{600 \times 102.5} \qquad = 0.03$ N/mm^2

Effective f_{kx} (*parallel to bed joint*) $= f_{ka}$

for inner leaf $= 0.25 + 2.1 \times 0.02$ $= 0.31 \text{ N/mm}^2$
and
for outer leaf $= 0.40 + 3.1 \times 0.03$ $= 0.49 \text{ N/mm}^2$

$\therefore$ Elastic design moment of resistance $= \dfrac{f_{ka} \cdot Z}{\gamma_m}$

for inner leaf $= \dfrac{0.31 \times 600 \times (100)^2 \times 10^{-6}}{3.1 \times 6}$

$= 0.10 \text{ kN-m}$
and

for outer leaf $= \dfrac{0.49 \times 600 \times (102.5)^2 \times 10^{-6}}{3.1 \times 6}$

$= 0.16 \text{ kN-m}$

assuming the inner leaf to behave in bending like a solid section wall.

Elastic design moment in span

$$= \frac{W \cdot L}{12.5} = \frac{3.33 \times 2.45}{12.5}$$

$$= 0.65 \text{ kN-m}$$

from Case (vi) in Table A4.1; the higher bending moment at the support acts over a larger width and is not as critical as the bending moment in the span.

Design moment

in inner leaf $= \dfrac{0.10}{0.26} \times 0.65$ $= 0.25 \text{ kN-m}$

in outer leaf $= \dfrac{0.16}{0.26} \times 0.65$ $= 0.40 \text{ kN-m}$

i.e. inadequate strength.

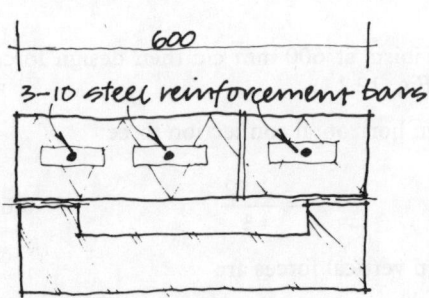

Fig. 12.11   *Reinforcing of masonry wall.*

The window frame not being able to provide additional strength in the critical area at mid-height, the most convenient way to give the necessary flexural strength to the wall is to reinforce the inner leaf with vertical steel bars, placed in the voids and concreted in position. It is advisable to use a stronger block, of about the same strength as the concrete infill. The calculation must now be repeated for the 600 mm length of wall at the ground floor to check its flexural strength. Alternatively, and perhaps more conveniently, the same steel bars may be used to reinforce the inner leaf of the blockwork from ground floor to roof. See Example A13.1.

Remarks: The partial safety factor for material strength has been taken as 3.1, the general figure for unsupervised small building work.

12.4   Connections

A section of the extension shows one of the sloping roof rafters (Fig. 12.12). These rafters, besides carrying roof load, have the vital role of stabilising the wall at the top in the same way that the first floor stabilises the wall at its mid-height. With suitable connection details there is no tendency for the sloping rafters to exert any significant horizontal forces on top of the wall when only vertical loads are present. However, the roof rafters and their connections must be able to transfer wind and other horizontal forces back to the existing building and to the side walls.

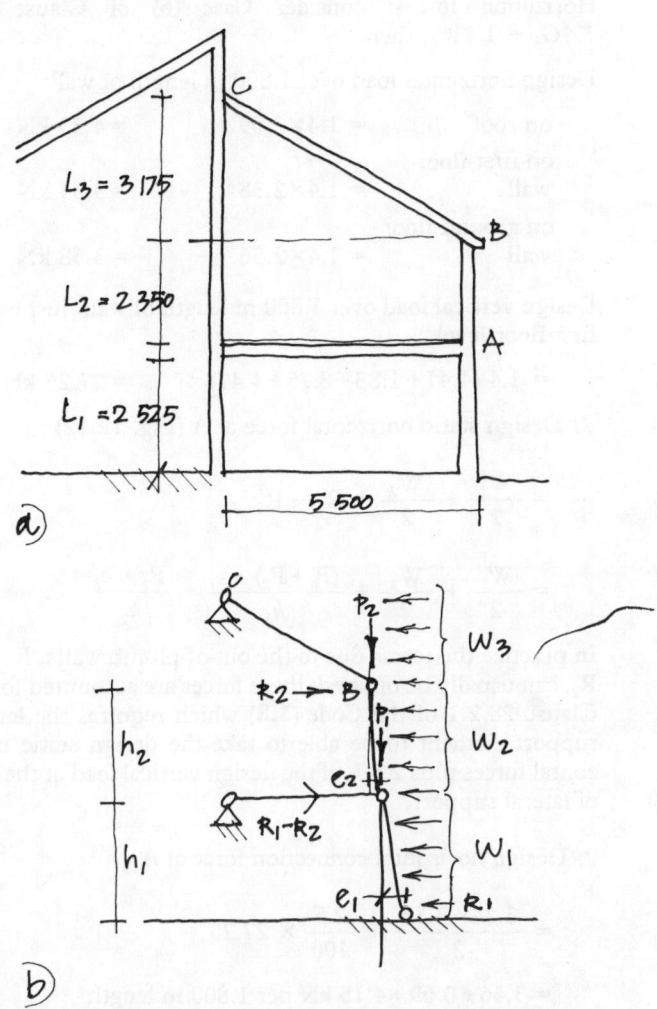

Fig. 12.12   *(a) Section and (b) model of wall under vertical and horizontal loads.*

Characteristic horizontal loads over 1.800 m length are:

on roof, $W_k = 1.800 \times 3.175 \times 0.54$ say

$= 3.09 \text{ kN}$

on first-floor wall, $W_k = 1.800 \times 2.450 \times 0.54$

$= 2.38 \text{ kN}$

on ground-floor wall, $W_k = 1.800 \times 2.625 \times 0.54$

$= 2.56 \text{ kN}$

Characteristic vertical loads over 1.800 m length of wall are:

from roof, G_k $= 4.41 \text{ kN}$
        Q_k $= 3.71 \text{ kN}$
from first floor, G_k $= 1.83 \text{ kN}$
        Q_k $= 7.42 \text{ kN}$

79

from wind
uplift
$$W_k = \frac{1.800 \times 5.500 \times -0.54}{2}$$
$$= -2.68 \text{ kN}$$

from self-weight of outer leaf of wall, at first-floor level
$$G_k = 3.885 \times 2.25 \qquad = 8.75 \text{ kN}$$

from self-weight of inner leaf of wall, at first-floor level
$$G_k = 3.885 \times 1.15 \qquad = 4.47 \text{ kN}$$

Connection forces at A – floor joists to wall

Horizontal forces: consider Case (b) of Clause 22: $1.4G_k + 1.4W_k$, then,

Design horizontal load over 1.800 m length of wall:

on roof	$= 1.4 \times 3.09$	$= 4.33$ kN
on first-floor wall	$= 1.4 \times 2.38$	$= 3.34$ kN
on ground-floor wall	$= 1.4 \times 2.56$	$= 3.58$ kN

Design vertical load over 1.800 m length of wall, just below first-floor level

$$= 1.4\,(4.41 + 1.83 + 8.75 + 4.47) \qquad = 27.25 \text{ kN}$$

∴ Design static horizontal force at A (Fig. 12.12)

$$= \frac{W_1}{2} + \frac{W_2}{2} + R_1 - R_2$$

$$= \frac{W_1}{2} + \frac{W_2}{2} + \frac{(P_1 + P_2) \cdot e_1}{h_1} - \frac{P_2 \cdot e_2}{h_2}$$

In practice the terms due to the out-of-plumb walls, R_1 and R_2, can usually be omitted; these forces are accounted for by Clause 28.2.1 of the Code (**5.3**) which requires the lateral support element to be able to take the design static horizontal forces plus 2.5% of the design vertical load at the line of lateral support,

∴ Design horizontal connection force at A

$$= \frac{3.34 + 3.58}{2} + \frac{2.5}{100} \times 27.25$$

$$= 3.46 + 0.69 = 4.15 \text{ kN per } 1.800 \text{ m length}$$

Vertical forces: consider Case (a) of Clause 22: $1.4G_k + 1.6Q_k$, then,

Design vertical force at A $= 1.4 \times 1.83 + 1.6 \times 7.42$
$$= 14.43 \text{ kN per } 1.800 \text{ m length}$$

with floor joists at 600 mm c/c then design forces on each joist are:

Design horizontal connection force $= \dfrac{4.15}{3}$
$$= 1.38 \text{ kN}$$

Design vertical force $= \dfrac{14.43}{3} = 4.81 \text{ kN}$

The Code assumes that, usually, for buildings of up to three storeys no straps are required. The horizontal forces are presumed to be taken in bond and friction. However, the friction force for Case (b) is small and, in some cases, straps may be required nevertheless; see App. **B4**.

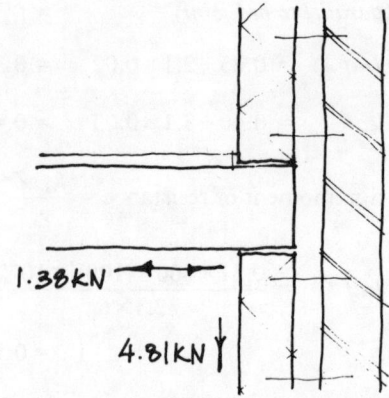

1.38KN

4.81KN

Fig. 12.13 Design vertical and horizontal forces on each joist spaced at 600 mm c/c.

Connection forces at B – roof joists to wall Horizontal forces: consider Case (b) of Clause 22 of the Code: $1.4G_k + 1.4W_k$, then,

Design vertical load over 1.800 m length of wall, at roof
$$= 1.4 \times 4.41 \qquad = 6.17 \text{ kN}$$

∴ Design horizontal connection force

$$= \frac{3.34}{2} + \frac{2.5}{100} \times 6.17$$

$$= 1.67 + 0.15 \quad = 1.82 \text{ kN per } 1.800 \text{ m length}$$

Vertical forces: consider Case (a) of Clause 22: $1.4G_k + 1.6Q_k$ for maximum force and Case (b) of Clause 22: $0.9G_k + 1.4W_k$ for minimum force.

∴ Design vertical forces at A

$$= 0.9 \times 4.41 + 1.4 \times -2.68$$
$$= -0.22 \text{ kN (min.) per } 1.800 \text{ m length}$$

and $= 1.4 \times 4.41 + 1.6 \times 3.71$
$$= 12.11 \text{ kN (max.) per } 1.800 \text{ m length}$$

with roof joists at 600 mm c/c then design forces on each joist are (Fig. 12.14):

Design horizontal connection force

$$= \frac{1.82}{3} \qquad = 0.61 \text{ kN}$$

Design vertical forces are

$$= \frac{-0.22}{3} \qquad = -0.08 \text{ kN}$$

or $= \dfrac{12.11}{3} \qquad = 4.04 \text{ kN}$

For cases where large wind uplift occurs straps will be needed to tie down the inner leaf. The Code allows direct tensile stress in the masonry to be relied on to resist wind uplift on roofs, using half the values given for f_{kx} in Table 3 of the Code (Table 2.3) (Clause 24.1).

∴ Design resistance to uplift at each joist
$$= \frac{0.25}{2} \times 100 \times \frac{600}{3.1} \times 10^{-3}$$

$$= 2.41 \text{ kN per } 600 \text{ mm length}$$

OK

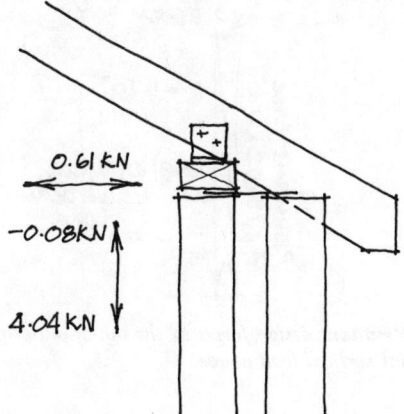

Fig. 12.14 *Design vertical and horizontal forces on each roof joist, showing wallplate and angle fixing bracket.*

Connection forces at C – roof joists to existing wall Horizontal wind forces on the roof will transfer to the side walls, which are in the same plane as the wind forces. However, it will be assumed for the purposes of calculating the connection force at C, that half the applied horizontal forces are taken by the connection at C (Fig. 12.15).

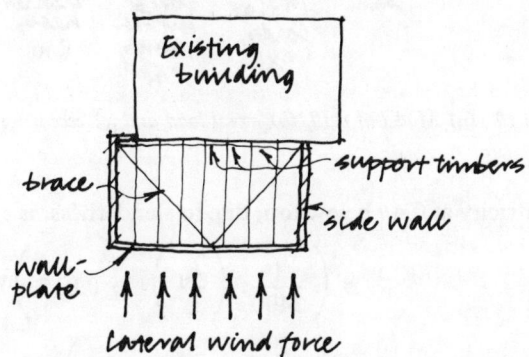

Fig. 12.15 *Plan on roof of extension showing roof joists and bracing.*

Horizontal forces:

Applied horizontal forces on roof: consider Case (b) of Clause 22: $1.4G_k + 1.4W_k$

Applied horizontal force on roof over 1.800 m

$$= W_3 + \frac{W_2}{2}$$

$$= 4.33 + \frac{3.34}{2} \qquad = 6.00 \text{ kN}$$

∴ Design horizontal connection force at C over 1.800 m length of wall

$$= \frac{6.00}{2} \text{ say} \qquad = 3.00 \text{ kN}$$

Vertical forces: consider Case (a) of Clause 22: $1.4G_k + 1.6Q_k$ for maximum force and Case (b) of Clause 22: $0.9G_k + 1.4W_k$ for minimum force

as at B, design vertical forces at C are:

$$= -0.22 \text{ kN (min.) per 1.800 m} \\ \text{length}$$

and $= 12.11 \text{ kN (max.) per 1.800 m}$
length

With roof joists at 600 mm c/c then design forces on each joists are,

Design horizontal connection force

$$= \frac{3.00}{3} \qquad = 1.00 \text{ kN}$$

Design vertical force $\qquad = -0.08 \text{ kN}$
or $\qquad = 4.04 \text{ kN}$

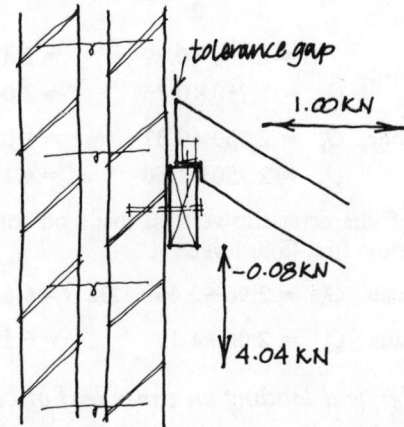

Fig. 12.16 *Design vertical and horizontal forces on support timber.*

Remarks: Sound connection details are essential to the stability of masonry buildings. Note that net wind uplift on a roof being the difference between roof dead load and wind suction is sensitive to any small change in these two quantities.

12.5 Bowed and out-of-plumb ground-floor wall – vertical and horizontal loads

The wall is re-checked because after construction a part of the ground-floor wall on the east elevation, where there are no window openings, is found to be out-of-plumb by 75 mm and bowed by 20 mm.

The design moment of resistance of the wall and the applied horizontal load is unchanged. However, the eccentricity of all the vertical loads has greatly increased.

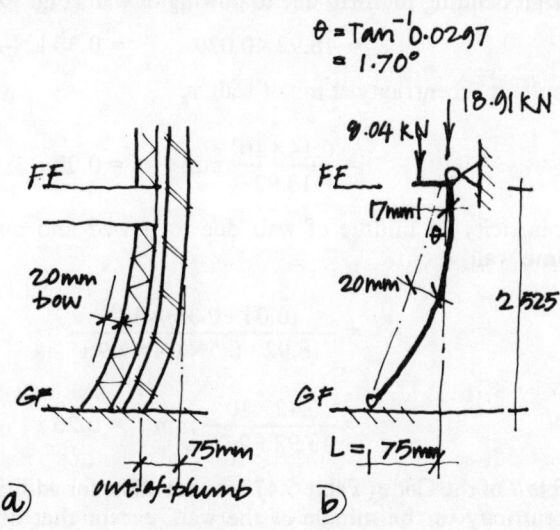

Fig. 12.17 *(a) Bowed wall and out-of-plumb wall and (b) model under design vertical loads.*

Vertical loading of out-of-plumb wall on east elevation
Consider a 1 m length of the inner leaf of blockwork.

Characteristic vertical loads on inner leaf of wall:

from self-weight of inner leaf of blockwork between ground- and first-floor level,

$$G_k = 1 \times 2.525 \times 1.15 = 2.90 \text{ kN/m}$$

from weight of first-floor wall,

$$G_k = 1 \times 2.575 \times 1.15 = 2.96 \text{ kN/m}$$

from roof,
$$G_k = 1 \times \frac{5.500}{2} \times 0.89$$

$$= 2.750 \times 0.89 = 2.45 \text{ kN/m}$$
$$Q_k = 2.750 \times 0.75 = 2.06 \text{ kN/m}$$

from first floor, $G_k = 2.750 \times 0.37 = 1.02 \text{ kN/m}$
$$Q_k = 2.750 \times 1.50 = 4.13 \text{ kN/m}$$

Summary of characteristic vertical loads on inner leaf of wall, just below first-floor level:

Dead loads $G_k = 2.96 + 2.45 + 1.02 = 6.43 \text{ kN/m}$

Live loads $Q_k = 2.06 + 4.13 = 6.19 \text{ kN/m}$

12.5.1 Vertical loading on inner leaf of bowed and out-of-plumb wall

Consider Case (a) of Clause 22: $1.4G_k + 1.6Q_k$

Design vertical load, at top of wall (Fig. 12.17)

$$= 1.4 \times 6.43 + 1.6 \times 6.19$$
$$= 18.91 \text{ kN/m}$$

Design bending moments due to eccentric first-floor loads,

$$= (1.4 \times 1.02 + 1.6 \times 4.13) \times 0.017$$
$$= 8.04 \times 0.017 = 0.14 \text{ kN-m/m}$$

Design bending moment due to self-weight of inner leaf

$$= \frac{W \cdot L}{8}$$

$$= \frac{1.4 \times 2.80 \times 0.075}{8} = 0.04 \text{ kN-m/m}$$

Design bending moment due to bowing of wall (Fig. 12.18)

$$= 18.92 \times 0.020 = 0.38 \text{ kN-m/m}$$

Resultant eccentricity at top of wall, e_x

$$= \frac{0.14 \times 10^3}{18.92} \text{ mm} = 0.08 t$$

Eccentricity in middle of wall due to bowed and out-of-plumb wall, e_z

$$= \frac{(0.04 + 0.38) \times 10^3}{18.92 + 0.5 \times 1.4 \times 2.96}$$

$$= \frac{0.42 \times 10^3}{18.92 + 2.07} \text{ mm} = 0.20 t$$

Table 7 of the Code (Table 5.4) does not allow for additional eccentricity in the middle of the wall, except that due to slenderness. Hence the capacity reduction factor will be calculated using Appendix B of the Code.

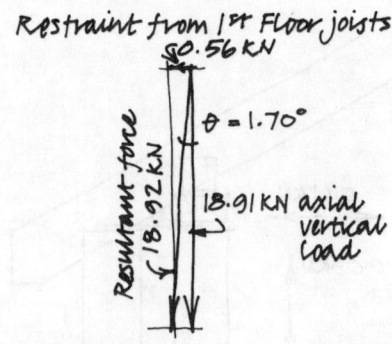

Fig. 12.18 Resultant design forces at the top of out-of-plumb wall due to an axial vertical load above.

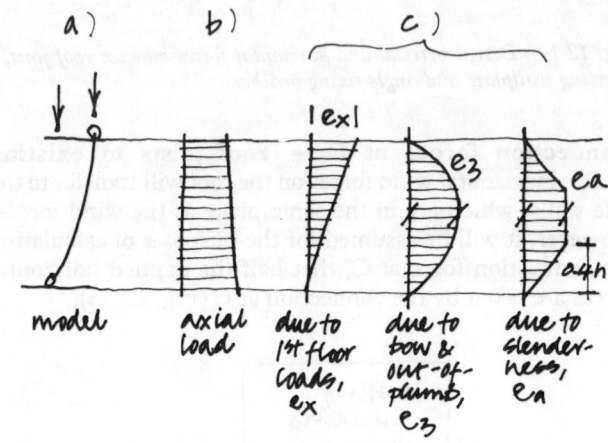

Fig. 12.19 (a) Model of wall, (b) axial load and (c) eccentricities in wall.

Eccentricity at $0.4h$ below top, due to slenderness, is e_a

$$= \left(\frac{1}{2400} \cdot \left(\frac{h_{ef}}{t_{ef}} \right)^2 - 0.015 \right) t$$

$$= \left(\frac{1}{2400} \cdot (14)^2 - 0.015 \right) t$$

$$= 0.067 t$$

$\therefore$ Total eccentricity at $0.4h$ below top, e_t (Fig. 12.19)

$$= 0.6e_x + e_z + e_a$$
$$= 0.6 \times 0.08t + 0.20t + 0.067t$$
$$= 0.32t$$

$\therefore e_t > e_x \therefore e_m = e_t = 0.32t$

This eccentricity can be assumed to be just acceptable being less than, say, $0.35t$.

$\therefore$ Capacity reduction factor, β

$$= 1.1 \left(1 - 2\frac{e_m}{t} \right)$$

$$= 1.1 (1 - 2 \times 0.32) = 0.40$$

$\therefore$ Design vertical load resistance of inner leaf of wall

$$= \frac{\beta \cdot t \cdot f_k}{\gamma_m} \text{ per unit length}$$

$$= \frac{0.40 \times 100 \times 1,000 \times 3.50 \times 10^{-3}}{3.1}$$

$$= 46 \text{ kN/m}$$

cf. design vertical load

$$= 18.91 + 2.03 \qquad = 21 \text{ kN/m}$$

OK

12.5.2 Horizontal loading on inner leaf of bowed and out-of-plumb wall

Consider Case (b) of Clause 22: $1.4G_k + 1.4W_k$

inner leaf of wall:

Design vertical load at top of wall, ignoring wind uplift,

$$= 1.4 \times 6.43 \qquad = 9.00 \text{ kN/m}$$

Design bending moment due to eccentric first-floor loads,

$$= 1.4 \times 1.02 \times 0.017 \quad = 0.03 \text{ kN-m/m}$$

Design bending moment due to self-weight of inner leaf of wall

$$= 0.04 \text{ kN-m/m}$$

Design bending moment due to bowing of wall,

$$= 9.00 \times 0.020 \qquad = 0.18 \text{ kN-m/m}$$

∴ Design bending moment in inner leaf of wall at middle

$$= (0.6 \times 0.03) + 0.04 + 0.18$$
$$= 0.24 \text{ kN-m/m}$$

Design bending moment due to wind, on both leaves of wall from, say, Case (v) in Table A4.1.

$$= \frac{W.L}{14}$$

$$= \frac{1.4 \times (1 \times 2.625 \times 0.54) \times 2.625}{14}$$

$$= 0.37 \text{ kN-m/m}$$

The inner leaf of the wall takes wind forces in proportion to its elastic design moment of resistance, hence total design bending moment on inner leaf of wall

$$= 0.24 + \frac{0.10}{0.26} \times 0.37 \quad = 0.38 \text{ kN-m/m}$$

Assuming, conservatively, that f_{ka} is the same as that in the top storey,

Design moment of resistance of inner leaf

$$= \frac{0.31 \times 10^3 \times (100)^2 \times 10^{-6}}{3.1 \times 6}$$

$$= 0.16 \text{ kN-m/m}$$

Hence there is inadequate flexural strength and the wall needs rebuilding.

Remarks: Note that the reduction in the design vertical load resistance is almost entirely due to bow in the wall, an out-of-plumb but straight wall having negligible effect on this. However, it should be noted that the out-of-plumb ground-floor wall has increased the design horizontal connection force at first-floor level (Fig. 12.18). In the present case, the design horizontal connection force at first-floor level, over 1.00 m length of wall, would be **(12.4)**

$$3.46/1.8 + 1.4(6.43 + 2.575 \times 2.25)0.075/2.525 = 2.43 \text{ kN}$$

if first-floor wall were plumb.

Hence with floor joists at 600 mm c/c,

Design horizontal connection force per joist (Fig. 12.20)

$$= 2.43 \times 0.600 \qquad = 1.46 \text{ kN}$$

In general, because of the required horizontal force, a wall should not be out-of-plumb by more than about 25 mm over a storey-height; in new work the wall should be built within 8 mm out-of-plumb over a storey-height of 2.4 m.

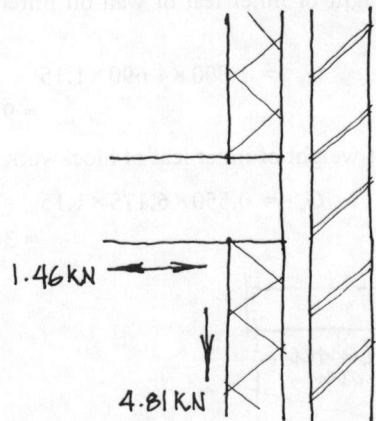

Fig. 12.20 *Design connection forces on inner leaf at first-floor level junction.*

12.6 Wall on south elevation – vertical and horizontal loads

12.6.1 Vertical loading of wall on south elevation

The position where the allowable compressive stress is most likely to be exceeded is in the inner leaf of the short wall, to one side of the garage doors (Figs 12.1 and 12.21).

Slenderness ratio of wall:

Effective thickness, t_{ef}

$$= \tfrac{2}{3}(102.5 + 100) \qquad = 135 \text{ mm}$$

Effective height, h_{ef}

$$= 1 \times 2,525 \qquad = 2,525 \text{ m}$$

(Simple resistance to lateral movement assuming strap connection to wall from floor, see Fig. 12.22).

∴ Slenderness ratio

$$(SR) \qquad = \frac{2,525}{135} \qquad = 18.70$$

Allowable SR $= 27$ OK

Length of wall $= 550 \text{ mm} > 4t_{ef} = 4 \times 135 = 540 \text{ mm}$, hence treat as wall not column.

Characteristic vertical load on inner leaf just below first-floor level:

from roof load and wind uplift on lintel over entrance door

$$G_k = 0.45 \times \frac{5.500}{2} \times 0.89$$

$$= 1.24 \times 0.89 \qquad = 1.10 \text{ kN}$$

$$Q_k = 1.24 \times 0.75 \qquad = 0.93 \text{ kN}$$

$$W_k = 1.24 \times -0.54 \qquad = -0.67 \text{ kN}$$

83

from floor load on lintel over entrance door

$$G_k = 1.24 \times 0.37 \qquad = 0.46 \text{ kN}$$

$$Q_k = 1.24 \times 1.50 \qquad = 1.86 \text{ kN}$$

from weight of inner leaf of wall on lintel over entrance door

$$G_k = 0.450 \times 6.175 \times 1.15$$
$$= 3.19 \text{ kN}$$

from weight of inner leaf of wall on lintel over garage doors

$$G_k = 1.800 \times 4.690 \times 1.15$$
$$= 9.71 \text{ kN}$$

from self-weight of inner leaf of blockwork

$$G_k = 0.550 \times 6.175 \times 1.15$$
$$= 3.90 \text{ kN}$$

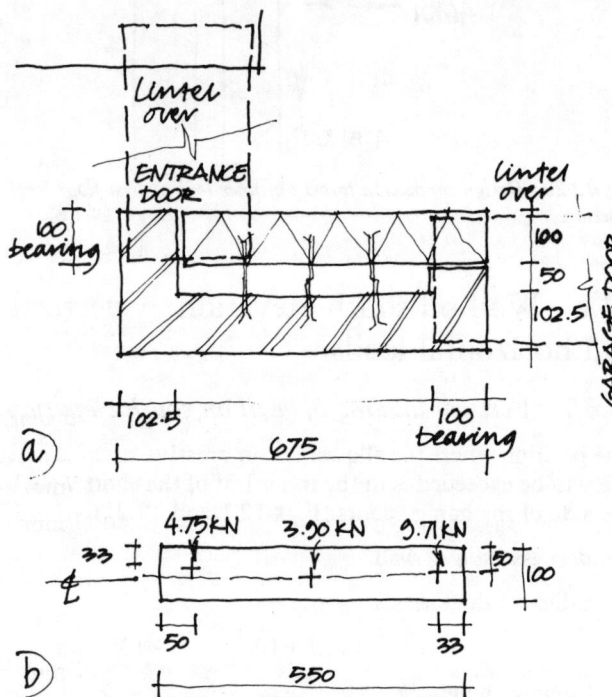

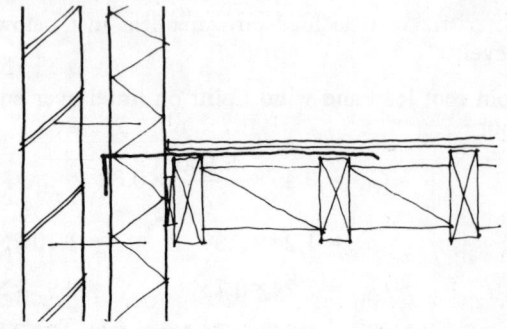

Fig. 12.21 (a) Plan on cavity wall and (b) plan on inner skin, showing characteristic dead loads.

Summary of characteristic vertical loads on inner leaf just below first-floor level:

Dead loads, $G_k = 1.10 + 0.46 + 3.19 + 9.71 + 3.90$
$$= 4.75 + 9.71 + 3.90 \quad = 18.36 \text{ kN}$$

Live loads, $Q_k = 0.93 + 1.86 \qquad = 2.79 \text{ kN}$

Wind loads, $W_k \qquad\qquad\qquad = -0.67 \text{ kN}$

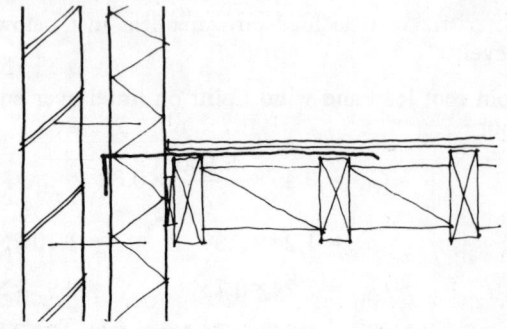

Fig. 12.22 Floor to wall detail.

Design vertical loads: Consider Case (a) of Clause 22: $1.4G_k + 1.6Q_k$

Check

(a) concentrated load under lintels:
Garage door lintel is worst case and this is the same as Case (d) in Fig. A8.1

∴ local design strength

$$= \frac{1.25 f_k}{\gamma_m}$$

∴ design local bearing resistance

$$= \frac{1.25 \times 3.50 \times (100 \times 100) \times 10^3}{3.10}$$
$$= 14.11 \text{ kN}$$

design local load $= 1.4 \times 9.71 \qquad = 13.59 \text{ kN}$

just OK

In practice some of the load will arch into the wall so that the actual local load on the lintel bearing will be slightly less than that calculated (Fig. 12.23); the lintel acts partly as a tie.

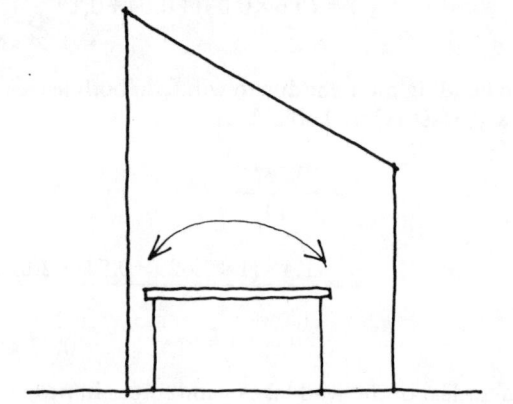

Fig. 12.23 Arching of part of the load over the garage opening.

(b) Wall 0.4h below top:

The wall has eccentricities about both the major and minor axes at the top. However, the element is classified as a 'wall', according to the Code, because $b > 4t_{ef}$ (Fig. 12.24).

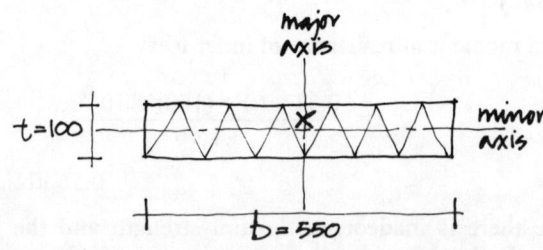

Fig. 12.24 Plan on wall showing position of resultant load.

Contrary to the case of a 'column', only eccentricity about the minor axis need be considered, in general, hence

Design vertical load $= 1.4 \times 18.36 + 1.6 \times 2.79$
$$= 30.17 \text{ kN}$$

(this load is assumed to be uniformly distributed along the 550 mm length of the wall at this level; otherwise see **5.2.1e**).

Design bending moment due to eccentric lintel load about

minor axis

$$= (1.4 \times 4.75 + 1.6 \times 2.79) \times 0.017$$

$$= 11.11 \times 0.017 \qquad = 0.19 \text{ kN-m}$$

∴ resultant eccentricity, e_x (Fig. 12.25)

$$= \frac{0.19 \times 10^3}{30.17} = 7 \text{ mm}$$

$$= 0.07t$$

Slenderness ratio (SR) $\qquad = 18.70$

∴ $\beta = 0.72$ by interpolation from Table 5.4.

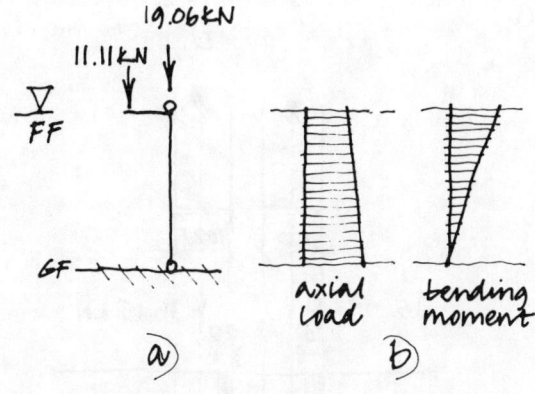

Fig. 12.25 (a) Design vertical loads and (b) axial load and bending moment pattern in wall.

Design vertical load resistance

$$= \frac{\beta . t . f_k}{\gamma_m} \text{ per unit length, where } f_k \text{ is multiplied by}$$

$$(0.70 + 1.5A)$$

$$= \frac{0.72 \times 100 \times 550 \times 3.50(0.70 + 1.5 \times 0.100 \times 0.550) \times 10^{-3}}{3.10}$$

$$= 34.98 \text{ kN}$$

cf. design vertical load $= 30.17$ kN

just OK

12.6.2 Horizontal loading on wall on south elevation

Short wall Check short wall to one side of garage doors at mid-height. Horizontal load resisted by both leaves of wall, which are 550 mm and 675 mm long.

Check on limiting dimensions:

Max. height allowed

$$= 40t_{ef} \text{ (Clause 36.3)}$$

$$= 40 \times 0.135 \qquad = 5.400 \text{ m}$$

Actual height $\qquad = 2.525$ m

OK

Consider Case (b) of Clause 22: $0.9G_k + 1.4W_k$

Characteristic vertical load on outer leaf just below first-floor level:

from weight of outer leaf on lintel over entrance door:

$$G_k = 0.450 \times 6.175 \times 2.25$$

$$= 6.25 \text{ kN}$$

from weight of outer leaf on lintel over garage doors:

$$G_k = 1.800 \times 4.690 \times 2.25$$

$$= 19.00 \text{ kN}$$

from self-weight of outer skin of brickwork above:

$$G_k = 0.675 \times 6.175 \times 2.25$$

$$= 9.37 \text{ kN}$$

Summary of vertical loads on outer leaf:

Dead loads, $G_k = 6.25 + 19.00 + 9.37 = 34.62$ kN

Horizontal load on both leaves of wall:

$$W_k = (0.675 + 1.800) \times 2.625 \times 0.54$$

$$= 3.51 \text{ kN}$$

Design moment of resistance of wall:

Consider Case (b) of Clause 22: $0.9G_k + 1.4W_k$

Design vertical load:

on inner leaf $\quad = 0.9 \times 18.36 - 1.4 \times 0.67$

$$= 15.59 \text{ kN}$$

on outer leaf $\quad = 0.9 \times 34.62 \quad = 31.15 \text{ kN}$

Design horizontal load

$$= 1.4 \times 3.51 \qquad = 4.92 \text{ kN}$$

'Elastic' design moment of resistance:

Elastic flexural strength of wall, f_{kx} (*parallel to bed joint*)
$= f_{ka}$

for inner leaf $\qquad = 0.25 \text{ N/mm}^2$

and *for outer leaf* $\qquad = 0.40 \text{ N/mm}^2$

from Table 3 of the Code (Table 2.3).

Design compressive stress, g_A

for inner leaf $\quad = \dfrac{15.59 \times 10^3}{550 \times 100} \quad = 0.28 \text{ N/mm}^2$

and *for outer leaf* $\quad = \dfrac{31.15 \times 10^3}{675 \times 102.5} \quad = 0.46 \text{ N/mm}^2$

Effective $f_{kx} \qquad = f_{ka}$

for inner leaf $\quad = 0.25 + 3.1 \times 0.28 \quad = 1.11 \text{ N/mm}^2$

and *for outer leaf* $= 0.40 + 3.1 \times 0.46 \quad = 1.82 \text{ N/mm}^2$

∴ Elastic design moment of resistance $= \dfrac{f_{ka} . Z}{\gamma_m}$ which

for inner leaf $\quad = \dfrac{1.11 \times 550 \times (100)^2 \times 10^{-6}}{3.1 \times 6}$

$$= 0.32 \text{ kN-m}$$

and *for outer leaf* $\quad = \dfrac{1.82 \times 675 \times (102.5)^2 \times 10^{-6}}{3.1 \times 6}$

$$= 0.69 \text{ kN-m}$$

and *for both leaves* $\qquad = 1.01$ kN-m

This 'elastic' design moment of resistance is smaller than any likely 'elastic' design moments; see Table A4.1. Therefore a 'plastic' method of design will be tried.

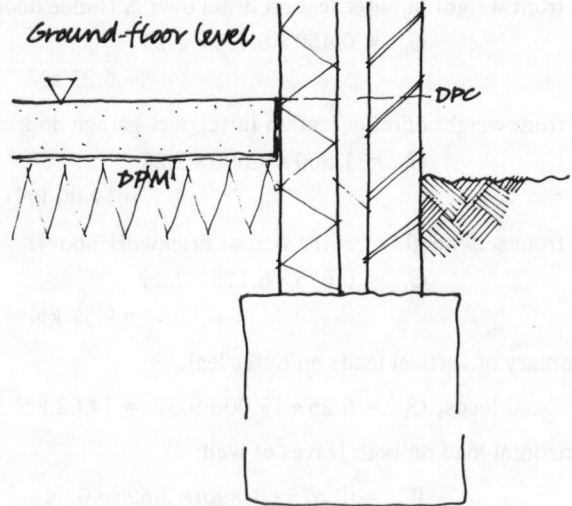

Fig. 12.26 Detail at ground-floor level.

'Plastic' design moment of resistance:

A flexible dpc usually only takes negligible tensile stress across it. Where it is important to maintain the tensile stress and therefore the elastic flexural strength of the wall, an engineering brick may be used as a dpc.

In this case a flexible dpc is used. Hence when sufficient bending moment is applied to the wall at dpc level it cracks relatively easily. However, as the wall hinges about one side the axial load in the wall comes into play and enables the wall to resist bending (Fig. 12.27). This bending resistance will be known as the 'plastic' flexural strength of the wall and will be higher when the axial load in the wall above is high. The 'plastic' flexural strength depends on the compressive strength of the wall but mainly on the axial load. By contrast the elastic flexural strength of a wall depends on its effective tensile strength which, however, is increased by axial load.

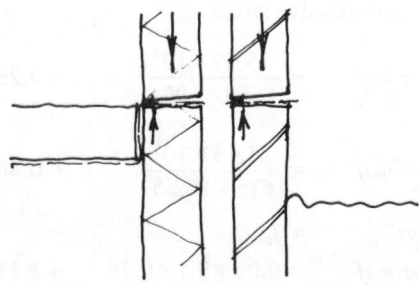

Fig. 12.27 'Plastic' resistance to bending due to axial load in wall.

Max. allowable compressive stress at 'hinge'

$$= \frac{f_k}{\gamma_m} \text{ which}$$

for inner leaf $= \dfrac{3.50}{3.10}$ $= 1.12 \text{ N/mm}^2$

and *for outer leaf* $= \dfrac{7.10}{3.10}$ $= 2.29 \text{ N/mm}^2$

Min. width of stress blocks, $s = \dfrac{P}{f_k/\gamma_m}$ per unit length (Fig. 12.28) which

for inner leaf $= \dfrac{15.59}{1.12 \times 0.550}$ $= 25 \text{ mm}$

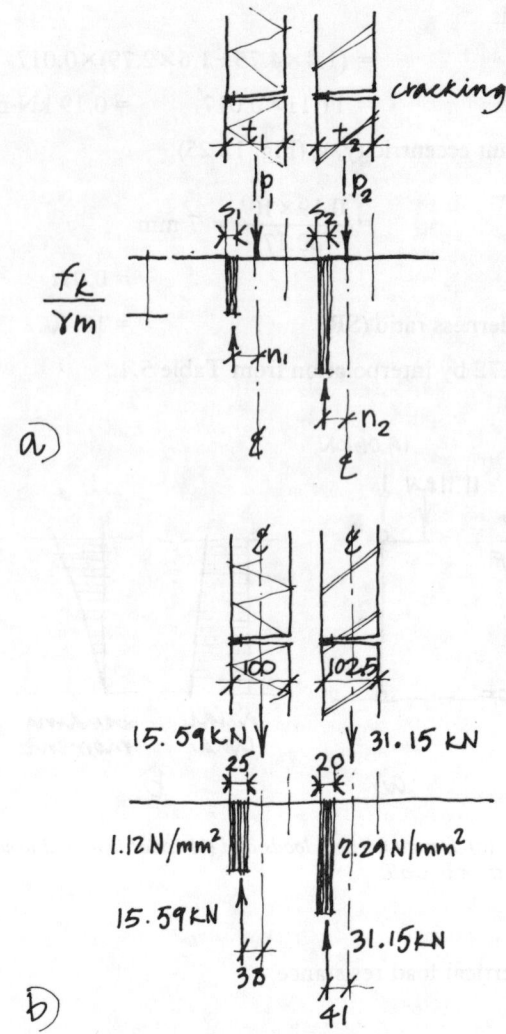

Fig. 12.28 'Plastic' flexural strength showing assumed compressive stress at edge of walls: (a) for general case and (b) for the particular design case.

and *for outer leaf* $= \dfrac{31.15}{2.29 \times 0.675}$ $= 20 \text{ mm}$

hence lever arm, $n = \dfrac{t}{2} - \dfrac{s}{2}$ which

for inner leaf $= 50 - 12$ $= 38 \text{ mm}$

and *for outer leaf* $= 51 - 10$ $= 41 \text{ mm}$

'Plastic' design moment of resistance $= P \cdot n$ which

for inner leaf $= 15.59 \times 38 \times 10^{-3}$ $= 0.59 \text{ kN-m}$

and *for outer leaf* $= 31.15 \times 41 \times 10^{-3}$ $= 1.27 \text{ kN-m}$

and *for both leaves of wall* $= 1.86 \text{ kN-m}$

Hence for the ground-floor cavity wall

$$i = \frac{\text{'plastic' design moment of resistance}}{\text{'elastic' design moment of resistance}} = \frac{1.86}{1.01}$$

$= 1.84$, say $= 1$

The 'elastic' and 'plastic' design moments of resistance will each have a constant value over the height of the ground-floor wall, as the axial load is approximately constant too.

Design moment in wall:

Take Case (iv) from Table A4.2(a) as the most appropriate

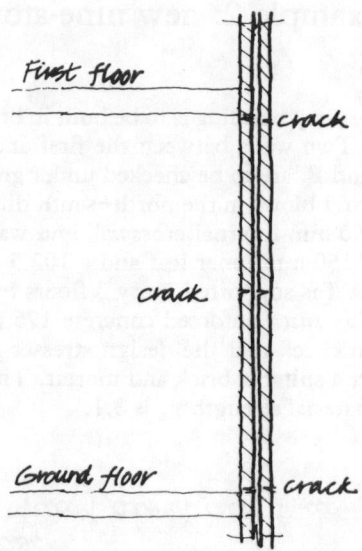

Fig. 12.29 *Cracked sections mobilising 'plastic' flexural strength of a wall in the span and at supports (i = 1).*

with 'plastic' hinges forming at dpc and first-floor level assuming $i = 1$ for both ground- and first-floor cavity walls.

'Plastic' design moment in wall at ground- and first-floor level

$$= \frac{W.L}{11.5}$$

$$= \frac{4.92 \times 2.625}{11.5} \quad = 1.12 \text{ kN-m}$$

'Elastic' design moment in ground-floor wall

$$= \frac{W.L}{25}$$

$$= \frac{4.92 \times 2.625}{25} \quad = 0.51 \text{ kN-m}$$

'Elastic' design moment in first-floor wall

$$= \frac{W.L}{11.5} \quad = 1.12 \text{ kN-m}$$

At ground floor,
'Plastic' design moment of resistance   = 1.86 kN-m
and 'Elastic' design moment of resistance = 1.01 kN-m

Because of the large area of wall, and the return wall above the entrance door, the 'elastic' design moment of resistance above first-floor level is clearly much greater than 1.12 kN-m. OK

Remarks: If, in a single-span wall fixed at both ends and under vertical load, the 'elastic' design moment of resistance in the span is exceeded, the wall cracks and forms 'plastic' hinges (Fig. 12.29); the design moment of resistance at supports and in the span are approximately equal and the bending moment pattern is the same as that for Case (iii) in Table A4.2(a). Such considerations also apply in this example, so that i is taken as equal to one, rather than a half, when estimating the design moment in the ground-floor cavity wall. Note that Case (iii) in Table A4.2(a) is, in effect, similar to the case treated in Clause 36.8 of the Code; however, Clause

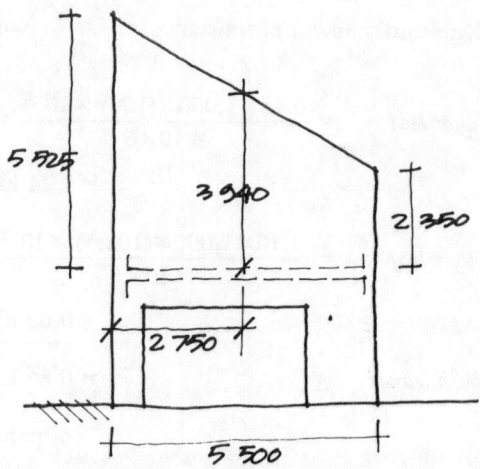

Fig. 12.30 *Elevation on south wall.*

36.8 uses a slightly higher overall factor of safety than the 'plastic' hinge method described; the Code recommends that the ratio, height divided by thickness, does not exceed 20 (or 25 for narrow brick walls).

Masonry panel at 1st-floor level

The first-floor wall carries only small vertical loads but will undergo bending due to wind forces. Check the wall as a wind panel (Fig. 12.30). The worst bending is assumed to take place on the vertical line 2.750 m from the end of the west wall and at mid-height on this line. Treat as a rectangular panel with sides 5.500 m say × 3.940 m having simple restraint on the top edge and moment restraint on the other three sides (Fig. 12.31).

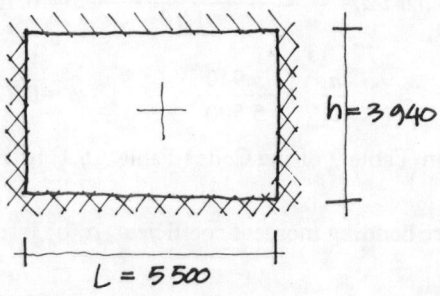

Fig. 12.31 *Assumed equivalent wall panel.*

Limiting dimensions of wall panel from Clause 36.3 of the Code (**6.2**) given by

height or length $\leqslant 50 \times 0.135$	= 6.750 m
height × length $\leqslant 2{,}250 \times (0.135)^2$	= 41.00 m²
cf. actual height × length = 3.940×5.500	= 21.67 m²

OK

Design moment of resistance:

Consider Case (b) of Clause 22: $0.9G_k + 1.4W_k$

Elastic flexural strength of wall panel, f_{kx} (*perpendicular to bed joint*) = f_{kb}, which

for inner leaf	= 0.45 N/mm²
and *for outer leaf*	= 1.10 N/mm²

from Table 3 of the Code (Table 2.3).

87

∴ Elastic design moment of resistance $= \frac{f_{kb} \cdot Z}{\gamma_m}$ which

$$for\ inner\ leaf = \frac{0.45 \times 1{,}000 \times (100)^2 \times 10^{-6}}{3.10 \times 6}$$

$$= 0.24\ \text{kN-m}$$

$$for\ outer\ leaf = \frac{1.10 \times 1{,}000 \times (102.5)^2 \times 10^{-6}}{3.10 \times 6}$$

$$= 0.62\ \text{kN-m}$$

and *for both leaves* $= 0.86$ kN-m
per m height
of panel

Design moment:

Design vertical stress at centre of panel, where bending moment is assumed to be greatest, g_d,

$$for\ inner\ leaf = 0.9 \times \frac{3.940}{2} \times \frac{1.15}{100}$$

$$= 0.02\ \text{N/mm}^2$$

and *for outer leaf* $= 0.9 \times \frac{3.940}{2} \times \frac{2.25}{102.5}$

$$= 0.03\ \text{N/mm}^2$$

Orthogonal strength ratio, $\mu = \frac{f_{ka}}{f_{kb}}$

$$for\ inner\ leaf = \frac{0.25 + 3.1 \times 0.02}{0.45} = 0.69$$

and *for outer leaf* $= \frac{0.40 + 3.1 \times 0.03}{1.10} = 0.45$

$$\frac{h}{L} = \frac{3.940}{5.500} = 0.72$$

Case H in Table 9 of the Code (Table A6.1) is the relevant case:

and hence bending moment coefficient, α, by interpolation, is

for inner leaf	$= 0.021$
and *for outer leaf*	$= 0.026$

Design moment $= \alpha(W_k \cdot \gamma_f)L^2$

∴ Maximum characteristic wind load, W_k

$$on\ inner\ leaf = \frac{0.24}{0.021 \times 1.4 \times 5.5^2} = 0.27\ \text{kN/m}^2$$

$$on\ outer\ leaf = \frac{0.62}{0.026 \times 1.4 \times 5.5^2} = 0.56\ \text{kN/m}^2$$

and *for both leaves* $= 0.83$ kN/m^2

OK

12.7 Example 2: new nine-storey building

A new nine-storey building is to be built in brick as shown in Fig. 12.32. Two walls between the first and second floor, marked A and B, are to be checked under gravity loads and when the wind blows in the north–south direction. Wall A is a solid 150 mm internal crosswall and wall B is a cavity wall with a 150 mm inner leaf and a 102.5 mm outer leaf. The outer leaf is supported every 3 floors by the floor slab which is of *in situ* reinforced concrete 175 mm thick. It is necessary to check that the design stresses are satisfactory and to select a suitable brick and mortar. The partial safety factor for material strength γ_m is 3.1.

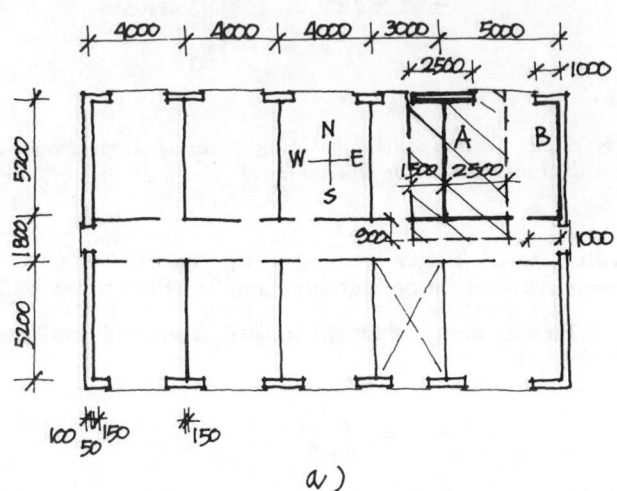

a)

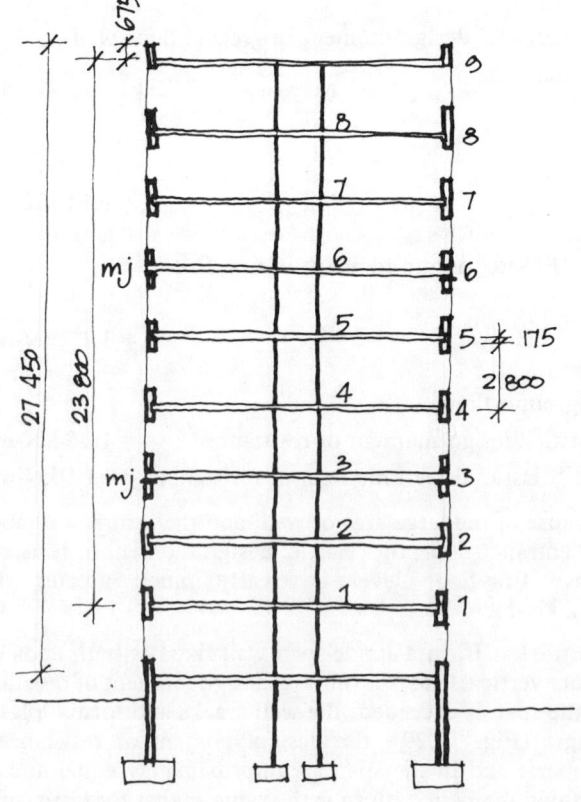

b)

Fig. 12.32 (a) Plan and (b) section on nine-storey building.

12.8 Characteristic loadings:

Roof:

Dead load, G_k = 3.6 kN/m²

Live load, Q_k = 1.5 kN/m²

Floor:

Dead load including weight of partitions, G_k
       = 4.8 kN/m²

Live load, Q_k = 1.5 kN/m²

Wall:

Dead load, G_k

outer skin of cavity wall:
102.5 mm brickwork   = 2.25 kN/m²

internal crosswalls and inner skin of cavity wall:
150 mm perforated brickwork and plaster
       = 3.80 kN/m²

Wind load, W_k

Dynamic wind pressure at top of building, q
       = 1.40 kN/m²

Take wind force on north or south face above first-floor level

$= C_f . q . A_e = 1.1 \times 1.40 \times 23.800 \times 20.400$

       = 748 kN

Dynamic wind pressure at 1st floor, q = 1.00 kN/m²

Wind suction on second-floor east wall with north/south wind

$= (C_{pe} - C_{pi}) . q$

$= -0.9 \times 1.00$ = 0.9 kN/m²

from CP3: Chapter V; Part 2: 1972 Clauses 7.2 and 7.3.

12.9 Vertical loads and bending moments on walls A and B just above first-floor level

For wall A,

Shaded part of floor (Fig. 12.32) has areas of $4.0 \times 6.1 = 24.4$ m² and length of load-bearing walls in shaded area = $2.5 + 5.2 + 4.0 = 11.7$ m, hence characteristic vertical loads on wall, assuming floor loads to be evenly distributed on the load-bearing walls, are

$G_k = (3.6 + 7 \times 4.8)24.4/11.7 + 8 \times 3.80 \times 2.800$

       = 163 kN/m

$Q_k = 8 \times 1.5 \times 0.6 \times 24.4/11.7$ = 15.0 kN/m

with 60% live load reduction from CP3: Chapter V: Part 1

and for wall B,

$G_k = (3.6 + 7 \times 4.8)15.25/7.2 + 8 \times 3.80 \times 2.800$
     $+ (0.675 + 6 \times 2.800)2.25$ = 203 kN/m

      including weight of outer leaf of wall above third floor.

$Q_k = 8 \times 1.5 \times 0.6 \times 15.25/7.2$ = 15.2 kN/m

Characteristic load on 3 m floor span is:

   $G_k = 14.4$ kN/m

   $Q_k = 4.5$ kN/m

and characteristic load on 5 m floor span is:

   $G_k = 24.0$ kN/m

   $Q_k = 7.5$ kN/m

Moment of inertia of walls A and B Assuming maximum flange widths on shear walls A and B to be the actual widths or $h/6$ and $h/12$ respectively (**8.3f**) – half that normally allowed in local bending – or the maximum outstand to be $6\,t$ ($=615$ mm) then the effective area of the wall in resisting overturning forces is as shown in Fig. 12.33.

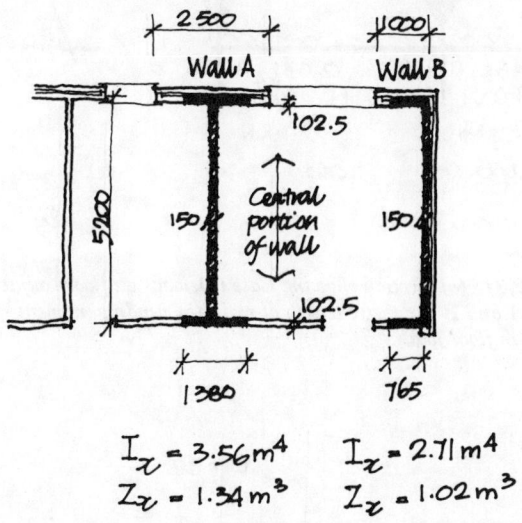

Fig. 12.33 *Plan and section on shear walls A and B with effective section area shown hatched.*

Assuming the sum of the moments of inertia of all the walls in the north–south direction is 35 m⁴ then, ignoring torsion effects which are small in this case and assuming each wall to behave as an independent cantilever, wind force carried by wall A at first-floor level

$= \dfrac{3.56}{35} \times 748$ = 76 kN

with bending moment at first-floor level

$= \dfrac{76 \times 23.800}{2}$ = 905 kN-m

and bending stress $= \dfrac{905 \times 10^{-3}}{1.34}$ = +0.68 N/mm²

and for wall B,

bending moment at first-floor level

$= \dfrac{58 \times 23.800}{2}$ = 691 kN-m

and bending stress $= \dfrac{691 \times 10^{-3}}{1.02}$ = +0.68 say +0.69 N/mm²

12.10 Interior wall A

Slenderness ratio of wall $= \dfrac{0.75 \times 2800}{150} = 14$

Consider Case (a) of Clause 22: $0.9 G_k / 1.4 G_k + 1.6 Q_k$

Assume checkerboard loading on first and second floors in order to give the most unfavourable curvature to wall A (Fig. 12.34).

Eccentricities in wall A may be found by assuming load on each side acts at an eccentricity $= t/3$ or by moment distribution (Fig. 12.35). In this case, using moment distribution

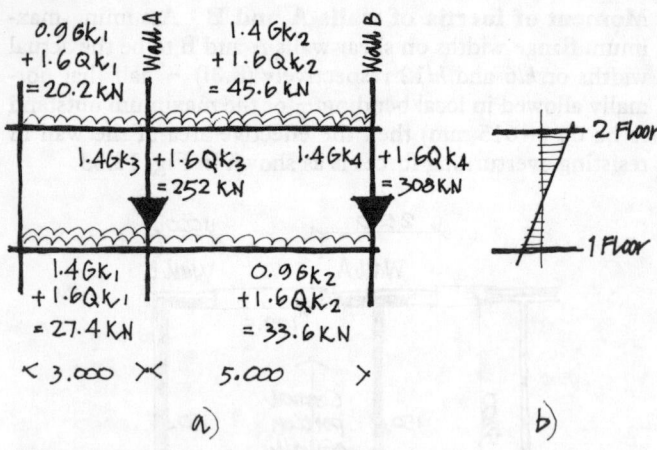

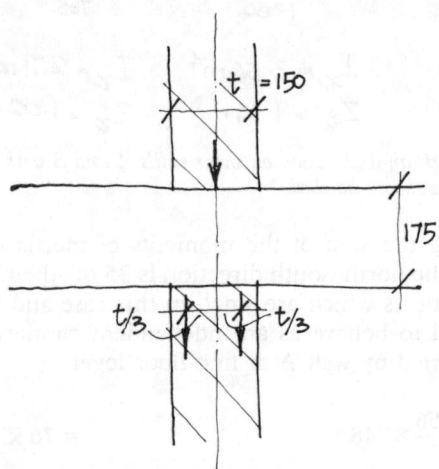

Fig. 12.34 (a) Section showing Case (a) loads on floors adjacent to walls A and B per metre length of wall; (b) bending moment in wall A due to floor loads.

Fig. 12.35 Section at joint of floor slab and wall.

and for the given spans and thicknesses (Figs 12.34 and 12.35) and taking the flexural rigidity, EI, of the masonry to be 20% of that for a similarly sized concrete element, then approximately 7.5% of the out-of-balance fixed end moment would go to each wall, above and below the joint, the remaining 85% being distributed back into the slabs. Hence,

bending moment at top of wall A

$$= 0.075 \left(\frac{45.6 \times 5}{12} - \frac{20.2 \times 3}{12} \right) = 1.05 \text{ kN-m/m}$$

and eccentricity $= \dfrac{1.05 \times 10^3}{252} = 4 \text{ mm} = 0.03 \, t$

and bending moment at bottom of wall A

$$= 0.075 \left(\frac{33.6 \times 5}{12} - \frac{27.4 \times 3}{12} \right) = 0.54 \text{ kN-m/m}$$

and eccentricity $= \dfrac{0.54 \times 10^3}{252} = 2 \text{ mm} = 0.02 \, t$

which acts in the opposite direction to that at the top of the wall and is conservatively taken as zero. Hence, from Table 5.4, $\beta = 0.89$.

The design vertical load resistance must be greater than the design load of 318 kN/m. Hence required characteristic strength of masonry,

$$f_k \geqslant \frac{252 \times 3.1}{0.89 \times 150 \times 1.15} = 5.09 \text{ N/mm}^2$$

using factor of 1.15 from Clause 23.1.2.

Combinations of load other than that shown in Fig. 12.34 should also be considered; however, note that Case (a) of Clause 22 does not include the possibility of there being no live load on one span.

Consider Case (b) of Clause 22: $0.9G_k/1.4G_k + 1.4W_k$

With bending about major axis,

stress at windward end of wall

$$= \frac{0.9 \times 163}{150} - 1.4 \times 0.68 = 0.03 \text{ N/mm}^2$$

and stress at leeward end of wall

$$= \frac{1.4 \times 163}{150} + 1.4 \times 0.68 = 2.47 \text{ N/mm}^2$$

Consider Case (c) of Clause 22: $1.2G_k + 1.2Q_k + 1.2W_k$

With bending about major axis,

stress at ends of wall

$$= \frac{1.2 \times 163 + 1.2 \times 15.0}{150} \pm 1.2 \times 0.68$$

$$= 1.42 \pm 0.82 = 2.24 \text{ and } 0.60 \text{ N/mm}^2$$

For both Cases (b) and (c) the eccentricity of vertical load about the minor axis in the central portion of wall A (Fig. 12.33), by a process similar to that for Case (a), is $0.02t$. Hence from Table 7 of the Code (Table 5.4), $\beta = 0.89$. Case (b) is the worst case. Assuming, conservatively, that there is the same eccentricity at the ends of the wall then the required characteristic strength of masonry,

$$f_k \geqslant \frac{2.47 \times 3.1}{0.89 \times 1.15} = 7.48 \text{ N/mm}^2$$

This is a worse case than Case (a); select a 35 N/mm² brick with a mortar designation (ii); see Table 2 of the Code (Table 2.2).

Design shear stress for Case (b),

$$= \frac{1.4 \times 76 \times 10^3}{150 \times 5200} = 0.14 \text{ N/mm}^2$$

which is less than the allowable shear stress,

$$= \left(0.35 + \frac{0.6 \times 0.9 \times 163}{150} \right) \frac{1}{2.5} = 0.37 \text{ N/mm}^2$$

OK

12.11 Exterior wall B

Slenderness ratio of wall $= \dfrac{0.75 \times 2800}{168}$

$$= 12.5$$

With wind in the north–south direction suction induces a bending moment in wall B about its minor axis. Only the inner leaf of the wall is considered to take this bending moment, being thicker and having very high axial load com-

pared to the outer leaf. The wall spans vertically and from Case (iii) in Table A4.2(a),
Bending moment at top, bottom and middle of wall due to characteristic loads

$$= \frac{W.L}{16}$$

$$= \frac{0.9 \times (2.800)^2}{16} \qquad = 0.44 \text{ kN-m/m}$$

Consider Case (a) of Clause 22: $0.9 \, G_k/1.4 \, G_k + 1.6 \, Q_k$

Assume floor load acts at eccentricity of $t/6$.
Hence resultant eccentricity (Fig. 12.34)

$$= \frac{0.5 \times 45.6 \, t/6}{308} = 0.02 \, t$$

Assuming that eccentricity at bottom of wall is zero then

$\beta = 0.92$ and required characteristic strength of masonry,

$$f_k \geqslant \frac{308 \times 3.1}{0.92 \times 150 \times 1.15} \qquad = 6.02 \text{ N/mm}^2$$

Note that calculation of the eccentricity by moment distribution, assuming 80% of the fully rigid joint moment is obtained in the walls (**A11.2**), would give a bending moment at top of wall B

$$= 0.80 \times 0.15 \left(\frac{45.6 \times 5}{12} \right)$$

$$= 2.28 \text{ kN-m/m}$$

and eccentricity $= \dfrac{2.28 \times 10^3}{308} = 7 \text{ mm} = 0.05 \, t$

This usually gives an upper limit for the eccentricity.

Consider Case (b) of Clause 22: $0.9G_k/1.4G_k + 1.4W_k$

Wind on north or south faces causes bending about major axis with stress at windward end of wall

$$= \frac{0.9 \times 203}{150} - 1.4 \times 0.69 \qquad = 0.25 \text{ N/mm}^2$$

and stress at leeward end of wall

$$= \frac{1.4 \times 203}{150} + 1.4 \times 0.69 \qquad = 2.86 \text{ N/mm}^2$$

Suction on east face, together with eccentric floor load, causes bending about minor axis (Fig. 12.36).

Worst position is at bottom of wall where moments are additive. Hence, bending moment at bottom of wall B

$$= 0.5 \times 1.4 \times 24 \times 0.150/6 + (1.4 \times 0.44)$$
$$= 0.42 + 0.62 \qquad = 1.04 \text{ kN-m/m}$$

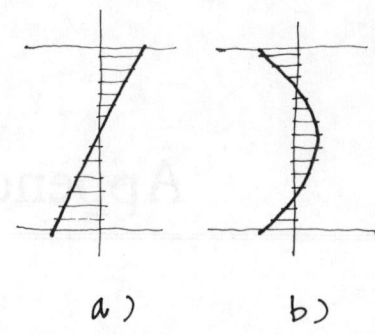

Fig. 12.36   Bending moments in wall B (a) due to floor loads and (b) due to wind loads.

and eccentricity $= \dfrac{1.04 \times 10^3}{285} = 4 \text{ mm} = 0.03 \, t$

with design vertical load in central portion of wall
$$= 285 \text{ kN/m.}$$

Hence $\beta = 0.92$, requiring that

$$f_k \geqslant \frac{2.86 \times 3.1}{0.92 \times 1.15} \qquad = 8.38 \text{ N/mm}^2$$

Consider Case (c) of Clause 22: $1.2 \, G_k + 1.2 \, Q_k + 1.2 \, W_k$

Wind on north or south faces causing bending about major axis with stress at ends of wall

$$= \frac{1.2 \times 203 + 1.2 \times 15.2}{150} \pm 1.2 \times 0.69$$

$$= 1.74 \pm 0.83 \qquad \begin{array}{l} = 2.57 \text{ or} \\ 0.91 \text{ N/mm}^2 \end{array}$$

Bending moment at bottom of wall

$$= 0.5 \, (1.2 \times 24 + 1.2 \times 7.5) \times 0.150/6 + (1.2 \times 0.44)$$
$$= 0.48 + 0.53 \qquad = 1.01 \text{ kN-m/m}$$

and eccentricity $= \dfrac{1.01 \times 10^3}{262} = 4 \text{ mm} = 0.03 \, t$

with design vertical load in middle of wall $= 262$ kN/m.
Hence $\beta = 0.92$, requiring that,

$$f_k \geqslant \frac{2.57 \times 3.1}{0.92 \times 1.15} \qquad = 7.53 \text{ N/mm}^2$$

Case (b) is the worst case; the brick and mortar will be the same as that for wall A. Note that there are no tensile stresses. The walls on the north and south faces should also be checked in a similar way. Finally the whole building should be checked with the wind blowing in the east–west direction. In practice the calculations would be done at each floor level and the results tabulated. Note that eccentricities in the upper floor levels would be greater than those calculated here.

Appendix A   Design data

A1   Definitions

A1.1   Centroids

The centre of gravity of a body is the one point in the body at which, in theory, it is possible for the body to balance about. The centroid is the analogous 'balance point' for a two-dimensional form, which has no thickness or weight, such as a line or an area. The area may be visualised as being a very thin card of uniform thickness and weight. If the area has an axis of symmetry then the centroid of the area will be somewhere on that line of symmetry. Hence if the area has two axes of symmetry then the centroid is at the intersection of the two axes of symmetry. In general the centroid of an area may be found by taking moments of areas about two different axes, or one axis if there is a line of symmetry. For an area of any shape the coordinates of the centroid are given by (Fig. A1.1),

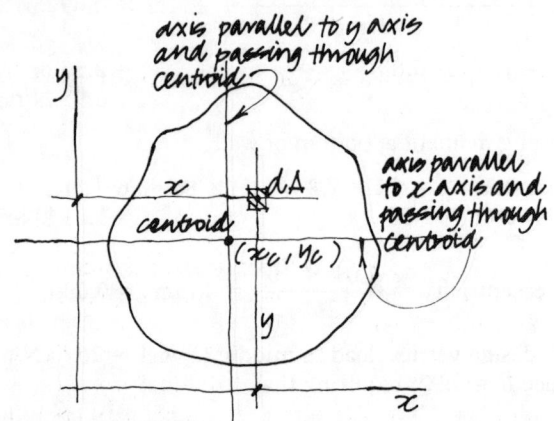

Fig. A1.1   General section.

$$x_c = 1/A. \int_A x.dA \quad \text{and} \quad y_c = 1/A. \int_A y.dA.$$

Example A1.1

Find the position of the centroid of the section shown in Fig. A1.2 which is symmetrical about the y-axis.

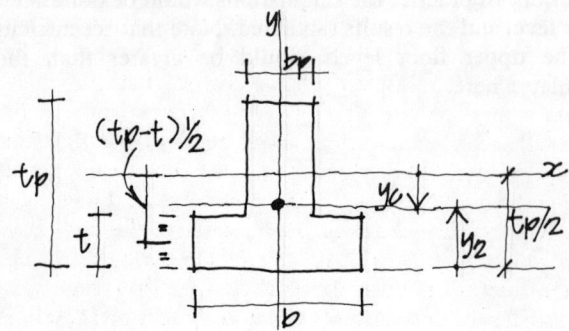

Fig. A1.2   Section.

Taking moments about the x-axis, placed $t_p/2$ from the end of the section,

$$y_c = 1/A. \int_A y.dA = [0.5t(b - b_p)(t_p - t)]/[t_p.b_p + t(b - b_p)]$$

and $x_c = 0$.

A1.2   Moments of inertia

The moment of inertia of a body is a measure of the resistance of the mass of that body to rotation about a particular axis. The moment of inertia, or strictly speaking the second moment of area, of a two-dimensional form such as a line or an area is given by (Fig. A1.1)

$$I_y = \int_A x^2 . dA \quad \text{and}$$

$$I_x = \int_A y^2 . dA$$

where I_x and I_y are the moments of inertia relative to the x- and y-axes respectively; these axes are in the same plane as the line or area considered. For a beam, wall or slab which has a cross-section with a large moment of inertia, the resistance to bending will be greater than for an element of the same material and weight with a small moment of inertia (Fig. A1.3). The moment of inertia of an area may be taken about an axis in any position or direction. However, for any particular axis direction, the minimum moment of inertia is that taken about the axis which passes through the centroid of the section (Fig. A1.1). For the purposes of calculation, the moment of inertia required is invariably that taken about an axis passing through the centroid of the section.

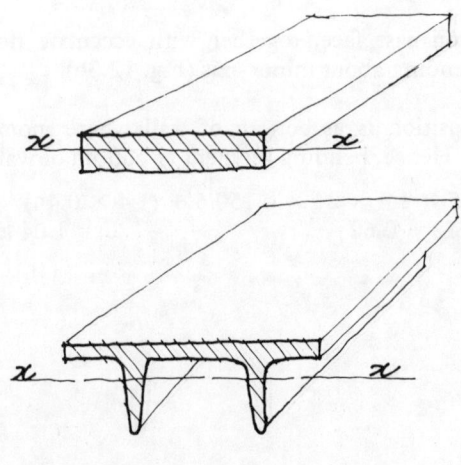

Fig. A1.3   Section through slab and slab with ribs.

Example A1.2

Find the moment of inertia of a beam with a rectangular cross-section about an axis which is parallel to the x-axis and passes through the centroid of the section (Fig. A1.4).

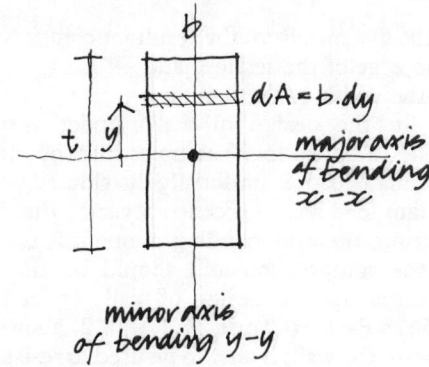

Fig. A1.4   Section.

$$I_x = \int_A y^2 . dA = \int_{-t/2}^{t/2} b.y^2.dy$$

$$= b \left| y^3/3 \right|_{-t/2}^{t/2} = b.t^3/12$$

There are two axes at right angles to each other, which pass through the centroid, about one of which the moment of inertia is a maximum, the so-called major axis of bending, and about the other of which the moment of inertia is a minimum, the so-called minor axis of bending (Fig. A1.4). These axes are known as principal axes of bending; bending moments about other axes should be resolved into moments about these principal axes before calculation of the bending stresses. The moments of inertia about the principal axes of bending are the figures usually quoted for the properties of the section; for a section which is symmetrical about at least one axis, in general, I_x and I_y are the quoted moments of inertia about the two principal axes. Sections which do not have at least one axis of symmetry, such as L- or Z-shaped sections, will have principal axes which do not coincide with the x and y coordinates (Fig. A1.5). However, masonry L- or Z-shaped sections may often be treated as having principal axes in the x- and y-directions because of the restraining effect of the masonry to which they are attached (App. **A3**).

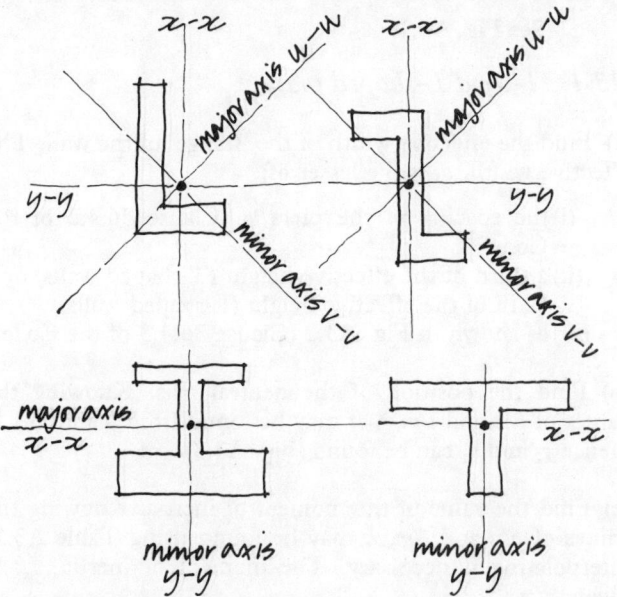

Fig. A1.5   Sections have principal axes, which may or may not coincide with the x- and y-axes.

A1.3   Bending stress and section modulus

The so-called engineers' theory of bending is usually used to calculate bending stresses. This theory assumes that when an element is bent there is a neutral surface in the element which is not stressed either in tension or compression (Fig. A1.6). The line of intersection of the neutral surface with a transverse section of the element is known as the neutral axis. The theory gives that this axis always passes through the centroid of the section considered and that

$$E/r = M/I = f/y$$

where f is the tensile or compressive stress at a point,

y is the distance from the neutral axis to the point,

I is the moment of inertia of the section considered,

E is the Young's Modulus of the material used,

M is the bending moment at the section considered, and

r is the radius of curvature of the element at the section considered due to bending.

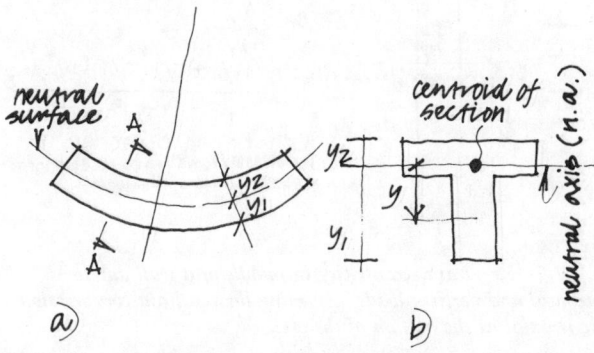

Fig. A1.6   (a) Curvature of beam under bending; (b) section A-A.

Hence $f_1 = M.y_1/I = M/Z_1$ and
$f_2 = M.y_2/I = M/Z_2$

where f_1 and f_2 are the bending stresses at the bottom and top of the element respectively and Z_1 and Z_2 are the section moduli of the element.

The section modulus is equal to the moment of inertia of the section divided by y_1 or y_2, the distances from the neutral axis to the bottom or top surface of the element respectively (Fig. A1.6). For an element which is symmetrical about the neutral axis there is only one section modulus, Z_1 being equal to Z_2.

A1.4   Core area

The core, or kern, area of a section is a property of the section, just as the position of the neutral axis is. It may be defined as that central area of the cross-section through which the resultant compression force must pass in order that there are no tensile stresses at that section, or for a material unable to accept tensile stresses, that no cracking occurs at that section. This is an important concept for a material such as masonry, which has little or no tensile strength. For an infinitely long wall the core area is a central strip with a width equal to a third of the wall thickness, the so-called middle third rule. For a column the core area is diamond-shaped (**1.1**k). An aim of the design may be to ensure that the resultant external compression force always falls within the core area of the running section (**1.1**c). The core area for various sections is given in App. **A2**.

A1.5 Eccentricity and stress due to bending and compression

Any perpendicular section through a masonry wall, column or arch is, in general, subject to a bending moment as well as an axial compression force. The eccentricity of the compression force about an axis through the centroid of the section considered, is equal to the total bending moment about that axis divided by the total axial compression force acting at right angles to the section, M/P. Note that the total bending moment about the chosen axis can be caused by vertical or horizontal forces or any other forces in the plane of the element (Fig. A1.7). The eccentricity is simply an alternative way of considering the effects, including the resulting stresses, caused by an axial force and a bending moment.

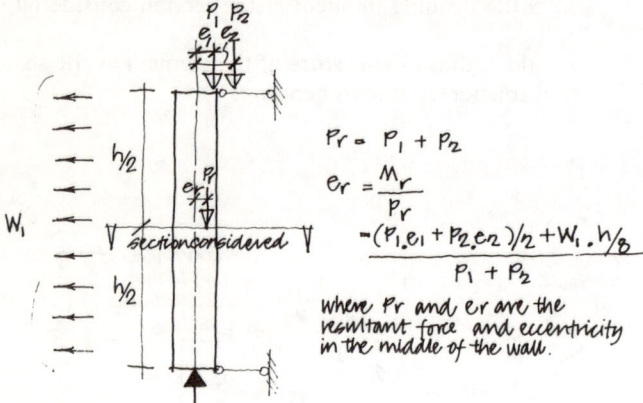

Fig. A1.7 Resultant eccentricity in middle of a wall due to horizontal and vertical loads, assuming the resultant compression force is axial at the bottom of the wall.

For an elastic material strong enough to take the calculated tensile stresses or for an elastic material in which the resultant compression force falls within the core or kern area, the maximum stresses are given by

$$P/A \pm M/Z = P/A(1 \pm e.A/Z)$$

where P is the resultant compression force,
  M is the resultant bending moment,
  e is the resultant eccentricity,
  A is the area of the section and
  Z is the section modulus.

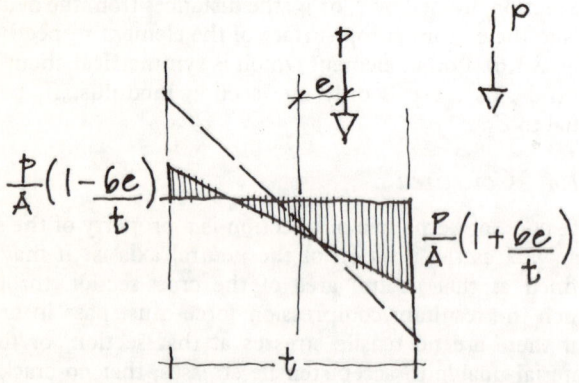

Fig. A1.8 Stresses due to a vertical load, with varying eccentricity, in a rectangular-shaped section able to accept tensile stresses.

For rectangular-shaped sections these stresses are given by (Fig. A1.8):

$$P/A(1 \pm 6e/t)$$

where t is the depth of the section. If the material has no tensile strength and the resultant compression force falls out-

side the core or kern area, cracking will occur and the maximum compressive stress for a rectangular-shaped section assuming an elastic distribution of stress would be (Fig. A1.9):

$$2P/(3x.w)$$

where x is the distance from the resultant compression force to the edge of the section, and
  w is the width of the section.

Note that for the design of walls under vertical load BS 5628:Part 1 (Appendix B) assumes that the compressive stress, at ultimate load, is uniformly distributed (Fig. A1.9). If the resultant load has an eccentricity such that it lies outside the section, then the bending moment is more significant than the compression and should be checked (Fig. A1.8). Note that for the design of walls under horizontal forces BS 5628:Part 1 (Clause 36.4.3) (**6.2**) allows the flexural strength of the wall, if any, to be used to resist the bending moments.

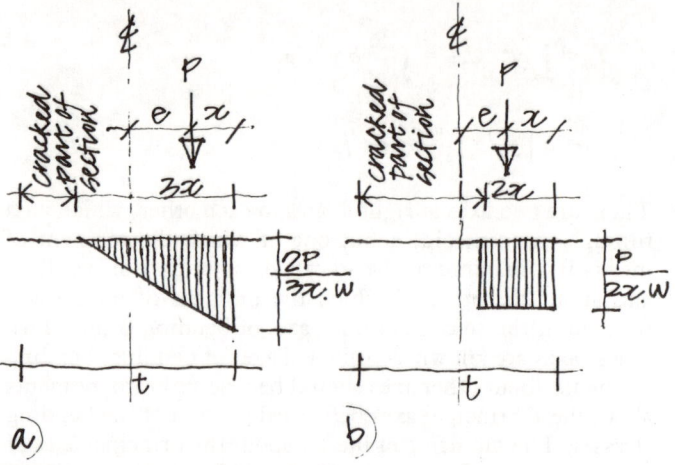

Fig. A1.9 Compressive stresses in wall assuming (a) an elastic distribution of stress or (b) a uniformly distributed stress.

A2 Properties of sections
  See Tables A2.1 and A2.2.

A3 The moment of inertia and section modulus of T-, I-, L- and Z-shaped sections in masonry walls
  See Fig. A3.1.

A3.1 T- and L-shaped walls

(a) Find the effective width of the 'flange' of the wall. The effective width, b, is the lesser of:

  (i) the spacing of the piers, s (Clause 36.4.3 of the Code);
or (ii) a third of the effective height (T-shaped walls) or a sixth of the effective height (L-shaped walls);
or (iii) as shown in Fig. A3.2 (Clause 36.4.3 of the Code).

(b) Find the position of the neutral axis. Knowing the values of b/n and t/m, y_2/t may be found from Table A3.1. Hence y_2 and y_1 can be found (Fig. A3.3).

(c) Find the value of the moment of inertia. Knowing the values of b/n and t/m, k may be found using Table A3.2, interpolating if necessary. The moment of inertia, I_x, is given by

$$I_x = k.n.t^3/12$$

(continued on p 99)

Table A2.1   Section properties

	Distance of extremities of section from $x-x$ axis through centroid, y_1 and y_2	Moments of inertia about $x-x$ and $y-y$ axis, I_x and I_y	Section moduli $Z\left(=\dfrac{I_x}{y}\right)$ about $x-x$ axis, Z_1 and Z_2	Area A	Radius of gyration about $x-x$ and $y-y$ axis, $k_x = \sqrt{\dfrac{I_x}{A}}$ and $k_y = \sqrt{\dfrac{I_y}{A}}$
a	$y_1 = y_2 = t/2$	$I_x = \dfrac{b.t^3}{12}$ $I_y = \dfrac{t.b^3}{12}$	$Z_1 = Z_2$ $= \dfrac{b.t^2}{6}$	$A = b.t$	$k_x = \dfrac{t}{\sqrt{12}}$ $k_y = \dfrac{b}{\sqrt{12}}$
b					
c	$y_1 = y_2 = t/2$	$I_x = \dfrac{1}{12}(b.t^3 - b_1.t_1^3)$ $I_y = \dfrac{1}{12}(t.b^3 - t_1.b_1^3)$ (case b only)	$Z_1 = Z_2$ $= \dfrac{b.t^3 - b_1.t_1^3}{6t}$	$A = b.t - b_1.t_1$	$k_x = \sqrt{\dfrac{b.t^3 - b_1.t_1^3}{12(b.t - b_1.t_1)}}$ $k_y = \sqrt{\dfrac{t.b^3 - t_1.b_1^3}{12(b.t - b_1.t_1)}}$ (case b only)
d (flanges restrained)					
e	$y_1 = y_2 = t/2$	$I_x = \dfrac{1}{12}(n.t^3 + b_1.m^3)$ $I_y = \dfrac{1}{12}(m.b^3 + t_1.n^3)$	$Z_1 = Z_2$ $= \dfrac{n.t^3 + b_1.m^3}{6t}$	$A = m.b_1 + n.t$	$k_x = \sqrt{\dfrac{n.t^3 + b_1.m^3}{12(m.b_1 + n.t)}}$ $k_y = \sqrt{\dfrac{m.b^3 + t_1.n^3}{12(m.b_1 + n.t)}}$
f	$y_1 = t - y_2$ y_2 (See Table A3.1)	I_x (see Table A3.2) $I_y = \dfrac{1}{12}(m.b^3 + t_1.n^3)$	$Z_1 = \dfrac{I_x}{y_1}$ $Z_2 = \dfrac{I_x}{y_2}$	$A = m.b + n.t_1$	$k_x = \sqrt{\dfrac{I_x}{A}}$ $k_y = \sqrt{\dfrac{m.b^3 + t_1.n^3}{12(m.b + n.t_1)}}$
g (flanges restrained)	$y_1 = t - y_2$ y_2 (See Table A3.1) $x_1 = b - x_2$ x_2 (See Table A3.1)	I_x (See Table A3.2) I_y (See Table A3.2)	$Z_1 = \dfrac{I_x}{y_1}$ $Z_2 = \dfrac{I_x}{y_2}$	$A = m.b + n.t_1$	$k_x = \sqrt{\dfrac{I_x}{A}}$ $k_y = \sqrt{\dfrac{I_y}{A}}$
h	$y_1 = y_2 = t/2$	$I_x = I_y = \dfrac{\pi.t^4}{64}$	$Z_1 = Z_2 = \dfrac{\pi.t^3}{32}$	$A = \dfrac{\pi.t^2}{4}$	$k_x = k_y = t/4$
i	$y_1 = y_2 = t/2$	$I_x = I_y = \dfrac{\pi}{64}(t^4 - t_1^4)$	$Z_1 = Z_2$ $= \dfrac{\pi}{32}\left(\dfrac{t^4 - t_1^4}{t}\right)$	$A = \dfrac{\pi}{4}(t^2 - t_1^2)$	$k_x = k_y = \dfrac{\sqrt{(t^2 + t_1^2)}}{4}$

Table A2.2   *Core (kern) area for sections, showing core dimensions, e_x and e_y*

	e_x	e_y
a	$\dfrac{b}{6}$	$\dfrac{t}{6}$
b	$\dfrac{t.b^3 - t_1.b_1^3}{6b(t.b - t_1.b_1)}$	$\dfrac{b.t^3 - b_1.t_1^3}{6t(b.t - b_1.t_1)}$
c	$\dfrac{t}{8}$	$\dfrac{t}{8}$
d	$\dfrac{t^2 + t_1^2}{8t}$	$\dfrac{t^2 + t_1^2}{8t}$
e	$e_{u1} = \dfrac{Z_{1(v-v)}}{A}$ $e_{u2} = \dfrac{Z_{2(v-v)}}{A}$	$e_{v1} = \dfrac{Z_{1(u-u)}}{A}$ $e_{v2} = \dfrac{Z_{2(u-u)}}{A}$

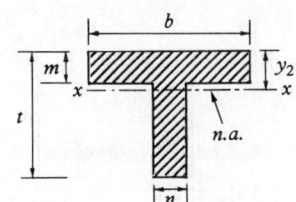

Table A3.1   Position of neutral axis from top of T- or L-shaped section, y_2

Values of $\dfrac{y_2}{t}$

$\dfrac{b}{n} =$	$\dfrac{t}{m} =$													
	25	20	16	13	10	8	6	5	4	3.5	3	2.5	2.25	2
2.00	0.482	0.477	0.472	0.467	0.459	0.451	0.440	0.433	0.425	0.421	0.417	0.414	0.415	0.417
2.20	0.478	0.473	0.467	0.461	0.452	0.443	0.431	0.423	0.413	0.409	0.405	0.403	0.403	0.406
2.40	0.475	0.469	0.462	0.455	0.445	0.435	0.421	0.413	0.403	0.398	0.394	0.392	0.393	0.397
2.60	0.471	0.465	0.457	0.449	0.438	0.427	0.412	0.403	0.393	0.388	0.384	0.383	0.385	0.389
2.80	0.468	0.461	0.453	0.444	0.431	0.420	0.404	0.394	0.384	0.379	0.375	0.374	0.377	0.382
3.00	0.464	0.457	0.448	0.438	0.425	0.412	0.396	0.386	0.375	0.370	0.367	0.367	0.369	0.375
3.20	0.461	0.453	0.443	0.433	0.419	0.406	0.388	0.378	0.367	0.362	0.359	0.360	0.363	0.369
3.40	0.458	0.449	0.439	0.428	0.413	0.399	0.381	0.370	0.359	0.355	0.352	0.353	0.357	0.364
3.60	0.455	0.445	0.434	0.423	0.407	0.393	0.374	0.363	0.352	0.348	0.345	0.347	0.351	0.359
3.80	0.452	0.442	0.430	0.418	0.402	0.387	0.367	0.356	0.346	0.341	0.339	0.342	0.346	0.354
4.00	0.449	0.438	0.426	0.413	0.396	0.381	0.361	0.350	0.339	0.335	0.333	0.336	0.341	0.350
4.20	0.446	0.434	0.422	0.409	0.391	0.375	0.355	0.344	0.333	0.329	0.328	0.332	0.337	0.346
4.40	0.443	0.431	0.418	0.404	0.386	0.370	0.349	0.338	0.328	0.324	0.323	0.327	0.333	0.343
4.60	0.440	0.428	0.414	0.400	0.381	0.364	0.344	0.333	0.322	0.319	0.318	0.323	0.329	0.339
4.80	0.437	0.424	0.410	0.396	0.376	0.359	0.338	0.327	0.317	0.314	0.314	0.319	0.326	0.336
5.00	0.434	0.421	0.406	0.391	0.371	0.354	0.333	0.322	0.312	0.310	0.310	0.315	0.322	0.333
5.20	0.431	0.418	0.403	0.387	0.367	0.349	0.328	0.317	0.308	0.305	0.306	0.312	0.319	0.331
5.40	0.428	0.414	0.399	0.383	0.362	0.345	0.324	0.313	0.304	0.301	0.302	0.309	0.316	0.328
5.60	0.425	0.411	0.395	0.379	0.358	0.340	0.319	0.308	0.299	0.297	0.298	0.306	0.313	0.326
5.80	0.423	0.408	0.392	0.376	0.354	0.336	0.315	0.304	0.295	0.293	0.295	0.303	0.311	0.324
6.00	0.420	0.405	0.388	0.372	0.350	0.332	0.311	0.300	0.292	0.290	.292	0.300	0.308	0.321
6.50	0.413	0.398	0.380	0.363	0.340	0.322	0.301	0.290	0.283	0.282	0.284	0.294	0.303	0.317
7.00	0.407	0.390	0.372	0.354	0.331	0.312	0.292	0.282	0.275	0.274	0.278	0.288	0.298	0.312
7.50	0.401	0.383	0.365	0.346	0.328	0.304	0.283	0.274	0.268	0.268	0.272	0.283	0.294	0.309
8.00	0.395	0.377	0.357	0.338	0.315	0.296	0.276	0.267	0.261	0.262	0.267	0.279	0.290	0.306
8.50	0.389	0.370	0.350	0.331	0.307	0.288	0.269	0.260	0.255	0.256	0.262	0.275	0.286	0.303
9.00	0.384	0.364	0.344	0.324	0.300	0.281	0.262	0.254	0.250	0.252	0.258	0.271	0.283	0.300
9.50	0.378	0.358	0.337	0.318	0.293	0.275	0.256	0.248	0.245	0.247	0.254	0.268	0.280	0.298
10.00	0.373	0.353	0.331	0.311	0.287	0.268	0.250	0.243	0.240	0.243	0.250	0.265	0.278	0.295
10.50	0.368	0.347	0.325	0.305	0.281	0.262	0.245	0.238	0.236	0.239	0.247	0.262	0.275	0.293
11.00	0.363	0.342	0.320	0.299	0.275	0.257	0.240	0.233	0.232	0.235	0.244	0.260	0.273	0.292
11.50	0.358	0.336	0.314	0.294	0.270	0.252	0.235	0.229	0.228	0.232	0.241	0.258	0.271	0.290
12.00	0.353	0.331	0.309	0.288	0.264	0.247	0.230	0.225	0.225	0.229	0.238	0.256	0.269	0.283

Table A3.2 Moment of inertia of T- or L-shaped section, I_x
Values of k

$$I_x = k \cdot \frac{n.t^3}{12}$$

$\dfrac{b}{n}=$	$\dfrac{t}{m}=$													
	25	20	16	13	10	8	6	5	4	3.5	3	2.5	2.25	2
2.00	1.106	1.129	1.156	1.183	1.222	1.257	1.302	1.328	1.353	1.363	1.370	1.373	1.373	1.375
2.20	1.127	1.153	1.184	1.217	1.262	1.302	1.353	1.381	1.408	1.419	1.425	1.427	1.427	1.431
2.40	1.147	1.177	1.212	1.249	1.300	1.345	1.401	1.431	1.459	1.470	1.476	1.477	1.478	1.484
2.60	1.166	1.201	1.240	1.281	1.337	1.386	1.446	1.478	1.507	1.518	1.523	1.524	1.525	1.533
2.80	1.186	1.224	1.267	1.312	1.372	1.425	1.489	1.523	1.552	1.562	1.567	1.567	1.570	1.580
3.00	1.205	1.246	1.293	1.342	1.407	1.463	1.530	1.565	1.594	1.603	1.607	1.608	1.611	1.625
3.20	1.224	1.269	1.319	1.371	1.440	1.500	1.569	1.604	1.633	1.642	1.646	1.646	1.651	1.668
3.40	1.242	1.290	1.345	1.399	1.473	1.535	1.606	1.642	1.670	1.679	1.681	1.683	1.689	1.709
3.60	1.261	1.312	1.369	1.427	1.504	1.568	1.642	1.678	1.705	1.713	1.715	1.717	1.725	1.749
3.80	1.279	1.333	1.393	1.454	1.534	1.601	1.676	1.712	1.739	1.746	1.747	1.750	1.759	1.787
4.00	1.296	1.354	1.417	1.481	1.564	1.632	1.708	1.744	1.770	1.776	1.778	1.781	1.792	1.825
4.20	1.314	1.374	1.440	1.506	1.592	1.662	1.739	1.775	1.800	1.806	1.807	1.811	1.825	1.861
4.40	1.331	1.394	1.463	1.531	1.620	1.692	1.769	1.804	1.828	1.834	1.834	1.840	1.856	1.897
4.60	1.348	1.413	1.485	1.556	1.647	1.720	1.798	1.833	1.856	1.860	1.861	1.868	1.886	1.932
4.80	1.365	1.433	1.507	1.580	1.673	1.747	1.825	1.859	1.881	1.885	1.886	1.895	1.915	1.966
5.00	1.382	1.452	1.528	1.603	1.698	1.773	1.852	1.885	1.906	1.910	1.910	1.921	1.944	2.000
5.20	1.398	1.470	1.549	1.626	1.723	1.799	1.877	1.910	1.930	1.933	1.933	1.946	1.972	2.033
5.40	1.414	1.489	1.570	1.648	1.747	1.824	1.902	1.934	1.953	1.955	1.956	1.970	1.999	2.066
5.60	1.430	1.507	1.590	1.670	1.770	1.848	1.925	1.957	1.974	1.977	1.977	1.994	2.026	2.098
5.80	1.446	1.525	1.610	1.692	1.793	1.871	1.948	1.979	1.995	1.997	1.998	2.017	2.052	2.129
6.00	1.461	1.542	1.629	1.712	1.815	1.893	1.970	2.000	2.016	2.017	2.019	2.040	2.078	2.161
6.50	1.499	1.585	1.676	1.762	1.868	1.947	2.022	2.050	2.063	2.064	1.066	2.094	2.140	2.237
7.00	1.536	1.626	1.721	1.810	1.917	1.996	2.069	2.095	2.106	2.107	2.111	2.146	2.200	2.312
7.50	1.571	1.665	1.763	1.855	1.964	2.042	2.133	2.137	2.146	2.147	2.153	2.196	2.258	2.386
8.00	1.605	1.703	1.804	1.898	2.008	2.086	2.154	2.176	2.183	2.184	2.193	2.244	2.315	2.458
8.50	1.639	1.739	1.843	1.939	2.049	2.126	2.192	2.212	2.218	2.219	2.230	2.290	2.371	2.530
9.00	1.671	1.775	1.881	1.977	2.088	2.164	2.228	2.246	2.250	2.251	2.266	2.335	2.425	2.600
9.50	1.702	1.809	1.917	2.014	2.125	2.200	2.261	2.277	2.280	2.282	2.300	2.379	2.478	2.670
10.00	1.732	1.841	1.951	2.050	2.160	2.234	2.292	2.306	2.309	2.312	2.333	2.421	2.531	2.739
10.50	1.762	1.873	1.985	2.084	2.193	2.265	2.321	2.334	2.336	2.340	2.365	2.463	2.583	2.807
11.00	1.791	1.904	2.017	2.116	2.225	2.296	2.348	2.360	2.362	2.367	2.396	2.504	2.634	2.875
11.50	1.818	1.933	2.047	2.147	2.255	2.324	2.374	2.385	2.386	2.393	2.426	2.544	2.684	2.942
12.00	1.846	1.962	2.077	2.177	2.284	2.351	2.399	2.408	2.409	2.418	2.455	2.584	2.734	3.010

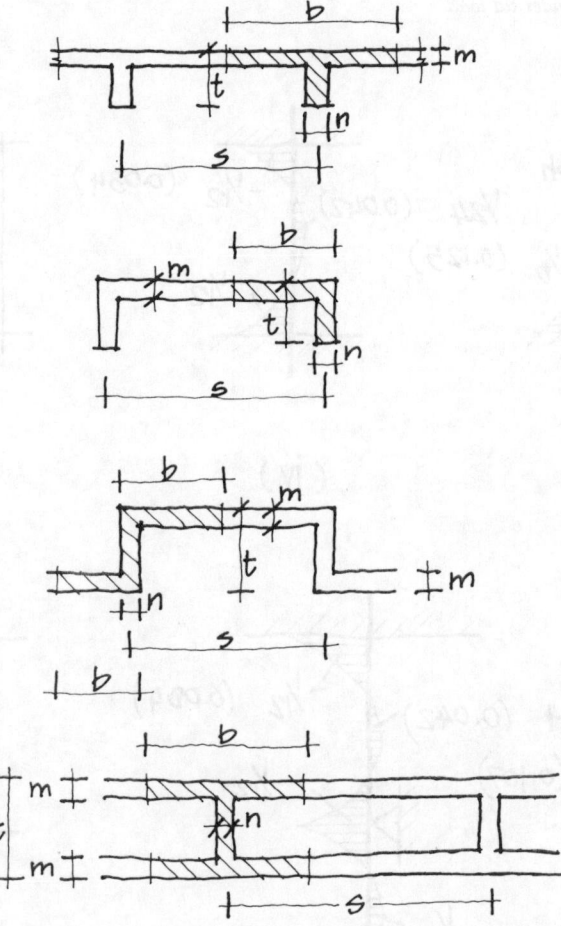

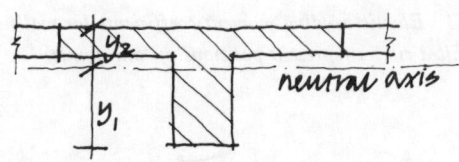

Fig. A3.3   Plan on T-section pier.

(d) Find the value of the section modulus. The maximum and minimum section moduli, Z_1 and Z_2, are given by

$$Z_1 = I_x/y_1 \quad \text{and} \quad Z_2 = I_x/y_2 \text{ (Fig. A3.3)}.$$

A3.2   I- and Z-shaped sections

(a) Find effective width of 'flange' of section, b, following same rules for I- and Z-shaped sections as those for T- and L-shaped sections respectively.

(b) Find value of moment of inertia, I_x, which is given by

$$I_x = (b.t^3 - b_1.t_1^3)/12 \quad \text{(Table A2.1)}.$$

(c) Find the value of the section modulus, Z, which is given by

$$Z_x = 2I_x/t.$$

A4   Bending moment coefficients for walls spanning one way under uniformly distributed (ud) load
See Tables A4.1 and A4.2.

Assumptions used in calculating Tables A4.1 and A4.2: W is total load on each span, L is span distance (all spans equal), i is the 'plastic' moment of resistance of wall at the support divided by the 'elastic' moment of resistance of wall in the span (**6.1**g); supports taken as knife-edged (App. **A5**).

A5   Reduction in bending moment at supports due to width of support
See Fig. A5.1 and Table A5.1.

If the bending moment at the support, assuming knife-edge supports, is $m.W.L$ then maximum bending moment in the wall at the supports is $(m-k)W.L$.

Example A5.1

The moment at the central support of a two-bay span is equal to 0.107 W.L (Case (vi) of Table A4.1) and the width of support is 0.12 L; therefore the maximum moment in the wall over the support is

$$0.107 \, W.L - 0.028 \, W.L = 0.079 \, W.L$$

Assumptions: Reduction is only valid on calculated values of the elastic bending moment at a knife-edge support (Table A4.1); the maximum moment occurs over the width of the pier or other support to which the wall is attached (Fig. A5.1); the bending moment pattern in the span is the same as that which would occur with knife-edges at the centre of each support.

A6   Bending moment coefficients for walls spanning two ways under uniformly distributed (ud) load

See Table A6.1.

Fig. A3.1   Plans on T-, L-, I- and Z-shaped walls.

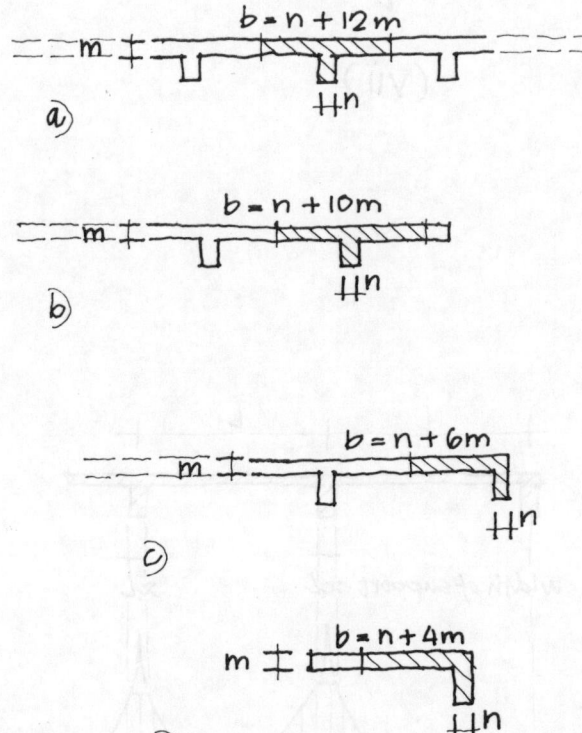

Fig. A3.2   Plans on T- and L-shaped walls showing effective width of (a) continuous T-shaped wall, (b) T-shaped wall with one end of flange unrestrained, (c) continuous L-shaped wall and (d) L-shaped wall with unrestrained flange.

Table A4.1 *Elastic bending moment coefficients for walls spanning one-way under ud load*
(Key: unfilled circles represent positions of 'pin' joints)

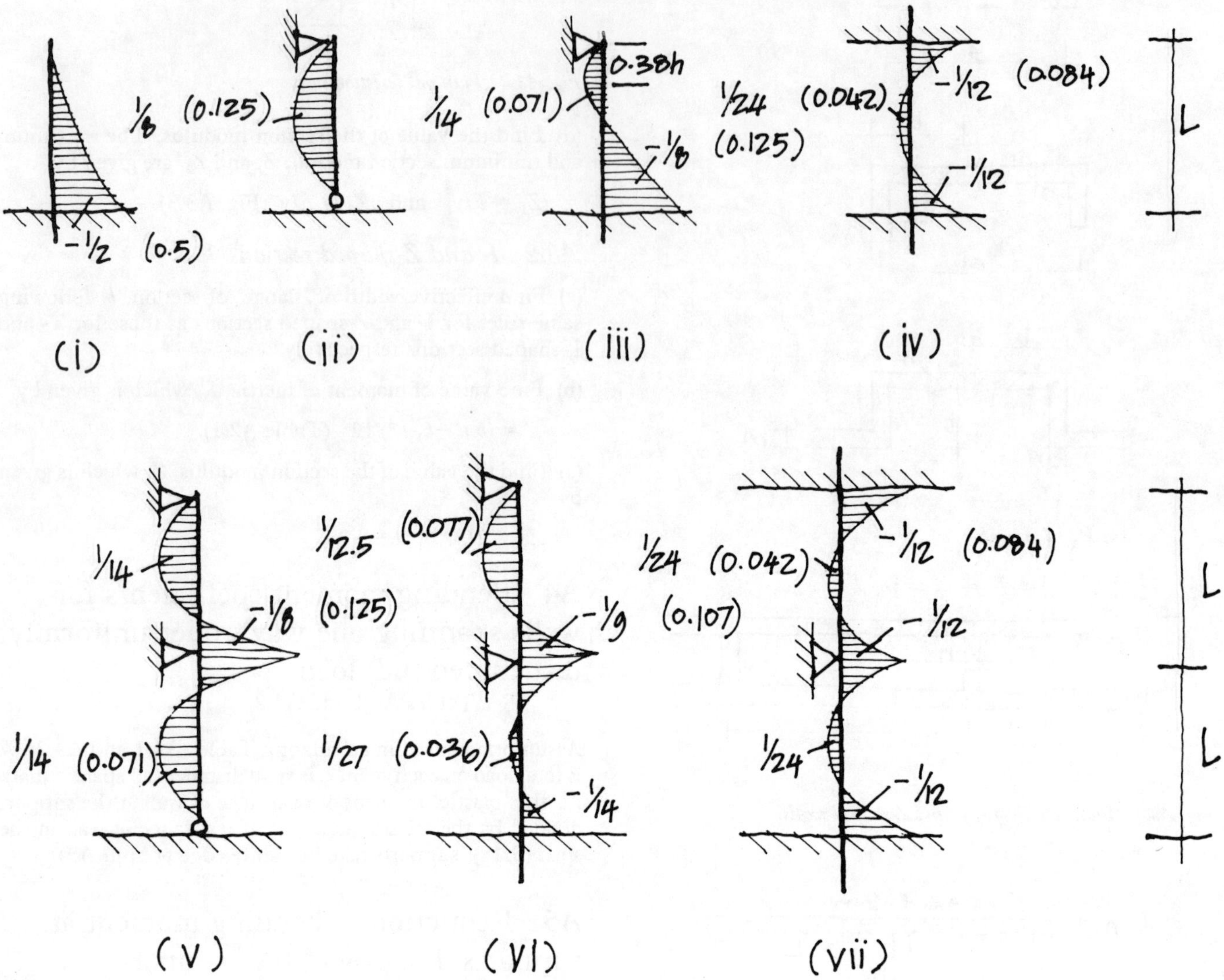

$\frac{1}{8}$ (0.125)

$-\frac{1}{2}$ (0.5)

(i)

$\frac{1}{14}$ (0.071)

(ii)

$\overline{0.38h}$

$-\frac{1}{8}$ (0.125)

(iii)

$\frac{1}{24}$ (0.042)

$-\frac{1}{12}$ (0.084)

$-\frac{1}{12}$

L

(iv)

$\frac{1}{14}$

$-\frac{1}{8}$ (0.125)

$\frac{1}{14}$ (0.071)

(v)

$\frac{1}{12.5}$ (0.077)

$-\frac{1}{9}$ (0.107)

$\frac{1}{27}$ (0.036)

$-\frac{1}{14}$

(vi)

$\frac{1}{24}$ (0.042)

$-\frac{1}{12}$ (0.084)

$-\frac{1}{12}$

$\frac{1}{24}$

$-\frac{1}{12}$

L

L

(vii)

Bending moment = (bending moment coefficient) × W × L

Table A5.1 *Reduction in bending moment due to width of support*

Width of support = $x.L$ $x =$	Reduction in bending moment at support due to width of support = $k.W.L$ $k =$
0.05	0.012
0.08	0.019
0.10	0.024
0.12	0.028
0.15	0.035
0.20	0.045

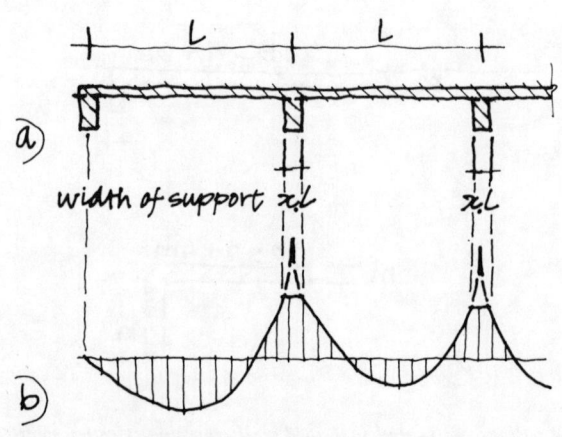

Fig. A5.1 (a) Plan on wall with piers; (b) moment diagram showing assumed reduction at supports.

Table A4.2   Elastic-plastic bending moment coefficients for walls spaning one-way under ud load (a) with i = 1.0 and (b) with i = 0.5
(Key: filled-in circles represent positions of 'plastic' hinges)

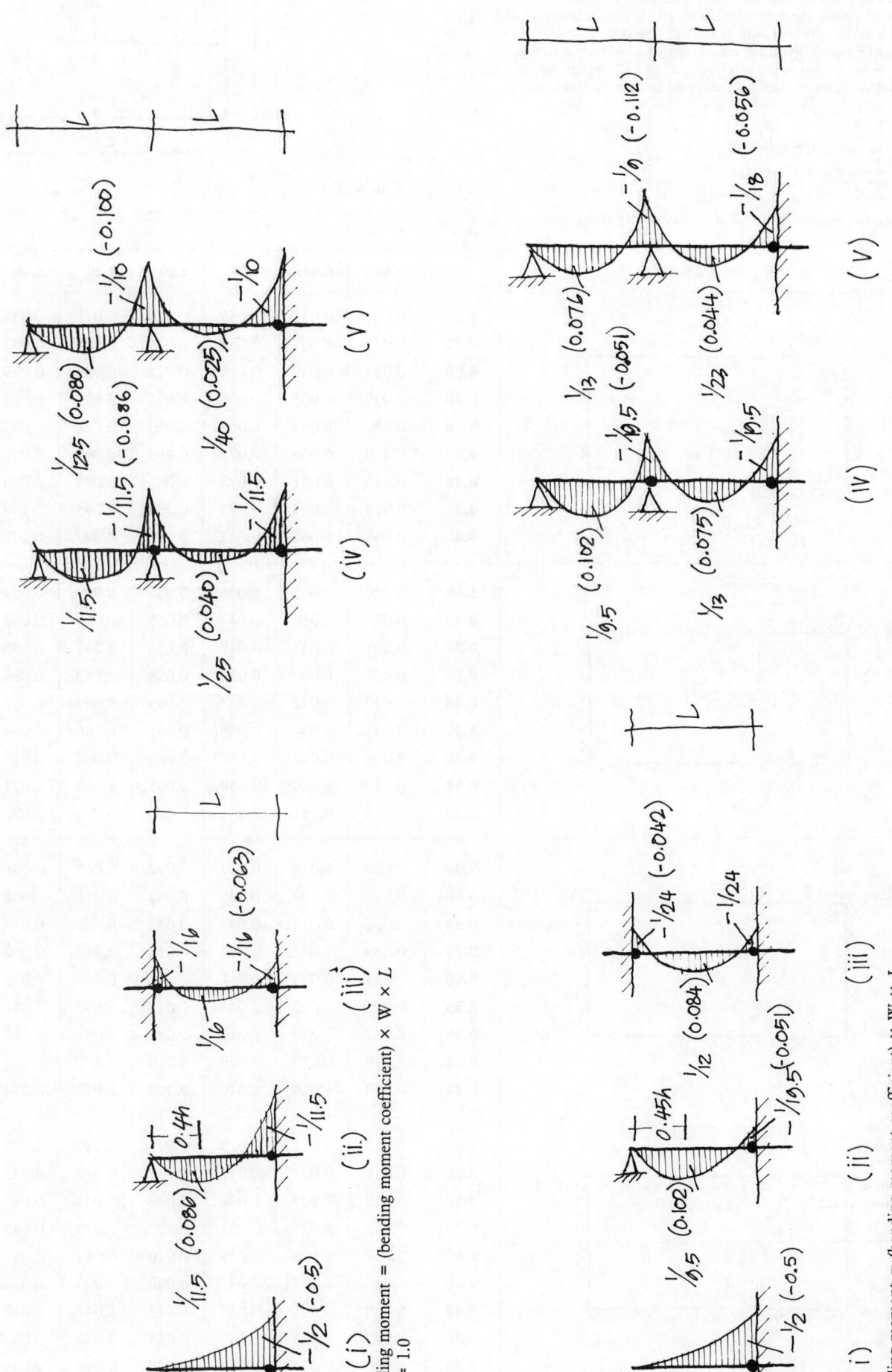

Bending moment = (bending moment coefficient) × W × L
a) $i = 1.0$

Bending moment = (bending moment coefficient) × W × L
b) $i = 0.5$

Table A6.1 *Bending moment coefficients, α, for walls spanning two ways under ud load (after Table 9 of BS 5628:Part 1)*

NOTE 1. Linear interpolation of μ and h/L is permitted.
NOTE 2. When the dimensions of a wall are outside the range of h/L given in this table, it will usually be sufficient to calculate the moments on the basis of a simple span. For example, a panel of type A having h/L less than 0.3 will tend to act as a freestanding wall, whilst the same panel having h/L greater than 1.75 will tend to span horizontally.

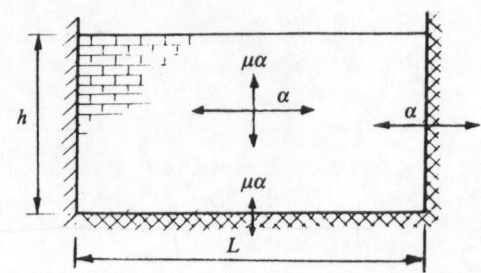

Key to support conditions

———— denotes free edge
////// simply supported edge
XXXXX an edge over which full continuity exists

Values of α

	$\mu =$	$h/L =$						
		0.30	0.50	0.75	1.00	1.25	1.50	1.75
A	1.00	0.031	0.045	0.059	0.071	0.079	0.085	0.090
	0.90	0.032	0.047	0.061	0.073	0.081	0.087	0.092
	0.80	0.034	0.049	0.064	0.075	0.083	0.089	0.093
	0.70	0.035	0.051	0.066	0.077	0.085	0.091	0.095
	0.60	0.038	0.053	0.069	0.080	0.088	0.093	0.097
	0.50	0.040	0.056	0.073	0.083	0.090	0.095	0.099
	0.40	0.043	0.061	0.077	0.087	0.093	0.098	0.101
	0.35	0.045	0.064	0.080	0.089	0.095	0.100	0.103
	0.30	0.048	0.067	0.082	0.091	0.097	0.101	0.104
B	1.00	0.024	0.035	0.046	0.053	0.059	0.062	0.065
	0.90	0.025	0.036	0.047	0.055	0.060	0.063	0.066
	0.80	0.027	0.037	0.049	0.056	0.061	0.065	0.067
	0.70	0.028	0.039	0.051	0.058	0.062	0.066	0.068
	0.60	0.030	0.042	0.053	0.059	0.064	0.067	0.069
	0.50	0.031	0.044	0.055	0.061	0.066	0.069	0.071
	0.40	0.034	0.047	0.057	0.063	0.067	0.070	0.072
	0.35	0.035	0.049	0.059	0.065	0.068	0.071	0.073
	0.30	0.037	0.051	0.061	0.066	0.070	0.072	0.074
C	1.00	0.020	0.028	0.037	0.042	0.045	0.048	0.050
	0.90	0.021	0.029	0.038	0.043	0.046	0.048	0.050
	0.80	0.022	0.031	0.039	0.043	0.047	0.049	0.051
	0.70	0.023	0.032	0.040	0.044	0.048	0.050	0.051
	0.60	0.024	0.034	0.041	0.046	0.049	0.051	0.052
	0.50	0.025	0.035	0.043	0.047	0.050	0.052	0.053
	0.40	0.027	0.038	0.044	0.048	0.051	0.053	0.054
	0.35	0.029	0.039	0.045	0.049	0.052	0.053	0.054
	0.30	0.030	0.040	0.046	0.050	0.052	0.054	0.055
D	1.00	0.013	0.021	0.029	0.035	0.040	0.043	0.045
	0.90	0.014	0.022	0.031	0.036	0.040	0.043	0.046
	0.80	0.015	0.023	0.032	0.038	0.041	0.044	0.047
	0.70	0.016	0.025	0.033	0.039	0.043	0.045	0.047
	0.60	0.017	0.026	0.035	0.040	0.044	0.046	0.048
	0.50	0.018	0.028	0.037	0.042	0.045	0.048	0.050
	0.40	0.020	0.031	0.039	0.043	0.047	0.049	0.051
	0.35	0.022	0.032	0.040	0.044	0.048	0.050	0.051
	0.30	0.023	0.034	0.041	0.046	0.049	0.051	0.052

Table A6.1 *Bending moment coefficients, α, for walls spanning two ways under ud load (after Table 9 of BS 5628:Part 1) (contd)*

Values of α

	$\mu =$	$h/L =$						
		0.30	**0.50**	**0.75**	**1.00**	**1.25**	**1.50**	**1.75**
E	**1.00**	0.008	0.018	0.030	0.042	0.051	0.059	0.066
	0.90	0.009	0.019	0.032	0.044	0.054	0.062	0.068
	0.80	0.010	0.021	0.035	0.046	0.056	0.064	0.071
	0.70	0.011	0.023	0.037	0.049	0.059	0.067	0.073
	0.60	0.012	0.025	0.040	0.053	0.062	0.070	0.076
	0.50	0.014	0.028	0.044	0.057	0.066	0.074	0.080
	0.40	0.017	0.032	0.049	0.062	0.071	0.078	0.084
	0.35	0.018	0.035	0.052	0.064	0.074	0.081	0.086
	0.30	0.020	0.038	0.055	0.068	0.077	0.083	0.089
F	**1.00**	0.008	0.016	0.026	0.034	0.041	0.046	0.051
	0.90	0.008	0.017	0.027	0.036	0.042	0.048	0.052
	0.80	0.009	0.018	0.029	0.037	0.044	0.049	0.054
	0.70	0.010	0.020	0.031	0.039	0.046	0.051	0.055
	0.60	0.011	0.022	0.033	0.042	0.048	0.053	0.057
	0.50	0.013	0.024	0.036	0.044	0.051	0.056	0.059
	0.40	0.015	0.027	0.039	0.048	0.054	0.058	0.062
	0.35	0.016	0.029	0.041	0.050	0.055	0.060	0.063
	0.30	0.018	0.031	0.044	0.052	0.057	0.062	0.065
G	**1.00**	0.007	0.014	0.022	0.028	0.033	0.037	0.040
	0.90	0.008	0.015	0.023	0.029	0.034	0.038	0.041
	0.80	0.008	0.016	0.024	0.031	0.035	0.039	0.042
	0.70	0.009	0.017	0.026	0.032	0.037	0.040	0.043
	0.60	0.010	0.019	0.028	0.034	0.038	0.042	0.044
	0.50	0.011	0.021	0.030	0.036	0.040	0.043	0.046
	0.40	0.013	0.023	0.032	0.038	0.042	0.045	0.047
	0.35	0.014	0.025	0.033	0.039	0.043	0.046	0.048
	0.30	0.016	0.026	0.035	0.041	0.044	0.047	0.049
H	**1.00**	0.005	0.011	0.018	0.024	0.029	0.033	0.036
	0.90	0.006	0.012	0.019	0.025	0.030	0.034	0.037
	0.80	0.006	0.013	0.020	0.027	0.032	0.035	0.038
	0.70	0.007	0.014	0.022	0.028	0.033	0.037	0.040
	0.60	0.008	0.015	0.024	0.030	0.035	0.038	0.041
	0.50	0.009	0.017	0.025	0.032	0.036	0.040	0.043
	0.40	0.010	0.019	0.028	0.034	0.039	0.042	0.045
	0.35	0.011	0.021	0.029	0.036	0.040	0.043	0.046
	0.30	0.013	0.022	0.031	0.037	0.041	0.044	0.047
I	**1.00**	0.004	0.009	0.015	0.021	0.026	0.030	0.033
	0.90	0.004	0.010	0.016	0.022	0.027	0.031	0.034
	0.80	0.005	0.010	0.017	0.023	0.028	0.032	0.035
	0.70	0.005	0.011	0.019	0.025	0.030	0.033	0.037
	0.60	0.006	0.013	0.020	0.026	0.031	0.035	0.038
	0.50	0.007	0.014	0.022	0.028	0.033	0.037	0.040
	0.40	0.008	0.016	0.024	0.031	0.035	0.039	0.042
	0.35	0.009	0.017	0.026	0.032	0.037	0.040	0.043
	0.30	0.010	0.019	0.028	0.034	0.038	0.042	0.044

Table A6.1 *Bending moment coefficients, α, for walls spanning two ways under ud load (after Roberts et al. (1986)) (contd)*

Values of α

	μ =	h/L =						
		0.30	0.50	0.75	1.00	1.25	1.50	1.75
A	0.25	0.050	0.071	0.085	0.094	0.099	0.103	0.106
	0.20	0.054	0.075	0.089	0.097	0.102	0.105	0.108
	0.15	0.060	0.080	0.093	0.100	0.104	0.108	0.110
	0.10	0.069	0.087	0.098	0.104	0.108	0.111	0.113
	0.05	0.082	0.097	0.105	0.110	0.113	0.115	0.116
B	0.25	0.039	0.053	0.062	0.068	0.071	0.073	0.075
	0.20	0.043	0.056	0.065	0.069	0.072	0.074	0.076
	0.15	0.047	0.059	0.067	0.071	0.074	0.076	0.077
	0.10	0.052	0.063	0.070	0.074	0.076	0.078	0.079
	0.05	0.060	0.069	0.074	0.077	0.079	0.080	0.081
C	0.25	0.032	0.042	0.048	0.051	0.053	0.054	0.056
	0.20	0.034	0.043	0.049	0.052	0.054	0.055	0.056
	0.15	0.037	0.046	0.051	0.053	0.055	0.056	0.057
	0.10	0.041	0.048	0.053	0.055	0.056	0.057	0.058
	0.05	0.046	0.052	0.055	0.057	0.058	0.059	0.059
D	0.25	0.025	0.035	0.043	0.047	0.050	0.052	0.053
	0.20	0.027	0.038	0.044	0.048	0.051	0.053	0.054
	0.15	0.030	0.040	0.046	0.050	0.052	0.054	0.055
	0.10	0.034	0.043	0.049	0.052	0.054	0.055	0.056
	0.05	0.041	0.048	0.053	0.055	0.056	0.057	0.058
E	0.25	0.023	0.042	0.059	0.071	0.080	0.087	0.091
	0.20	0.026	0.046	0.064	0.076	0.084	0.090	0.095
	0.15	0.032	0.053	0.070	0.081	0.089	0.094	0.098
	0.10	0.039	0.062	0.078	0.088	0.095	0.100	0.103
	0.05	0.054	0.076	0.090	0.098	0.103	0.107	0.109
F	0.25	0.020	0.034	0.046	0.054	0.060	0.063	0.066
	0.20	0.023	0.037	0.049	0.057	0.062	0.066	0.068
	0.15	0.027	0.042	0.053	0.060	0.065	0.068	0.070
	0.10	0.032	0.048	0.058	0.064	0.068	0.071	0.073
	0.05	0.043	0.057	0.066	0.070	0.073	0.075	0.077
G	0.25	0.018	0.028	0.037	0.042	0.046	0.048	0.050
	0.20	0.020	0.031	0.039	0.044	0.047	0.050	0.052
	0.15	0.023	0.034	0.042	0.046	0.049	0.051	0.053
	0.10	0.027	0.038	0.045	0.049	0.052	0.053	0.055
	0.05	0.035	0.044	0.050	0.053	0.055	0.056	0.057
H	0.25	0.014	0.024	0.033	0.039	0.043	0.046	0.048
	0.20	0.016	0.027	0.035	0.041	0.045	0.047	0.049
	0.15	0.019	0.030	0.038	0.043	0.047	0.049	0.051
	0.10	0.023	0.034	0.042	0.047	0.050	0.052	0.053
	0.05	0.031	0.041	0.047	0.051	0.053	0.055	0.056
I	0.25	0.011	0.021	0.030	0.036	0.040	0.043	0.046
	0.20	0.013	0.023	0.032	0.038	0.042	0.045	0.047
	0.15	0.016	0.026	0.035	0.041	0.044	0.047	0.049
	0.10	0.020	0.031	0.039	0.044	0.047	0.050	0.052
	0.05	0.027	0.038	0.045	0.049	0.052	0.053	0.055

Table A6.1 Bending moment coefficients, α, for walls spanning two ways under ud load (after Table 9 of BS 5628:Part 1) (contd)

Key to support conditions
——— denotes free edge
///// simply supported edge
✕✕✕✕✕ an edge over which full continuity exists

Values of α

μ =	h/L =						
	0.30	**0.50**	**0.75**	**1.00**	**1.25**	**1.50**	**1.75**
J							
1.00	0.009	0.023	0.046	0.071	0.096	0.122	0.151
0.90	0.010	0.026	0.050	0.076	0.103	0.131	0.162
0.80	0.012	0.028	0.054	0.083	0.111	0.142	0.175
0.70	0.013	0.032	0.060	0.091	0.121	0.156	0.191
0.60	0.015	0.036	0.067	0.100	0.135	0.173	0.211
0.50	0.018	0.042	0.077	0.113	0.153	0.195	0.237
0.40	0.021	0.050	0.090	0.131	0.177	0.225	0.272
0.35	0.024	0.055	0.098	0.144	0.194	0.244	0.296
0.30	0.027	0.062	0.108	0.160	0.214	0.269	0.325
K							
1.00	0.009	0.021	0.038	0.056	0.074	0.091	0.108
0.90	0.010	0.023	0.041	0.060	0.079	0.097	0.113
0.80	0.011	0.025	0.045	0.065	0.084	0.103	0.120
0.70	0.012	0.028	0.049	0.070	0.091	0.110	0.128
0.60	0.014	0.031	0.054	0.077	0.099	0.119	0.138
0.50	0.016	0.035	0.061	0.085	0.109	0.130	0.149
0.40	0.019	0.041	0.069	0.097	0.121	0.144	0.164
0.35	0.021	0.045	0.075	0.104	0.129	0.152	0.173
0.30	0.024	0.050	0.082	0.112	0.139	0.162	0.183
L							
1.00	0.006	0.015	0.029	0.044	0.059	0.073	0.088
0.90	0.007	0.017	0.032	0.047	0.063	0.078	0.093
0.80	0.008	0.018	0.034	0.051	0.067	0.084	0.099
0.70	0.009	0.021	0.038	0.056	0.073	0.090	0.106
0.60	0.010	0.023	0.042	0.061	0.080	0.098	0.115
0.50	0.012	0.027	0.048	0.068	0.089	0.108	0.126
0.40	0.014	0.032	0.055	0.078	0.100	0.121	0.139
0.35	0.016	0.035	0.060	0.084	0.108	0.129	0.148
0.30	0.018	0.039	0.066	0.092	0.116	0.138	0.158

Table A6.1 Bending moment coefficients, α, for walls spanning two ways under ud load (after Roberts et al. (1986)) (contd)

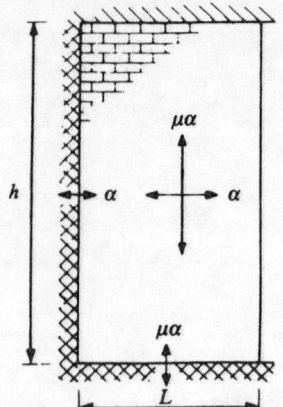

Key to support conditions

——————— denotes free edge

///// simply supported edge

XXXXX an edge over which full continuity exists

Values of α

	μ =	*h/L* =						
		0.30	**0.50**	**0.75**	**1.00**	**1.25**	**1.50**	**1.75**
J	0.25	0.032	0.071	0.122	0.180	0.240	0.300	0.362
	0.20	0.038	0.083	0.142	0.208	0.276	0.344	0.413
	0.15	0.048	0.100	0.173	0.250	0.329	0.408	0.488
	0.10	0.065	0.131	0.224	0.321	0.418	0.515	0.613
	0.05	0.106	0.208	0.344	0.482	0.620	0.759	0.898
K	0.25	0.028	0.056	0.091	0.123	0.150	0.174	0.196
	0.20	0.033	0.064	0.103	0.136	0.165	0.190	0.211
	0.15	0.040	0.077	0.119	0.155	0.184	0.210	0.231
	0.10	0.053	0.096	0.144	0.182	0.213	0.238	0.260
	0.05	0.080	0.136	0.190	0.230	0.260	0.286	0.306
L	0.25	0.021	0.044	0.073	0.101	0.127	0.150	0.170
	0.20	0.025	0.052	0.084	0.114	0.141	0.165	0.185
	0.15	0.031	0.061	0.098	0.131	0.159	0.184	0.205
	0.10	0.041	0.078	0.121	0.156	0.186	0.212	0.233
	0.05	0.064	0.114	0.164	0.204	0.235	0.260	0.281

A7 Bending moments in walls with openings

A7.1.1 Simplified method for walls with openings of any size

According to Appendix D of BS 5628:Part 1, a design bending moment for a wall with openings at the edge may be calculated by assuming the wall to be split into sub-panels with line loads applied to some of the edges (Fig. A7.1). In the case of wind loading, the line loads can be assumed to be uniformly distributed (ud). Each sub-panel is checked as a separate entity; the large sub-panels are usually the more critical, but the small ones, for example B in Case (a) of Fig. A7.1, may also need checking in some instances. Any, but usually the most favourable, pattern of splitting is chosen; in the case of panels like those drawn in Fig. A7.2 it is often advantageous to have a sub-panel the same length as the main panel, as shown, bonded masonry being very much stronger spanning horizontally than vertically.

On the lines separating the sub-panels, the masonry may be thought of as being able to transmit shear forces but not bending, although this is not usually true in a real case. For panels with openings at corners, the panel may be split as shown in Fig. A7.3. With the wall split into sub-panels, which have assumed lines of support along some edges, the bending moment in each sub-panel is calculated in turn, Table A6.1 being used to find moment coefficients. Normally the highest bending moment found is taken as the design moment.

It is a great convenience in analysis to be able to apply an equivalent ud load to a wall panel instead of a ud load and line load, if the latter is acting on the panel too. A yield-line approach developed by Johansen (1972) for ductile slabs gives an equivalent ud load, w_e for the case shown in Fig. A7.4(a) of

$$w_e = w(1 + 2\beta) = w(1 + c/h)$$

taking the worst case, where w is the actual ud load and β is the line load ratio, the ratio of the line load along an edge to the total ud load on the panel.

The bending moment coefficients in Table A6.1 have been obtained by correlating a yield-line method with test results, and these coefficients, in combination with the equivalent ud load, can be used to obtain the design bending moments in wall panels with openings. This simplified method can be used with any of the cases illustrated in Table A6.1.

It is suggested that, in general, line loads are not applied to

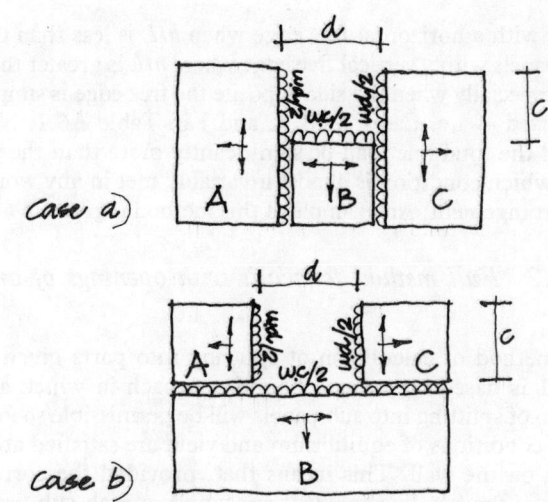

Fig. A7.1 *Wall panel split in two different ways.*

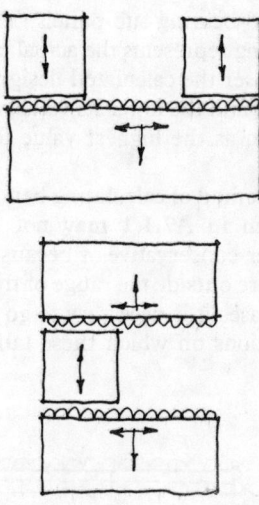

Fig. A7.2 *Wall panels split so that the principal parts span horizontally.*

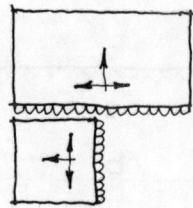

Fig. A7.3 *Wall panel, with corner opening, split into parts.*

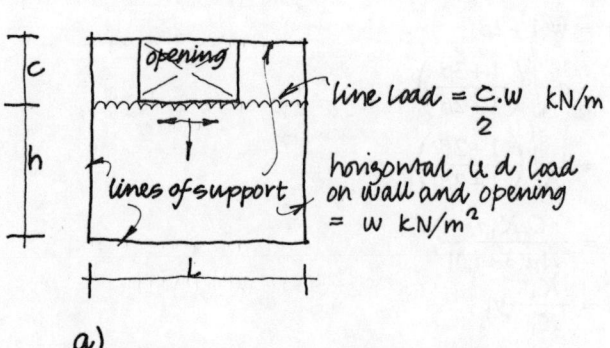

a)

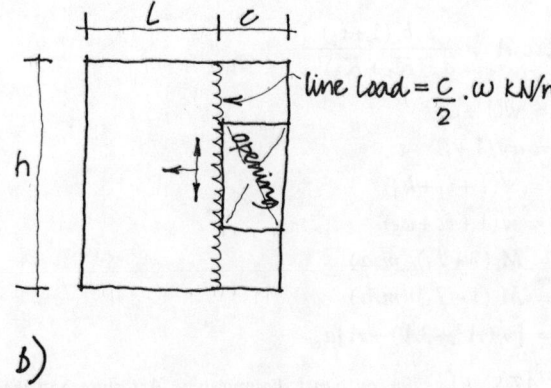

b)

Fig. A7.4 *Elevation on wall (a) with horizontal opening and (b) with vertical opening.*

panels with a horizontal free edge when h/L is less than 0.3 or to panels with a vertical free edge where h/L is greater than 1.75, especially when the side opposite the free edge is simply supported — i.e. Cases A, B, C and J in Table A6.1. Nor should the total line load be significantly more than the ud load, which condition is almost invariably met in any workable arrangement. An example of this method is given in **6.3**.

A7.1.2 Full method for walls with openings of any size

The method of calculation of splitting into parts given in **A7.1.1** is based on a lower bound approach in which any pattern of splitting into sub-panels will be permissible so long as the conditions of equilibrium and yield are satisfied at all points on the wall. This means that, provided the correct values of the bending moment are found in each sub-panel for the load considered and that these moments are always less than the design moment of resistance of the panel at each point, then the result will be safe whatever pattern of splitting is adopted and however the load is assumed to be distributed between the load-bearing sub-panels. The more closely the pattern of splitting represents the actual behaviour of the wall at failure the closer the calculated design moment will be to the true moment and the lower its value, assuming the design moment is taken as the highest value found.

However, the method of calculating bending moments in the sub-panels given in **A7.1.1** may not be adequate either because it is over-conservative or because the proportions of the sub-panels are outside the range of those treated in Table A6.1. In such cases it is necessary to go back to the original yield line equations on which these tables are based.

Figures A7.5 and A7.6 treat the two basic yield-line patterns for calculating bending moments in panels supported on three sides; Figure A7.7 treats another pattern for panels with high aspect ratios assuming the long side opposite the free edge to be simply supported. Also given are the positions of the yield lines and the equivalent values (with an 'e' subscript) for all those parameters affected when a wall with a ud load and a line load along its free edge is treated as a panel with an equivalent ud load only. In these figures, the letter i stands for the degree of restraint at the edge of the slab and the letter I for the degree of restraint at the ends of the discrete beam; m is the yield moment of the slab and M is the yield moment of the discrete beam. Note that the cases given can be applied to panels supported on four sides or two adjacent sides by using equivalent dimensions. It is important to remember that yield-line theory only claims to provide upper-bound solutions, that is ones that give an over-estimate or the correct value of the collapse load, so that that yield-line formula which gives the highest design moment must always be the one used; see Jones and Wood (1967) for an account of yield-line theory. It is assumed here that line loads do not alter the yield-line patterns significantly.

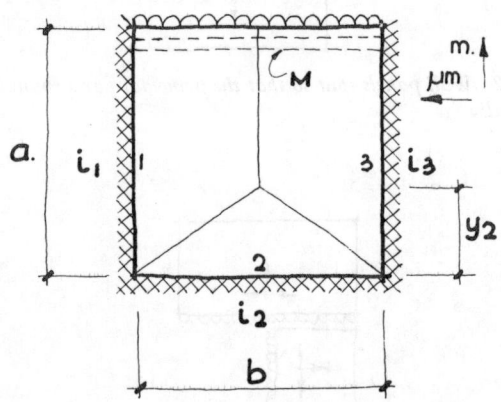

$$m = \frac{w_e \cdot b^2}{6\mu(K_1+K_3)^2}[\sqrt{(A+3)}-\sqrt{A}]^2$$

where $A = \dfrac{\mu.b^2(1+i_2)}{a_e^2(K_1+K_3)^2}$

$w_e = w(1+\beta)$

$a_e = a\sqrt{(1+\beta)}$

$K_1 = \sqrt{(1+i_1+k_1)}$

$K_3 = \sqrt{(1+i_3+k_3)}$

$k_1 = M(1+I_1)/(m.a)$

$k_3 = M(1+I_3)/(m.a)$

$y_2 = [\sqrt{(A^2+3A)}-A]a_e$

Fig. A7.5 Elevation on panel showing type A failure and yield-line data.

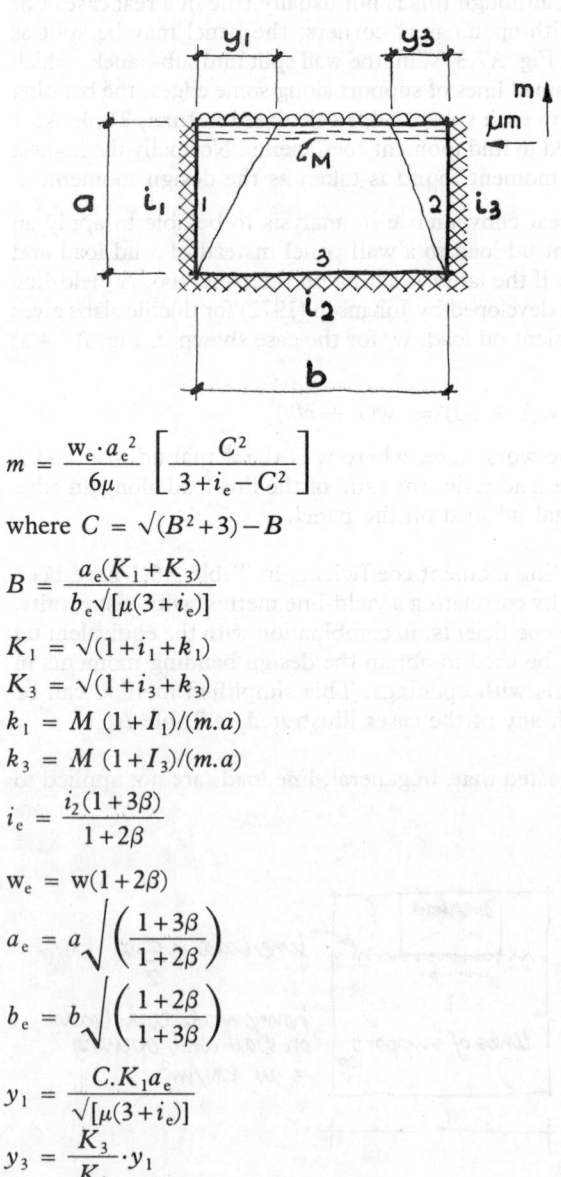

$$m = \frac{w_e \cdot a_e^2}{6\mu}\left[\frac{C^2}{3+i_e-C^2}\right]$$

where $C = \sqrt{(B^2+3)}-B$

$B = \dfrac{a_e(K_1+K_3)}{b_e\sqrt{[\mu(3+i_e)]}}$

$K_1 = \sqrt{(1+i_1+k_1)}$

$K_3 = \sqrt{(1+i_3+k_3)}$

$k_1 = M(1+I_1)/(m.a)$

$k_3 = M(1+I_3)/(m.a)$

$i_e = \dfrac{i_2(1+3\beta)}{1+2\beta}$

$w_e = w(1+2\beta)$

$a_e = a\sqrt{\left(\dfrac{1+3\beta}{1+2\beta}\right)}$

$b_e = b\sqrt{\left(\dfrac{1+2\beta}{1+3\beta}\right)}$

$y_1 = \dfrac{C.K_1a_e}{\sqrt{[\mu(3+i_e)]}}$

$y_3 = \dfrac{K_3}{K_1}.y_1$

Fig. A7.6 Elevation on panel showing type B failure and yield-line data.

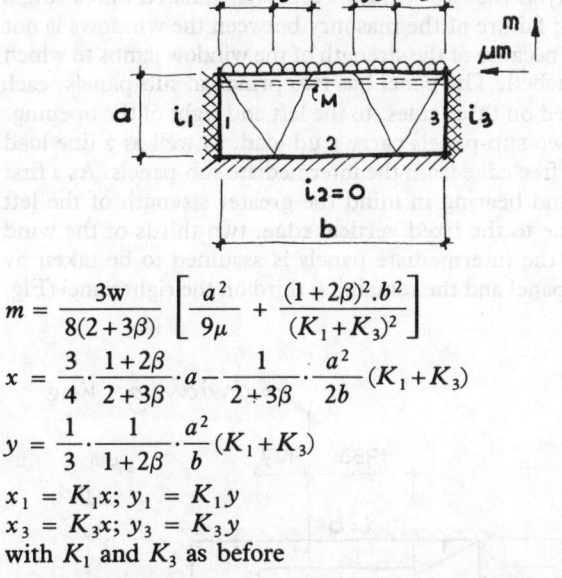

$$m = \frac{3w}{8(2+3\beta)}\left[\frac{a^2}{9\mu} + \frac{(1+2\beta)^2 \cdot b^2}{(K_1+K_3)^2}\right]$$

$$x = \frac{3}{4}\cdot\frac{1+2\beta}{2+3\beta}\cdot a - \frac{1}{2+3\beta}\cdot\frac{a^2}{2b}(K_1+K_3)$$

$$y = \frac{1}{3}\cdot\frac{1}{1+2\beta}\cdot\frac{a^2}{b}(K_1+K_3)$$

$$x_1 = K_1 x; \quad y_1 = K_1 y$$
$$x_3 = K_3 x; \quad y_3 = K_3 y$$

with K_1 and K_3 as before

Fig. A7.7 *Elevation on panel with simply supported bottom edge showing type C failure and yield-line data.*

A7.2 Walls with medium- or small-sized openings

The methods of calculation given in **A7.1** tend to ignore the two-way spanning action of the panel and could be over-conservative where the openings in the wall are only of moderate size. A good insight into the behaviour of such panels is given by yield-line theory. Johansen (1972, pp.6–13) gives a safe value for bending moments in slabs with such holes by transforming the actual slab to an equivalent slab which is simply supported at the edges and without holes. This transformation is explained in Fig. A7.8. The transformation gives the dimensions h_t and L_t of the equivalent simply supported slab. The letter i stands for the degree of restraint given by each of the four edges being equal to one if there is continuity at the edge, zero if the edge is only simply supported, or any value between these extremes which seems appropriate, depending on the edge details.

The bending moments may then be found by using Case E, the one for simply supported slabs, in Table A6.1. It is suggested that, as a general rule only, this method is applied to masonry wall panels where the opening has both a vertical dimension of less than $0.6h$ and a horizontal dimension of less than $0.35L$, these proportions reflecting the different strengths of masonry in the horizontal and vertical directions. These different strengths do not affect the use of the formula given in Fig. A7.8. This method becomes more conservative the larger the size of the opening and the more eccentrically it is placed, assuming other kinds of failure can be prevented (see below), but the method is recommended for a variety of panels on account of its simplicity. A wall panel with

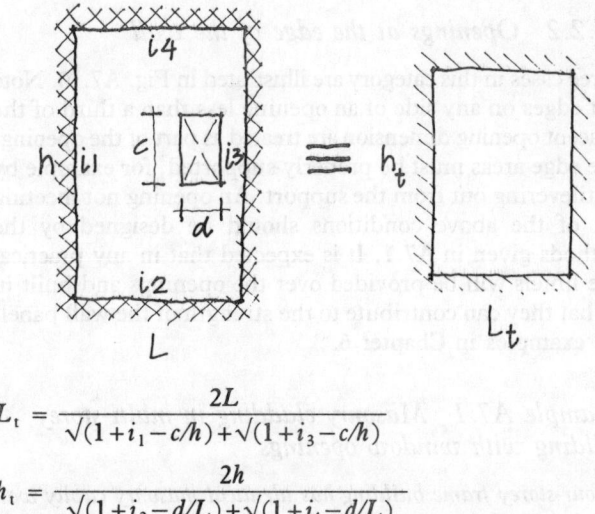

$$L_t = \frac{2L}{\sqrt{(1+i_1-c/h)}+\sqrt{(1+i_3-c/h)}}$$

$$h_t = \frac{2h}{\sqrt{(1+i_2-d/L)}+\sqrt{(1+i_4-d/L)}}$$

Fig. A7.8 *Transformations of a panel with an opening and of varying fixity at edges to an equivalent simply-supported slab without an opening.*

two or more openings may be aggregated into an equivalent single opening with minimum dimensions of $c_{eq} \times d_{eq}$: adding up the c dimensions (Fig. A7.8) of each individual opening, but not counting more than once those portions which overlap in the vertical direction, gives c_{eq} ; d_{eq} is calculated similarly in the horizontal direction.

Looking at the formulae in Fig. A7.8, it is clear that an opening placed near an edge reduces the effective fixity of that edge (and all the other edges it is assumed for the present); the last two terms in the square root brackets, e.g. $i_1 - c/h$, etc., then denote the effective fixity of each of the four edges, taking account of the effect of the opening; see Example A7.4. Walls in this category may be divided into two cases, those with openings in the central area and those with openings near the edge of the wall.

A.7.2.1 Openings in the central area of the wall

To avoid the occurrence of long, unsupported edges adjacent to the opening which could cause type C failure patterns (A7.1.2), it is suggested that no edge distance surrounding an opening be less than a third of the adjacent opening dimension (Fig. A7.9). This provides a further limitation on the dimensions of the opening. The maximum area of opening suggested previously, placed as near to one corner as possible, has been illustrated in Fig. A7.9. Openings nearer to the edge than this would need to be checked (A7.1.2) or reduced in size or treated as shown in Fig. A7.10.

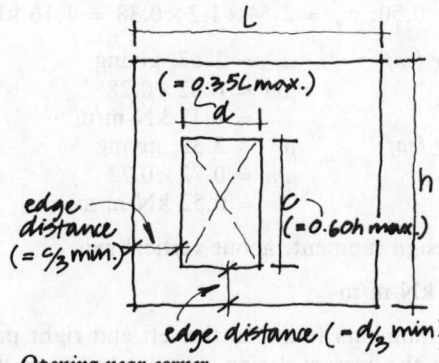

Fig. A7.9 *Opening near corner.*

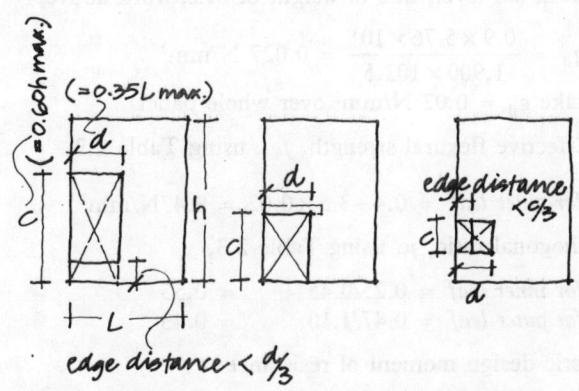

Fig. A7.10 *Dimensions c and d to be taken for calculation purposes.*

A7.2.2 Openings at the edge of the wall

Three cases in this category are illustrated in Fig. A7.10. Note that edges on any side of an opening less than a third of the adjacent opening dimension are treated as part of the opening. The edge areas must be properly supported, for example by cantilevering out from the support. An opening not meeting any of the above conditions should be designed by the methods given in **A7.1**. It is expected that in any practical case lintels will be provided over the openings and built in so that they can contribute to the strength of the wall panel. See examples in Chapter 6.

Example A7.1 Masonry cladding in multi-storey building with window openings

A four-storey frame building has identical masonry cavity wall panels in each bay at each level consisting of an inner 100 mm thick blockwork leaf, supported by the slab at each level, and a continuous outer leaf of brickwork 102.5 mm thick; there are two central window openings (Fig. A7.11). A mortar of designation (iii) is used. The brick has a water absorption of 9% and the blockwork has a compressive strength of 3.5 N/mm². The factor of safety for material strength, $\gamma_m = 3.5$. The horizontal wind load at the top is 0.38 kN/m². Check.

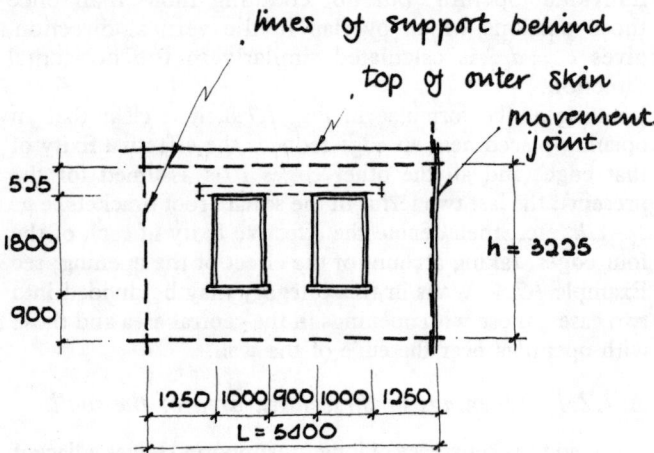

Fig. A7.11 Elevation on wall at top storey.

Horizontal loading. The panel checked is the critical one against a movement joint and at the top storey, where the axial load is least.

By inspection panel meets Code limiting dimensions. Select Case (b) of clause 22 of the Code: $0.9\,G_k + 1.2\,W_k$.

Design vertical compressive stress near centre of brickwork skin, at sill level, due to weight of brickwork above,

$$g_d = \frac{0.9 \times 5.76 \times 10^3}{1,900 \times 102.5} = 0.027 \text{ N/mm}^2$$

take $g_d = 0.02$ N/mm² over whole panel

$\therefore$ Effective flexural strength, f_{ka}, using Table 2.3,

for outer leaf $= 0.4 + 3.5 \times 0.02 = 0.47$ N/mm²

Orthogonal ratio, μ, using Table 2.3,

for inner leaf $= 0.25/0.45$ $= 0.55$
for outer leaf $= 0.47/1.10$ $= 0.43$

Elastic design moment of resistance

for inner leaf $= 0.45 \times 1.66/3.5 = 0.21$ kN-m/m
for outer leaf $= 1.10 \times 1.75/3.5 = 0.55$ kN-m/m

For analysis the two windows are amalgamated into a single opening; failure of the masonry between the windows is not possible because of the strength of the window jambs to which it is attached. The panel has two principal sub-panels, each supported on three sides, to the left and right of the opening. These two sub-panels carry a ud load, as well as a line load on their free edge from the intermediate sub-panels. As a first guess, and bearing in mind the greater strength of the left panel due to the fixed vertical edge, two-thirds of the wind load on the intermediate panels is assumed to be taken by the left panel and the remaining third on the right panel (Fig. A7.12).

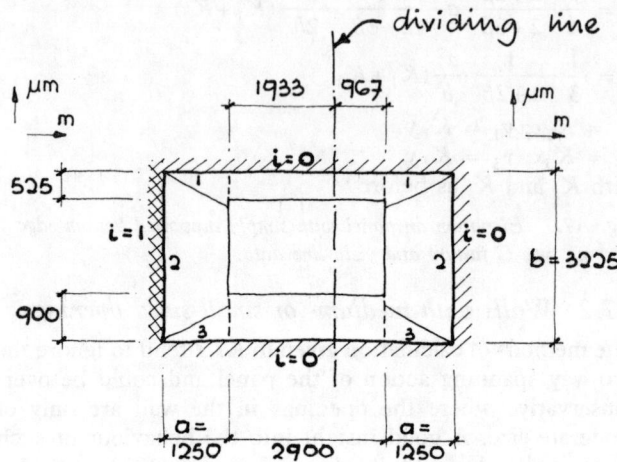

Fig. A7.12 Panel showing edge conditions and assumed load split between left and right sub-panels.

Taking a type B failure pattern (a type A failure pattern giving lower design moments), then

for left panel,

$\beta = 1933/1250 = 1.54$; $K_1 = K_3 = 1$; $i_2 = 1$;
$i_e = 1.38$; $a_e/b_e = 0.53$; $w_e = 4.08 \times 1.2 \times 0.38$
                    $= 1.86$ kN/m²;

then with proportion of total wind load taken by inner leaf $= 0.21/0.76 = 0.28$, then with moment key line directions as indicated in Fig. A7.12,

for inner leaf $\mu = 1/0.55 = 1.82$ giving
                  $\mu m = 0.53 \times 0.28 = 0.15$ kN/-m/m
for outer leaf $\mu = 1/0.43 = 2.32$ giving
                  $\mu m = 0.58 \times 0.72$
                      $= 0.42$ kN-m/m

$\therefore$ total design moment, about vertical axis
    $= 0.57$ kN-m/m

for right panel

$\beta = 967/1250 = 0.77$; $K_1 = K_3 = 1$; $i_2 = 0$;
$a_e/b_e = 0.50$; $w_e = 2.54 \times 1.2 \times 0.38 = 1.16$ kN/m²;

for inner leaf $\mu = 1.82$, giving
                  $\mu m = 0.62 \times 0.28$
                      $= 0.17$ kN-m/m
for outer leaf $\mu = 2.32$, giving
                    $\mu m = 0.72 \times 0.72$
                      $= 0.52$ kN-m/m

$\therefore$ total design moment, about vertical axis

    $= 0.69$ kN-m/m

When the moments found in the left and right panels are equal, then the correct design moment has been found; in the present case it would be sufficient to take an average value of 0.63 kN-m/m which is less than the total design moment of resistance of 0.76 kN-m/m, it being sufficiently accurate to add together the moments from the two leaves.

Alternatively the dividing line can be shifted slightly to the right so that the design moment increases on the left panel and decreases on the right panel; the design moment is sensitive to the value of β and this fact can be used to find a new approximate position of the dividing line. Repeating the calculation for a new load split (Fig. A7.13) gives

for left panel, total moment = 0.61 kN-m/m, and
for right panel, total moment = 0.62 kN-m/m

Looking at the intermediate sub-panels, it is assumed that the sub-panel above the opening is prevented from a type C or other type of failure by a lintel which is continuous across the tops of the two windows (Fig. A7.11). For the sub-panel below the opening, because of the two window sills, a type B failure is precluded (Fig. A7.14); a type C failure gives a higher design moment than type A.

For type C failure pattern, assuming, conservatively, that sub-panel takes half of the load on windows and all edges are simply supported,

$K_1 = K_3 = 1$; $\beta = 900/900 = 1$; $b = 2.9$ m; $a = 0.9$ m

ignoring first term in brackets, which is small, design moment, about vertical axis,

$m = 0.65$ kN-m/m < 0.76 kN-m/m   OK

It would be permissible, and more realistic, to assume that the two vertical edges of the sub-panel were fixed, in which case the design moment is almost halved. Note that type C failure patterns of the kind shown in Fig. A7.15 are not taken to be critical either because they give a lower moment than that already considered, in the case of the pattern shown on

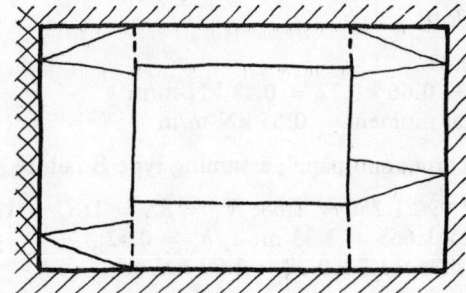

Fig. A7.15   *Type C failure patterns.*

the right, or because they would do so if the strength of the surrounding masonry and window frames in the real structure were taken into account, in the case of the pattern shown on the left. In this example, it is assumed that a valid mechanism would have yield lines intersecting the opening near the corners or near the middle of the opening at the bottom, where the wall does not benefit from the additional strength provided by a window frame. It is possible to include the bending strength of the window frame in the calculation by treating it as a discrete beam, which thus increases the 'K' factor. It may help to draw out the positions of the yield lines for each assumed failure pattern to see whether these give a plausible mechanism.

Example A7.2   Unreinforced masonry panel with opening

Repeat Example A7.1 *but with the wall split into a different arrangement of sub-panels, as shown in* Fig. A7.16, *with a bottom sub-panel supported on three sides and two top corner panels each supported on two sides.*

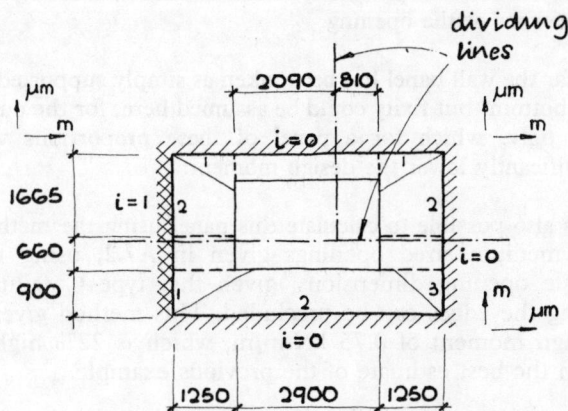

Fig. A7.16   *Elevation on panel and assumed load split.*

Assume to start with that the vertical dividing line, giving the load split between corner panels, is the same as that found in the previous example, for that sub-panel arrangement; it is assumed that the corner panels take a line load on the vertical edge only.

The horizontal dividing line, giving the line load on the horizontal edge of the bottom sub-panel, may be found by trial and error. For bottom sub-panel, assuming type B failure pattern, take

$\beta = 660/900 = 0.73$; $K_1 = \sqrt{2}$; $K_3 = 1$; $i_2 = 0$;
$a_e/b_e = 0.22$; $w_e = 2.47 \times 1.2 \times 0.38 = 1.12$ kN/m^2

for inner leaf,

$\mu = 0.55$
$\therefore m = 0.60 \times 0.28 = 0.17$ kN-m/m

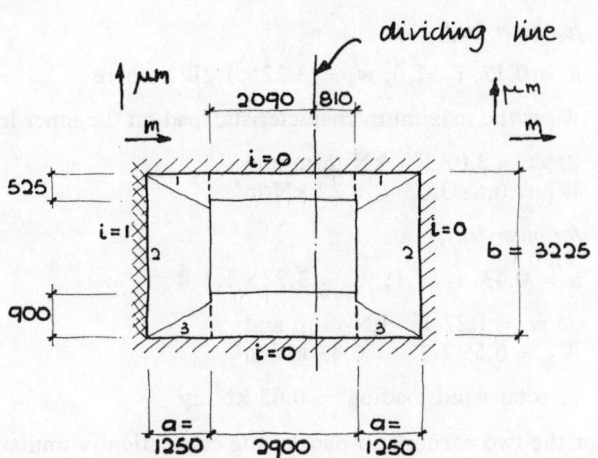

Fig. A7.13   *Panel with revised load split between sub-panels.*

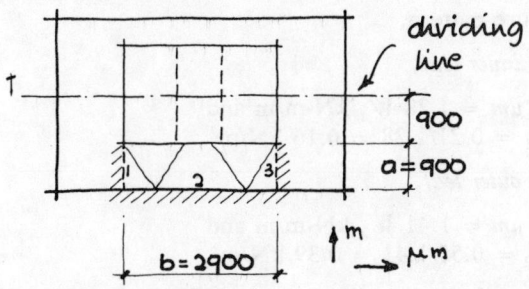

Fig. A7.14   *Sub-panel below opening showing assumed yield lines.*

for outer leaf,

$\mu = 0.43$

$\therefore m = 0.66 \times 0.72 = 0.48$ kN-m/m

$\therefore$ total moment $= 0.65$ kN-m/m

For left corner sub-panel, assuming type B failure pattern,

$\beta = 2\,090/1\,250 = 1.68$; $K_1 = K_3 = 1$; $i_e = 1.38$;
$b = 2 \times 1.665 = 3.33$ m; $a_e/b_e = 0.52$;
$w_e = 4.36 \times 1.2 \times 0.38 = 1.99$ kN/m^2

for inner leaf,

$\mu = 1.82 \therefore \mu m = 0.58 \times 0.28 = 0.16$ kN-m/m

for outer leaf,

$\mu = 2.32 \therefore \mu m = 0.64 \times 0.72 = 0.46$ kN-m/m

$\therefore$ total moment $= 0.62$ kN-m/m

For right corner sub-panel, assuming type B failure pattern,

$\beta = 810/1250 = 0.65$; $K_1 = K_3 = 1$; $i_2 = 0$;
$b = 2 \times 1.665 = 3.33$ m; $a_e/b_e = 0.48$;
$w_e = 2.30 \times 1.2 \times 0.38 = 1.05$ kN/m^2

for inner leaf,

$\mu = 1.82 \therefore \mu m = 0.58 \times 0.28 = 0.16$ kN-m/m

for outer leaf,

$\mu = 2.32 \therefore \mu m = 0.67 \times 0.72 = 0.48$ kN-m/m

$\therefore$ total moment $= 0.64$ kN-m/m

This subdivision of the panel gives good agreement, with the dividing lines only needing to be moved slightly to the right and towards the bottom for the optimum position. The design moment is slightly higher than that found in the previous example; the yield lines intersect the same sides of the opening in the two arrangements, except at the bottom right hand, where in an actual case the yield line would go directly into the corner of the opening.

So far the wall panel has been taken as simply supported at the bottom, but fixity could be assumed here, for the outer leaf only, which for a panel of these proportions will significantly lower the design moment.

It is also possible to calculate this panel using the method for medium sized openings given in **A7.2,** using the single opening dimensions, given that type C failures along the edges can be precluded. This method gives a design moment of 0.75 kN-m/m, which is 22% higher than the best estimate of the previous example.

Example A7.3 Reinforced masonry panel with opening

Repeat Example A7.1 *but with wind load of* 0.55 kN/m^2 *instead of* 0.38 kN/m^2.

The maximum characteristic wind load that can be carried by the unreinforced panel is approximately

$W_k = (0.76/0.62) \times 0.38 = 0.46$ kN/m$^2 < 0.55$ kN/m^2

Therefore reinforcement is provided in the top two courses of the inner leaf of the bottom sub-panel only, where it is better protected and gives a greater proportionate increase in strength than in the outer leaf. The reinforcement, consisting of two wires 55 mm apart horizontally, each having an area of 10 mm^2, has an average vertical spacing of 450 mm over the panel depth, this also being the maximum spacing. Hence from **A13.2.1,** for inner leaf, with $f_k = 3.5$

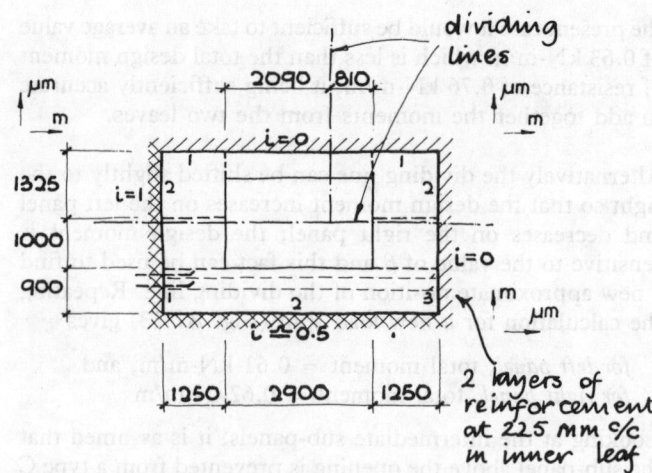

Fig. A7.17 *Elevation on panel and assumed load split.*

N/mm^2

$z = \left(1 - \dfrac{0.5 \times 10 \times 485 \times 3.5}{450 \times 77 \times 3.5 \times 1.15}\right) d = 0.94\,d < 0.95\,d$

$\therefore M_d = \dfrac{10 \times 485 \times 0.94 \times 77 \times 10^{-6}}{1.15 \times 0.450} = 0.68$ kN-m/m

and moment of resistance about horizontal axis is

$M_a = 0.25 \times 1.66/3.5 = 0.12$ kN-m/m

$\therefore \mu = 0.12/0.68 = 0.17$

It will be assumed, in this example, that the outer leaf of the wall is fixed at the bottom, the inner leaf being pinned, and that, at ultimate load, the full moment of resistance of each leaf is mobilised. For this case, it is most convenient to work in terms of the maximum allowable wind loads on each sub-panel rather than the design moments.

For bottom sub-panel with type B failure pattern, take

$\beta = 1000/900 = 1.11$; $K_1 = \sqrt{2}$; $K_3 = 1$; $a_e/b_e = 0.22$;

for inner leaf,

$\mu = 0.17$; $i_2 = 0$; $w_e = 3.22 \times 1.2 W_1$ where

W_1 is the maximum characteristic load on the inner leaf,

$\therefore m = 3.07\,W_1$ kN-m/m and
$W_1 = 0.68/3.07 = 0.22$ kN/m^2

for outer leaf,

$\mu = 0.43$; $i_2 = 1$; $w_e = 3.22 \times 1.2\,W_2$

$\therefore m = 1.27\,W_2$ kN-m/m and
$W_2 = 0.55/1.27 = 0.43$ kN/m^2

$\therefore$ total wind loading $= 0.65$ kN/m^2

For the two corner sub-panels, the calculation is similar to that of the previous example with all input values the same except that the characteristic wind load is W_k and $b = 2 \times 1.325 = 2.65$ m; for left corner sub-panel with type B failure pattern,

for inner leaf,

$\therefore \mu m = 1.28\,W_1$ kN-m/m and
$W_1 = 0.21/1.28 = 0.16$ kN/m^2

for outer leaf,

$\therefore \mu m = 1.41\,W_2$ kN-m/m and
$W_2 = 0.55/1.41 = 0.39$ kN/m^2

$\therefore$ total wind loading $= 0.55$ kN/m^2

Similarly for right corner sub-panel with type B failure pattern,

for inner leaf,

$\therefore \mu m = 1.15 \ W_1$ kN-m/m and $W_1 = 0.21/1.15 = 0.18$ kN/m^2

for outer leaf,

$\therefore \mu m = 1.34 \ W_2$ kN-m/m and $W_2 = 0.55/1.34 = 0.41$ kN/m^2

$\therefore$ total wind loading $= 0.59$ kN/m^2

The horizontal dividing line should be moved up slightly. Nevertheless the values of the wind loadings on each sub-panel are sufficiently close for design purposes; taking an average, the maximum allowable characteristic wind load would be 0.60 kN/m^2 which is satisfactory. Note that the reinforcement must be effectively continuous over the left support in order for the fixity of the inner leaf, i_2, to maintain a value of unity as assumed; otherwise i_2 should be reduced accordingly. The ties or other bars used to maintain continuity at the support should have a minimum lap of 225 mm with the bed joint reinforcement.

Example A7.4   Solid panel adjacent to panel with opening

Find moment coefficient for panel shown in Fig. 6.9, *which is continuous with another panel of the same height along its right-hand edge. This latter panel, however, contains an opening on its left-hand edge, like those in* Fig. A7.10, *for which c/h = 0.5.*

Fixity of right-hand edge of panel in Fig. 6.9,

$i = i_3 - c/h = 1 - 0.5 = 0.5$, as shown

$$\therefore L_t = \frac{2 \times 5.400}{\sqrt{(1+1)} + \sqrt{(1+0.5)}} = 4.092 \text{ m}$$

and $h_t = h = 6.000$ m

Considering Case E of Table A6.1, gives $\alpha = 0.040$, which is a better estimate than that given in Example 6.4.

A8   Concentrated loads on walls

A8.1   Local design strength of walls under beams

The allowable local design strength of a wall under a bearing depends on the wall/beam detail adopted (Figs A8.1, A8.2, A8.3 and A8.4); for any particular bearing type the end of the beam or the bearing pad must fall within the shaded area shown and the minimum dimensions, d and w, must be respected.

A8.2   Methods of calculating local design stress under a concentrated load

For a bearing type 1 or a bearing type 2 (Clause 34 of BS 5628:Part 1) the stress is assumed to be uniform immediately under the bearing and at lower levels, spreading at an angle of 45° (Fig. A8.5). For a bearing type 3 (Clause 34 of BS 5628:Part 1) the load is assumed not to be uniform immediately under the bearing but to be uniform at lower levels, spreading at an angle of 45° (Fig. A8.6). Note, in Fig. A8.6(b), that in theory the spreader beam should be built into the wall so as to prevent uplift.

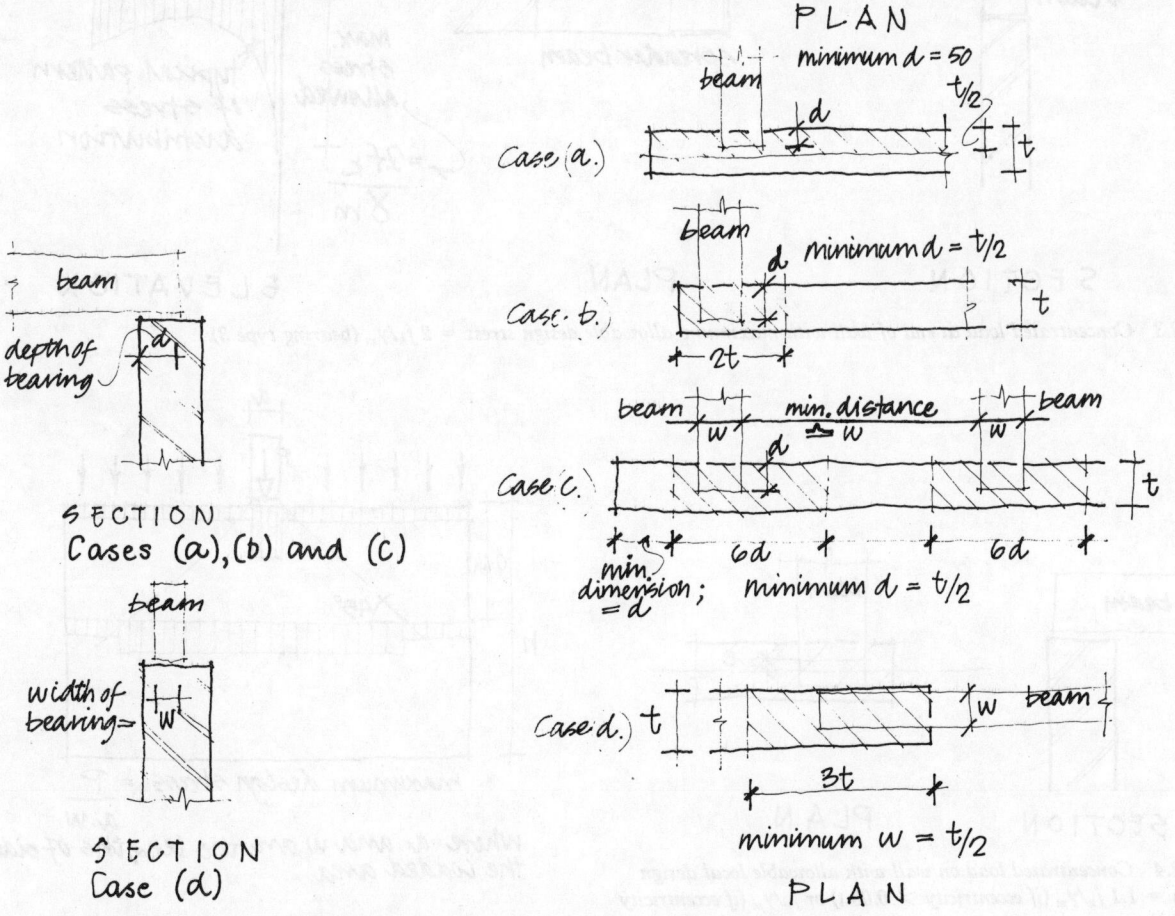

Fig. A8.1   *Concentrated loads on walls with allowable local design strength* $= 1.25 f_k/\gamma_m$ *(bearing type 1).*

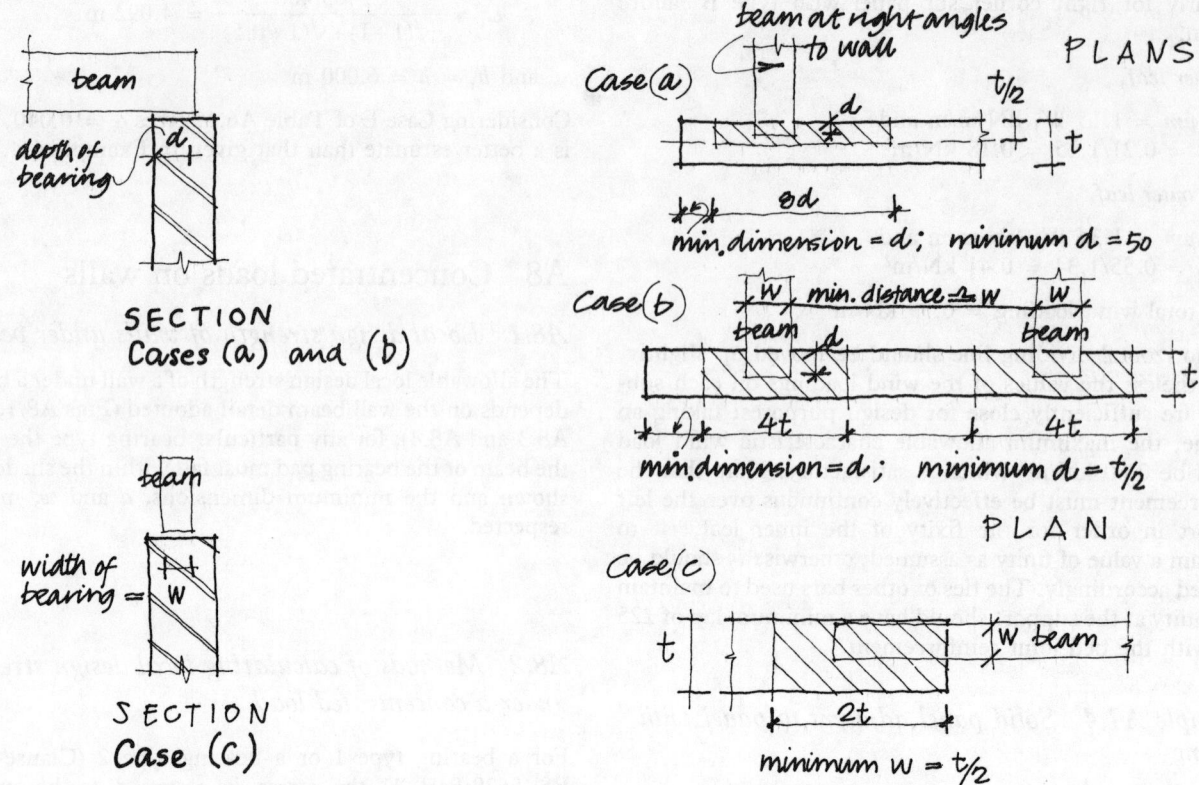

Fig. A8.2 *Concentrated loads on walls with allowable local design strength = $1.5 f_k/\gamma_m$ (bearing type 2).*

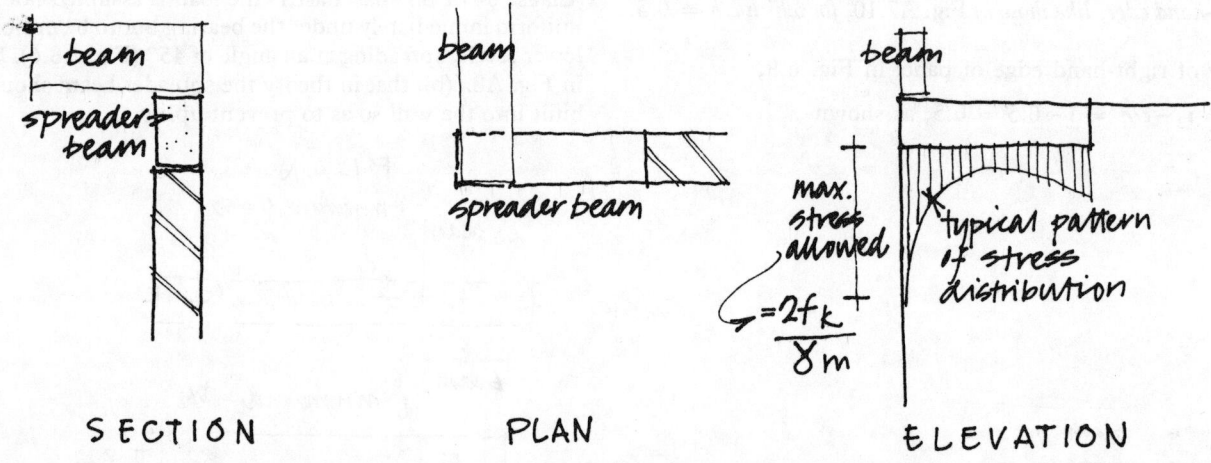

Fig. A8.3 *Concentrated load at end of wall with maximum allowable design stress = $2 f_k/\gamma_m$ (bearing type 3).*

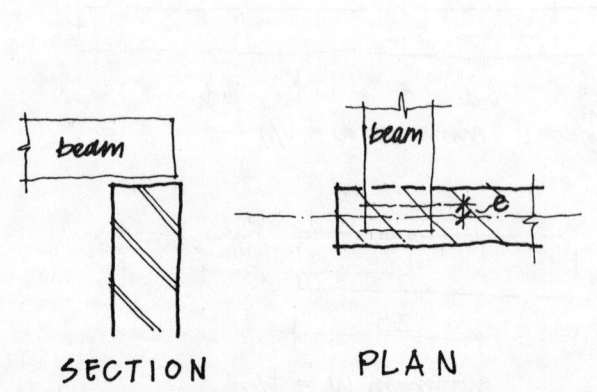

Fig. A8.4 *Concentrated load on wall with allowable local design strength = $1.1 f_k/\gamma_m$ (if eccentricity > 0.05 t) or f_k/γ_m (if eccentricity ≤ 0.05 t).*

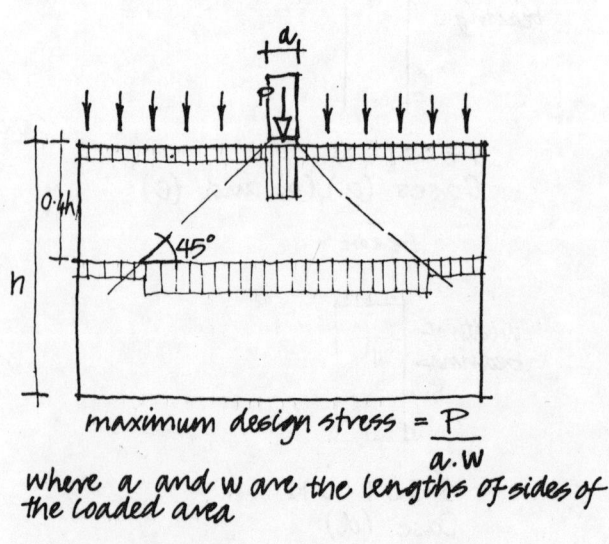

maximum design stress = $\dfrac{P}{a \cdot w}$

where a and w are the lengths of sides of the loaded area

Fig. A8.5 *Dispersion of concentrated load on wall.*

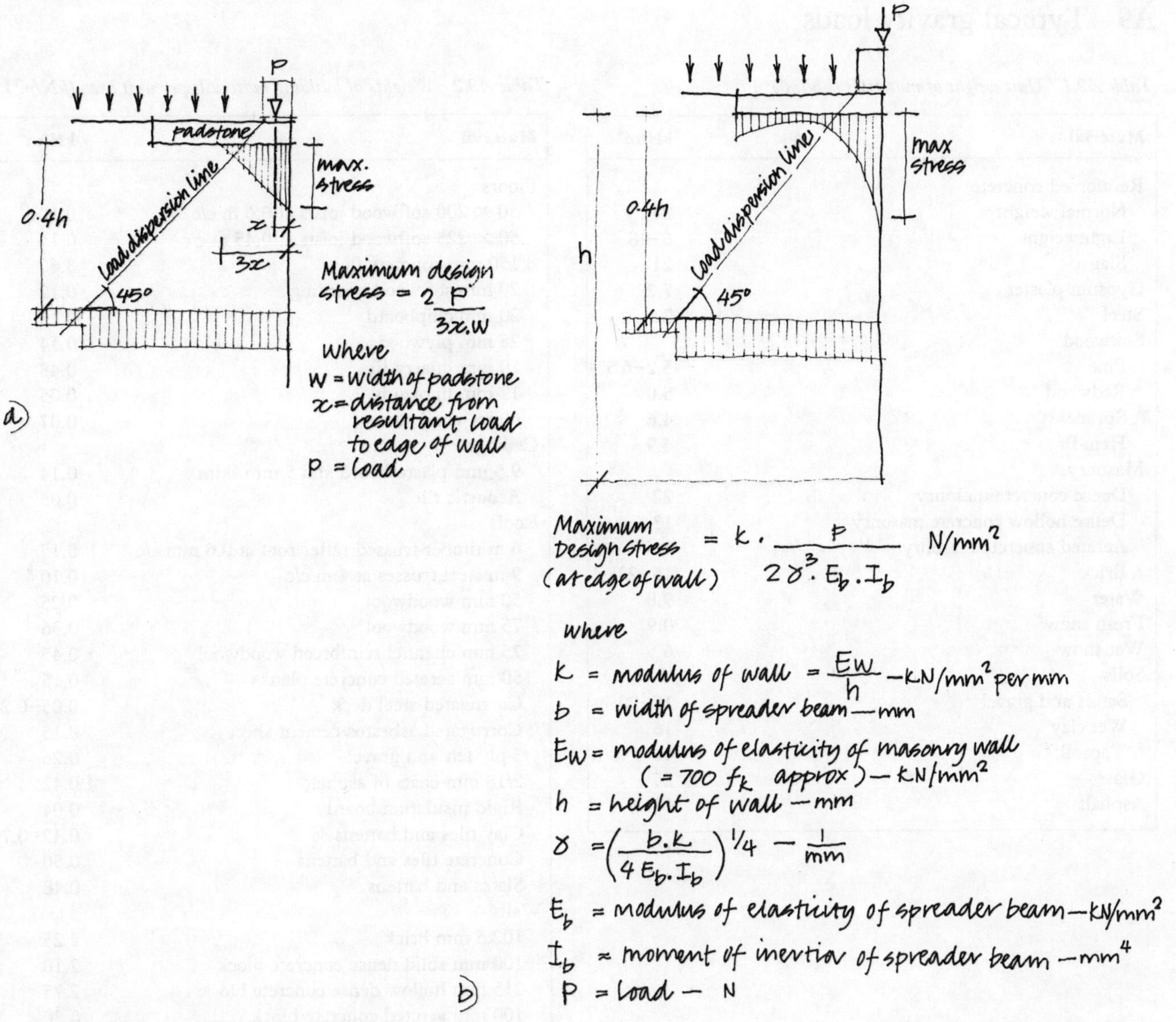
P
padstone
max.
stress
0.4h
Load dispersion line
x
3x
45°
Maximum design
stress = 2 P / 3x.W
where
W = width of padstone
x = distance from resultant load to edge of wall
P = load
a)
P
max
stress
0.4h
h
Load dispersion line
45°
Maximum Design stress (at edge of wall) = K . P / 2γ³ E_b . I_b N/mm²
where
K = modulus of wall = Ew / h – kN/mm² per mm
b = width of spreader beam – mm
Ew = modulus of elasticity of masonry wall (= 700 f_k approx) – kN/mm²
h = height of wall – mm
γ = (b.k / 4 E_b.I_b)^1/4 – 1/mm
E_b = modulus of elasticity of spreader beam – kN/mm²
I_b = moment of inertia of spreader beam – mm⁴
P = load – N
b)

Fig. A8.6 Maximum design stress at end of wall assuming (a) triangular stress distribution and (b) stress similar to that of semi-infinite beam on elastic foundation.

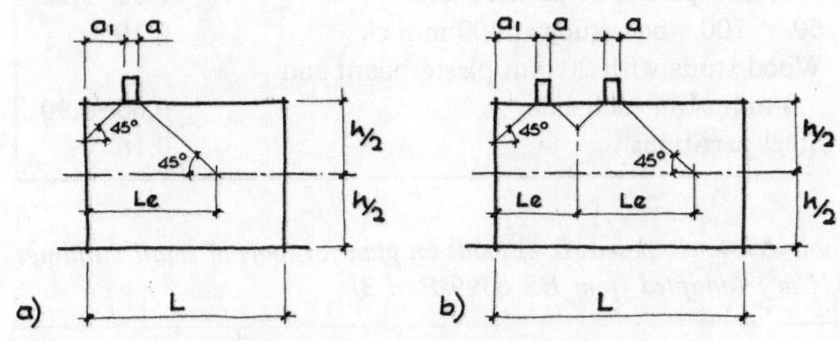
a₁
a
45°
45°
Le
h/2
h/2
L
a)
a₁
a
a
45°
45°
Le
Le
h/2
h/2
L
b)

Fig. A8.7 Elevation on wall (a) with single load and (b) with two parallel loads illustrating terms used in formula.

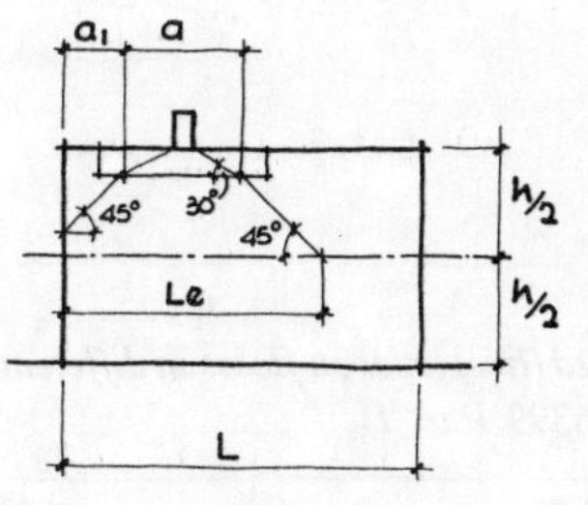
a₁
a
45°
30°
45°
Le
h/2
h/2
L

Fig. A8.8 Elevation on wall with spreader beam illustrating terms used in formula.

A8.3 Local design strength of walls under beams considering overall dimensions of wall

The empirical formula devised by Page and Hendry (1988) takes the overall dimensions of the wall as parameters, as well as dimensions local to the bearing. According to this, the strength enhancement factor for concentrated loads, i.e. the factor by which f_k/γ_m is multiplied, is equal to

$$\frac{0.55\ (1+0.5\ a_1/L)}{(A_b/A_e)^{0.33}}$$

but is not less than 1 or greater than $1.5 + a_1/L$ where A_b is the bearing area on the masonry $= a.w$; A_e is the effective area of the wall $= L_e.t$; a_1 is the distance from the edge of the bearing to the nearest end of the wall; L_e is the effective length of the wall which is the same as the actual length, L, for a short wall or column and is the dispersed length at mid-height for a long wall (Fig. A8.7) in which the load disperses at 45° to the horizontal. If a stiff spreader beam is used the load dispersion line can be assumed to run at 30° to the horizontal through the beam (Fig. A8.8).

A9 Typical gravity loads

Table A9.1 Unit weight of materials (kN/m³)

Material	kN/m³
Reinforced concrete	
Normal weight	24
Lightweight	6–18
Slag	21
Gypsum plaster	7.2
Steel	77
Softwood	
Pine	5.2–6.5
Redwood	5.0
Spruce	4.6
Hem-fir	4.9
Masonry	
Dense concrete masonry	22
Dense hollow concrete masonry	13.5–16
Aerated concrete masonry	9.5
Brick	16–22
Water	9.8
Fresh snow	0.9
Wet snow	6.5
Soils	
Sands and gravel	15–20
Wet clay	16
Topsoil	12
Glass	25
Asphalt	14

Table A9.2 Weights of building materials per unit area (kN/m²)

Material	kN/m²
Floors	
50 × 200 softwood joists at 0.6 m c/c	0.10
50 × 225 softwood joists at 0.45 m c/c	0.15
150 mm concrete floor	3.60
20 mm softwood boarding	0.12
20 mm chipboard	0.15
25 mm plywood	0.14
20 mm quarry tile	0.48
12 mm linoleum	0.05
6 mm vinyl tiles	0.07
Ceilings	
9.5 mm plasterboard and 5 mm skim	0.14
Acoustic tile	0.05
Roofs	
6 m timber-trussed rafter roof at 0.6 mm c/c	0.15
9 m steel trusses at 4 m c/c	0.10
50 mm woodwool	0.25
75 mm woodwool	0.36
75 mm channel reinforced woodwool	0.45
50 mm aerated concrete planks	0.15
Corrugated steel deck	0.05–0.20
Corrugated asbestos cement sheet	0.15
3-ply felt and gravel	0.26
2/18 mm coats of asphalt	0.42
Rigid insulation board	0.04
Clay tiles and battens	0.42–0.70
Concrete tiles and battens	0.50
Slates and battens	0.48
Walls	
102.5 mm brick	2.25
100 mm solid dense concrete block	2.10
215 mm hollow dense concrete block	2.75
100 mm aerated concrete block	0.80
Windows – glass and frame	0.40
Partitions	
75 mm lightweight concrete block	0.65
12.5 mm plaster or plasterboard	0.11–0.22
50 × 100 wood studs at 400 mm c/c	0.10
Wood studs with 20 mm plasterboard and	
5 mm skim each side	0.60–0.90
Steel partitions	0.18

Table A9.3 Imposed (live) loads on floors in different types of building (adapted from BS 6399:Part 1)

Type of building	Room or area	Uniformly distributed load kN/m²	Concentrated load on 300 mm square kN
Domestic	all rooms	1.5	1.4
Hotels	bedrooms	2.0	1.8
Schools	classrooms	3.0	2.7
	assembly areas	4.0–5.0	3.6
	corridors	4.0	4.5
Shops	sale areas	4.0	3.6
Offices	general areas	2.5	2.7
	storage areas	5.0	4.5
	computer rooms	3.5	4.5
	public areas	4.0	4.5

Table A9.4 Imposed (live) loads on plan for roofs of small buildings (kN/m²) (adapted from BS 6399:Part 3)

	Uniformly distributed load kN/m²	Concentrated load on 125 mm square kN
Roofs with access (slope ⩽ 10°)	1.50	1.8
Roofs without access (slope ⩽ 30°)	0.75†	0.9
Roofs without access (slope ⩾ 60°)	0	0.9
Roofs without access (slope = A°) for 30° < A < 60°	(60-A)/40†	0.9

†Stated ud load is assumed to be greater than snow load given in BS 6399: Part 3.

A10 Preliminary information on soils and foundations

Table A10.1 Soil identification (after Table 1 of BRE Digest 64)

Soil type	Field identification	Field assessment of structure and strength	Possible foundation difficulties
Gravels	Retained on No. 7 BS sieve and up to 76.2 mm. Some dry strength indicates presence of clay.	Loose – easily removed by shovel. 50 mm stakes can be driven well in.	Loss of fine particles in water-bearing ground.
Sands	Pass No. 7 and retained on No. 200 BS sieve. Clean sands break down completely when dry. Individual particles visible to the naked eye and gritty to fingers.	Compact – requires pick for excavation. Stakes will penetrate only a little way.	Frost heave, especially on fine sands. Excavation below water table causes runs and local collapse, especially in fine sands.
Silts	Pass No. 200 BS sieve. Particles not normally distinguishable with naked eye. Slightly gritty; most lumps can be moulded with the fingers but not rolled into threads. Shaking a small moist lump in the hand brings water to the surface. Silts dry rapidly; fairly easily powdered.	Soft – easily moulded with the fingers. Firm – can be moulded with strong finger pressure.	As for fine sands.
Clays	Smooth, plastic to the touch. Sticky when moist. Hold together when dry. Wet lumps immersed in water soften without disintegrating. Soft clays either uniform or show horizontal laminations. Harder clays frequently fissured, the fissures opening slightly when the overburden is removed or a vertical surface is revealed by a trial pit.	Very soft – exudes between fingers when squeezed. Soft – easily moulded with the fingers. Firm – can be moulded with strong finger pressure. Stiff – cannot be moulded with fingers. Hard – brittle or tough.	Shrinkage and swelling caused by vegetation. Long-term settlement by consolidation. Sulphate-bearing clays may attack concrete and corrode pipes. Poor drainage. Movement down slopes: most soft clays lose strength when disturbed.
Peat	Fibrous, black or brown. Often smelly. Very compressible and water retentive.	Soft – very compressible and spongy. Firm – compact.	Very low bearing capacity: large settlement caused by high compressibility. Shrinkage and swelling – foundations should be on firm strata below.
Chalk	White – readily identified.	Plastic – shattered, damp and slightly compressible or crumbly. Solid – needing a pick for removal.	Frost heave. Floor slabs on chalk fill particularly vulnerable during construction in cold weather. Swallow holes.
Fill	Miscellaneous material, e.g. rubble, mineral, waste, decaying wood.		To be avoided unless carefully compacted in thin layers and well consolidated. May ignite or contain injurious chemicals.

Table A10.2   *Choice of foundation (after Table 1 of BRE Digest 67)*

Soil type and site condition	Foundation	Details	Remarks
Rock, solid chalk, sands and gravels or sands and gravels with only small proportions of clay, dense silty sands.	Shallow strip or pad footings as appropriate to the load-bearing members of the building.	Breadth of strip footings to be related to soil density and loading (see Table A10.3). Pad footings should be designed for bearing pressures tabled in CP 101 : 1972. For higher pressures the depth should be increased and CP 2004 : 1972 'Foundations' consulted.	Keep above water wherever possible. Slopes on sand liable to erosion. Foundations 0.5 m deep should be adequate on ground susceptible to frost heave although in cold areas or in unheated buildings the depth may have to be increased. Beware of swallow holes in chalk.
Uniform, firm and stiff clays; (1) Where vegetation is insignificant;	Bored piles and ground beams, or strip foundations at least 1 m deep.	Deep strip footings of the narrow widths shown in Table A10.3 can conveniently be formed of concrete up to the ground surface.	
(2) Where trees and shrubs are growing or to be planted close to the site;	Bored piles and ground beams.	Bored piles dimensions as in Table A10.4.	Downhill creep may occur on slopes greater than 1 in 10. Unreinforced piles have been broken by slowly moving slopes.
(3) Where trees are felled to clear the site and construction is due to start soon afterward.	Reinforced bored piles of sufficient length with the top 3 m sleeved from the surrounding ground and with suspended floors, or thin reinforced rafts supporting flexible buildings, or basement rafts.		
Soft clays, soft silty clays.	Strip footings up to 1 m wide if bearing capacity is sufficient, or rafts.	See Table A10.3 and CP 101 : 1972.	Settlement of strips or rafts must be expected. Services entering building must be sufficiently flexible. In soft soils of variable thickness it is better to pile to firmer strata (See Peat and Fill below).
Peat, fill.	Bored piles with temporary steel lining or precast or *in situ* piles driven to firm strata below.	Design with large safety factor on end resistance of piles only as peat or fill consolidating may cause a downward load on pile (see BRE *Digest 63*). Field tests for bearing capacity of deep strata or pile loading tests will be required.	If fill is sound, carefully placed and compacted in thin layers, strip footings are adequate. Fills containing combustible or chemical wastes should be avoided.
Mining and other subsidence areas.	Thin reinforced rafts for individual houses with load-bearing walls and for flexible buildings.	Rafts must be designed to resist tensile forces as the ground surface stretches in front of a subsidence. A layer of granular material should be placed between the ground surface and the raft to permit relative horizontal movement.	Building dimensions at right angles to the front of long-wall mining should be as small as possible.

Table A10.3   Minimum width for strip foundations (after Table 2 of BRE Digest 67)

Soil type	Field assessment of structure and strength	Minimum width (mm) for total load (kN/m) of not more than:						
		16	24	32	40	48	56	64
Gravels and sands	Loose	300	450	600				
	Compact	225	225	300	375	450	525	600
Silts	Soft	450	675	900	Silts are frequently combined with			
	Firm	300	450	600	sands or clays. Values for composite			
Clays	Very soft	450	675	900	types are given in the Building			
	Soft	360	525	675	Regulations.			
	Firm	260	325	375	450	560	675	750
	Stiff or hard	225	225	300	375	450	525	600
Peat	Soft	Footings should be in firm ground beneath the peat. Rarely does						
	Firm	investigation show that the peat itself will support foundations.						
Chalk	Plastic	Assess as clay above.						
	Solid	Equal to the width of the wall.						
Fill		To be determined after investigation.						

Table A10.4   Load-carrying capacity of bored piles in kN (after Table 3 of BRE Digest 67)

NOTE: The figures are for clay which increases in strength with depth to the 'stiff' and 'hard' classifications near the bottom of the piles. The figures should not be applied to piles in other situations.

Strength classification	Diameter of pile (mm)	Length of pile (m)				
		2.5	3	3.5	4	4.5
Stiff	250	40	48	56	64	72
(Unconfined shear strength	300	50	60	70	80	90
more than 70 kN/m^2)	350	65	77	90	102	115
Hard	250	55	65	75	85	95
(Unconfined shear strength	300	70	82	94	106	118
more than 140 kN/m^2)	350	95	108	120	132	145

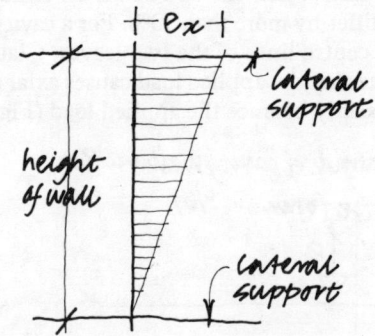

Fig. A11.1   Assumed variation of eccentricity due to vertical load over the height of a wall.

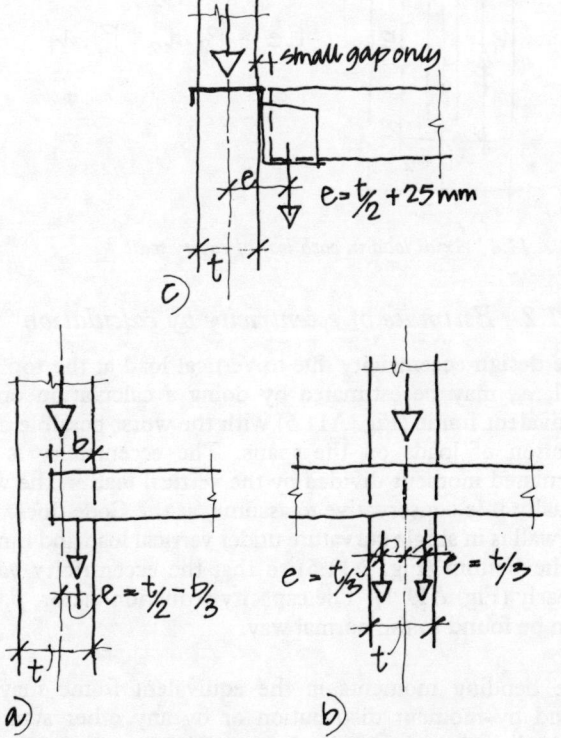

Fig. A11.2   Normal eccentricity (a) of slab from one side of wall, (b) of continuous or discontinuous slab loads on full width of wall (c) of joist hanger load from one side of wall.

A11   Eccentricity of load

The eccentricity of the vertical load at right angles to the line of a wall may be estimated either by rules of thumb, based on experience, or by calculation. Where a concrete slab is supported by a load-bearing wall, it is important that the slab has no upstand or downstand where it meets the wall otherwise the vertical load may become unnecessarily eccentric, and even unsafe.

A11.1   Estimate of eccentricity by rule of thumb

Usually only the eccentricity of the load at the top of the wall, e_x, is required. However, when e_x is known and with the assumption that any vertical load above a lateral support is axial (Clause 31 of the Code), the resultant eccentricity may be calculated at any height in the wall (Fig. A11.1). Clause 31 of the Code assumes as a general rule that the load transmitted to a wall by a floor or a roof on it acts at one third the depth of the bearing area from the face of the wall or, for a joist hanger/beam detail which is properly constructed, at 25 mm from the face of the wall (Fig. A11.2). Similarly for the cavity wall shown in Fig. A11.4, the eccentricity is often taken to be equal to $t/6$. However, in cases where the floor or

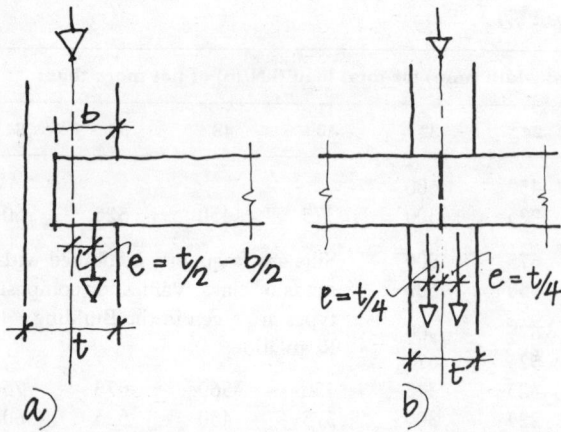

Fig. A11.3   *Eccentricity, with short spans, (a) of slab from one side of wall and (b) of continuous or discontinuous slab loads over full width of wall.*

roof structure is stiff or the span is short (span < 30 × wall thickness), it can be assumed that the load transmitted to the wall acts in the middle of the bearing area (Fig. A11.3). For the case shown in Fig. A11.2(b), it is usual to assume the total load is axial when the two spans on either side of the wall do not differ by more than 50%. For a cavity wall loaded between the centre lines of the two leaves, Clause 31 of the Code assumes that the applied load causes axial loads in each leaf which exactly balance the applied load (Fig. A11.4).

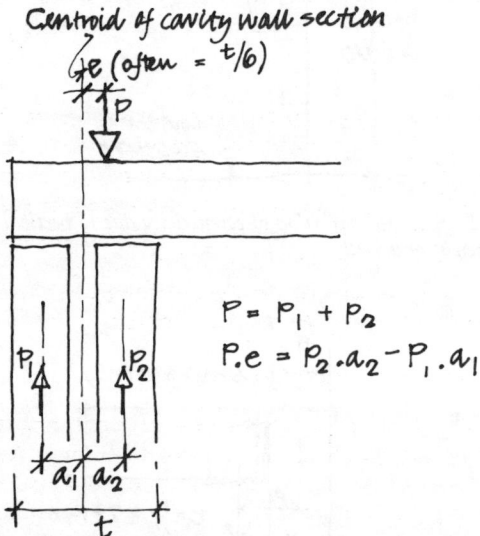

Fig. A.11.4   *Axial load in each leaf of cavity wall.*

A11.2   Estimate of eccentricity by calculation

The design eccentricity due to vertical load at the top of a wall, e_x, may be estimated by doing a calculation on an equivalent frame (Fig. A11.5) with the worst possible combination of loads on the spans. The eccentricity is the calculated moment divided by the vertical load in the wall. Usually it is conservative to assume, as the Code does, that the wall is in single curvature under vertical load and hinged at the bottom (Fig. A11.5) so that the eccentricity varies linearly (Fig. A11.1). The capacity reduction factor, β, can then be found in the normal way.

The bending moments in the equivalent frame may be found by moment distribution or by any other standard method. If there is rotation between the floor and the wall, the joint cannot be assumed to be rigid and the normal frame analysis is no longer valid; see Hendry (1990, pp. 121–131) for a general review. Hendry (1986) provides an approximate

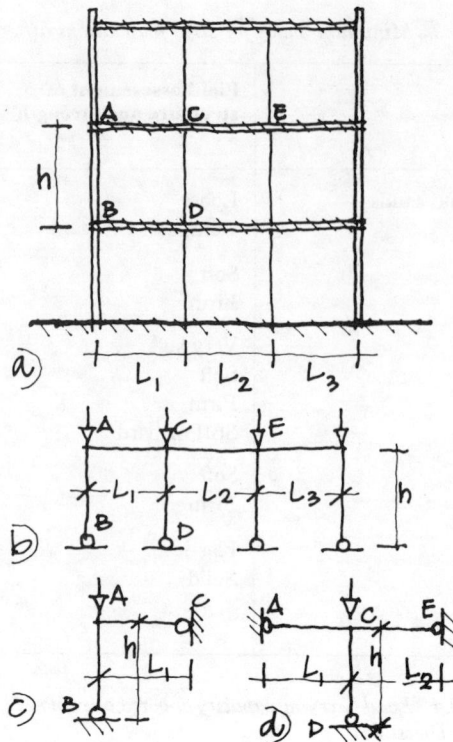

Fig. A11.5   *(a) Three-storey building; (b) using equivalent frame for one complete intermediate floor; or alternatively using an equivalent frame separately (c) for exterior wall and (d) for interior wall.*

method for calculating eccentricity in load-bearing walls with concrete floor slabs where the compressive stress in the walls is greater than 0.25N/mm^2. The method is based on use of a joint fixity factor, F, for the wall/slab joint, which is the factor by which the calculated moment in the wall is reduced after an analysis of the equivalent frame (Fig. A11.5) which assumes the joint to be rigid. The factor F is equal to $1/(0.44\alpha + 1.1)$; α is the slab/wall stiffness ratio in the structure which is equal to $(h/L).(E_s I_s / E_w I_w)$, the last term in the bracket being the ratio of the flexural rigidities of the slab and the wall. The element stiffnesses may have different values in the equivalent frame from those in the actual structure, depending on what equivalent frame is chosen: the equivalent frames in Fig. A11.5 only have walls below the slab and their stiffness must therefore be adjusted. Hendry suggests that a normal 250 mm thick cavity wall be treated as a single wall if the compressive stress in the wall exceeds 0.3 N/mm^2. An example follows in which the wall is not taken as hinged at the bottom.

Example A11.1   Wall in single curvature

Find the design eccentricity, e_m, in the wall AB which for the critical case is in single curvature having moments of 1.6 kN-m/m and 1.4 kN-m/m at A and B respectively. The design vertical load in the wall, ignoring self-weight, is 90 kN/m and the wall has a thickness of 150 mm and a slenderness ratio of 16 (Fig. A11.6).

The eccentricity due to slenderness ratio, e_a, in the middle of the wall (Appendix B of the Code) is

$$= \left(\frac{(16)^2}{2400} - 0.015\right)t \qquad\qquad = 0.092\,t$$

The eccentricity due to the load varies linearly being $1.6 \times 10^3/90 = 18$ mm $= 0.119\,t$ at the top and $0.104\,t$ at the bottom (Fig. A11.7).

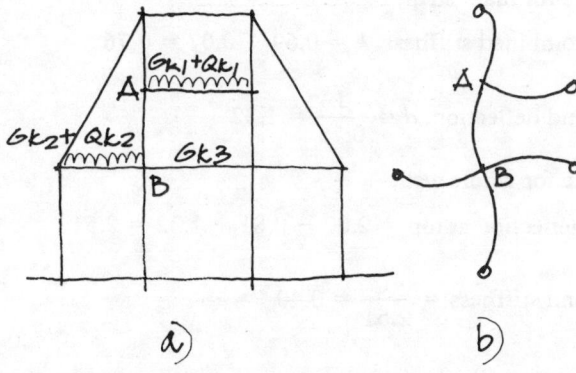

Fig. A11.6 (a) Building with load combination; (b) equivalent frame for calculation, shown in its deflected state.

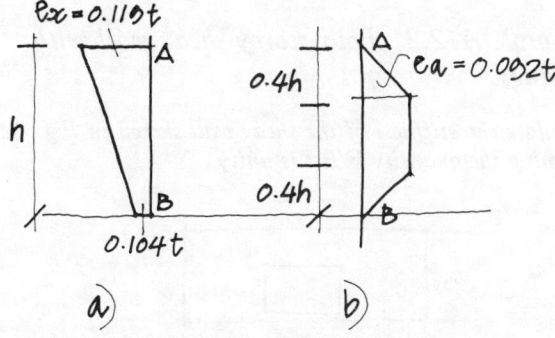

Fig. A11.7 Eccentricities (a) due to vertical load and (b) due to slenderness.

Hence eccentricity in middle,

$$e_t = 0.104\,t + 0.6 \times 0.015\,t + 0.092\,t \qquad = 0.205\,t$$
$$\therefore e_t > e_x \text{ and } e_m = e_t \qquad\qquad\qquad = 0.205\,t$$

A12 Stiffness of shear walls with and without openings

A12.1 Shear walls without openings

The deflection and bending moment in a shear wall will depend on the deflected shape of the wall. For example, if the shear wall deflects in the shape of a cantilever column then the maximum bending moment and deflection will be much larger than if the wall deflects in the shape of a frame column in which the stiffness of the floor or wall above gives the column fixity at the top (Fig. A12.1). It is normally accurate enough to assume that the shear wall deflects in one

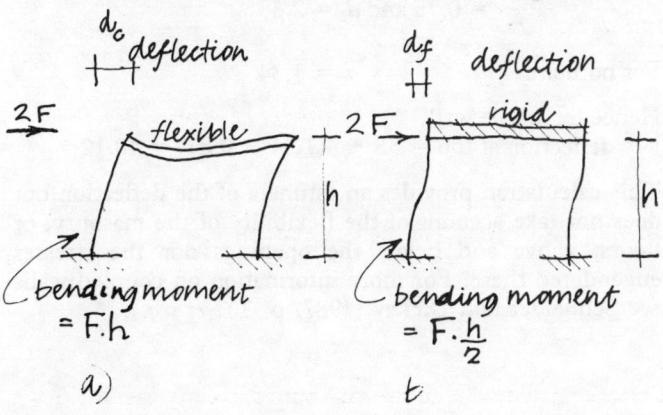

Fig. A12.1 Deflected shape of (a) cantilever columns and (b) frame columns.

of these two ways. The deflection at the top of a wall that deflects like a cantilever column is

$$d_c = F.h^3/(3E.I) + 1.2F.h/(G.A)$$
$$= F.h\,[\,4(h/L)^2 + 3\,]/(E.t.L)$$

and the deflection at the top of a wall that deflects like a frame column is

$$d_f = F.h^3/(12E.I) + 1.2F.h/(G.A)$$
$$= F.h\,[\,(h/L)^2 + 3\,]/(E.t.L)$$

where F is the horizontal force,
h is the height of the shear wall,
E is the modulus of elasticity of the masonry = 700 f_k approximately,
G is the shear modulus of the masonry, taken as 0.4 E,
I is the moment of inertia taken about the axis of bending, and
h, L and t are the height, length and thickness of the shear wall respectively.

The first term, in each case, is the deflection due to bending and the second term is that due to shear which is significant for walls with an h/L ratio less than five. The deflection of a simple shear wall is not often calculated but the formulae are useful in estimating the relative stiffness of a number of shear walls.

The stiffness of each wall, k, is the reciprocal of the deflection, d, when F, the horizontal force, is equal to one, i.e.

$$k = 1/d$$

If the shear deflection is ignored then for a number of shear walls having the same height and modulus of elasticity the relative stiffness depends only on the moment of inertia, I, and this is the value for relative stiffness used in **8.2**c and **8.2**d. In cases where this is not true, a calculated value of the stiffness, k, should replace I in the formulae given in **8.2**c and **8.2**d. In calculating the relative stiffness of a shear wall, the deflection due to shear is often ignored.

A12.2 Shear walls with openings

A method of estimating the stiffness of a shear wall with openings was proposed by the Concrete Masonry Association of California and is reproduced by Schneider and Dickey (1987, p. 314 *et seq.*). The method consists of calculating the deflection of the wall under a unit horizontal force, assuming the wall to be solid, and then subtracting the deflection of a solid strip having a height equal to that of the highest opening in the wall. To this total is added the deflection of the portions of masonry between the openings acting together; these portions of masonry may be subdivided in a similar way if there are yet further openings within these portions. The total deflection at the top of the wall due to a unit horizontal force is calculated and the reciprocal of this is the stiffness of the wall.

Example A12.1 Shear wall with door and window opening

Calculate the relative stiffness of the shear wall with openings shown in Fig. A12.2 compared with the same solid wall. As only relative deflections are required the quantity F/E.t is taken as unity.

The solid walls can be assumed to deflect like cantilever columns but portions of masonry adjacent to the openings can be assumed to deflect like frame columns.

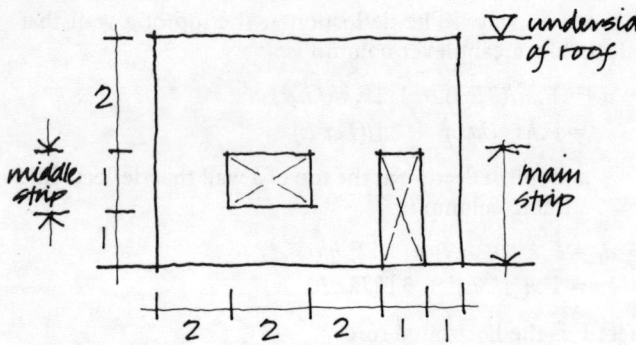

Fig. A12.2 *Elevation on single-storey shear wall with door and window opening.*

For the whole wall,

$$\frac{h}{L} = \frac{4}{8} = 0.5 \text{ and } d_c \text{ (solid)} = 2.00$$

For the main strip,

$$\frac{h}{L} = \frac{2}{8} = 0.25 \text{ and } d_c \text{ (solid)} = 0.81$$

For area to left of door,

$$\frac{h}{L} = \frac{2}{6} = 0.33 \text{ and } d_c \text{ (solid)} = 1.15$$

For middle strip,

$$\frac{h}{L} = \frac{1}{6} = 0.167 \text{ and } d_c \text{ (solid)} = 0.52$$

For each area either side of window,

$$\frac{h}{L} = \frac{1}{2} = 0.5 \text{ and } d_f = 1.63$$

$$\therefore \text{ stiffness, } k = \frac{1}{d_f} = 0.61$$

For both areas,
    combined stiffness $= 2\,k = 1.22$
    and deflection, $d_f = 1/1.22 = 0.82$

Hence for area to left of door,
    deflection, $d = 1.15 - 0.52 + 0.82 = 1.45$

and stiffness, $k = \frac{1}{1.45} = 0.69$

For area to right of door,

$$\frac{h}{L} = \frac{2}{1} = 2 \text{ and } d_f = 14$$

and stiffness, $k = \frac{1}{14} = 0.07$

Hence for main strip,

    combined stiffness, $k = 0.69 + 0.07 = 0.76$

and deflection, $d = \frac{1}{0.76} = 1.32$

Hence for whole wall,

    deflection at top $= 2.00 - 0.81 + 1.32 = 2.51$

and stiffness $= \frac{1}{2.51} = 0.40$

which is $\frac{0.40}{0.50} \times 100 = 80\%$ of the stiffness of the whole solid wall.

Example A12.2   Four-storey shear wall with openings

Calculate the stiffness of the shear wall shown in Fig. A12.3, assuming the quantity F/E.t is unity.

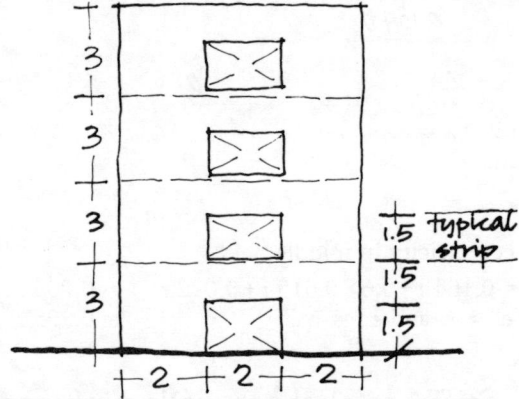

Fig. A12.3 *Elevation on four-storey shear wall with openings.*

For the whole wall,

$$\frac{h}{L} = \frac{12}{6} = 2 \text{ and } d_c \text{ (solid)} = 38$$

For typical strip,

$$\frac{h}{L} = \frac{1.5}{6} = 0.25 \text{ and } d_c \text{ (solid)} = 0.81$$

For area either side of opening,

$$\frac{h}{L} = \frac{1.5}{2} = 0.75 \text{ and } d_f = 2.68$$

For both areas, $d_f = 1.34$

Hence for whole wall,
    deflection at top $= 38 + 4(1.34 - 0.81) = 40.12$

This calculation provides an estimate of the deflection but does not take account of the flexibility of the masonry, or floors, above and below the openings nor the stresses engendered there. For more information on coupled walls see Schneider and Dickey (1987, p. 302 *et seq.*).

A13 Vertically reinforced walls in bending

A13.1 Introduction

The range of uses of masonry can be considerably extended by making use of reinforcement to act compositely with the masonry. Vertically reinforced masonry generally takes the form of small reinforced columns incorporated into masonry construction (Fig. B1.12), pocket-type walls (Fig. B1.11), where bricks are omitted to form the column shapes or reinforced hollow blockwork (Fig. B1.19). A variant is the grouted cavity wall (Fig. 2.21), in which reinforcing bars or mesh are concreted into the cavity.

In general there is only a limited advantage in using reinforcement to increase the axial compressive strength of masonry elements; reinforcement is most often used to increase bending strength. Where the axial load on an element is less than $0.1 f_k.A_m$, BS 5628:Part 2 allows the element to be designed for bending only, which provision greatly reduces the calculation necessary. For the generality of cases in this last category, the element need only be designed against maximum span to effective depth ratios and the required minimum bending and shear strengths. Various rules about the detailing of reinforcement are given in Clause 26 of BS 5628:Part 2, principally concerning anchorage and lap lengths. A section of this clause stipulates that, where secondary reinforcement is required, it should have a minimum area of at least 0.05% of the effective depth times the breadth of a section. This is relevant for grouted cavity walls; for pocket-type walls, in which it may be assumed all integral column arrangements are included, there is no obligation to provide secondary reinforcement.

A13.2 Structural design

The procedure for the design, according to Code, of vertically reinforced masonry closely follows that for unreinforced masonry: the same partial safety factors on load, γ_f, are used (**5.3**), the same rules govern the effective thickness of walls (**5.1**) and the same characteristic compressive strengths of masonry, f_k, are used (Table 2.2). Because it is desirable to equalise the strength of the masonry with the strength of the infilling concrete, or mortar, as far as possible, the minimum compressive strength of the masonry unit in reinforced work should be 7 N/mm² and the mortar at least a designation (ii); this limits the range compared with unreinforced work. Table 3 of BS 5628:Part 2, which tabulates compressive strengths, contains data for stronger masonry units only; also the section tabulation is for solid concrete blocks with a height to least horizontal dimension of 1.0, instead of 0.6 as given in Table 2 of BS 5628:Part 1 (Table 2.2). However, the two tables are similar in principle. The partial safety factors for strengths of materials are, in general, different for unreinforced and reinforced masonry, the partial safety factor, γ_{mm}, for strength of reinforced masonry in direct compression and bending being 2.0 or 2.3 for Special or Normal manufacturing control respectively.

A13.2.1 Procedure for checking vertically reinforced masonry elements in simple bending under horizontal load

(a) Check span to effective depth ratio is less than that given in Table A13.1 (Table 8 of BS 5628:Part 2); for walls subject mainly to wind loading, and not part of a building, Clause

Table A13.1 Limiting ratios of span to effective depth for reinforced laterally-loaded walls (after Table 8 of BS 5628:Part 2)

End condition	Ratio
Simply supported	35
Continuous or spanning in two directions	45
Cantilever with values of ρ up to and including 0.005†	18

†$\rho = A_s/b.d$

22.3 of BS 5628:Part 2 allows the given ratios to be increased by 30%. Failing this check, deflection and crack widths may need calculation in some cases.

(b) Check that design vertical load is less than $0.1f_k.A_m$ where f_k is the characteristic strength of the masonry (from Table 2.2 or Table 3 of BS 5628:Part 2) and A_m is the area of the masonry.

(c) Check bending strength. The bending strength should be greater than the design moment given by the worst of the load combinations; see **5.3**.

The bending resistance of a reinforced masonry wall should be governed by the tensile strength of the reinforcement rather than the compressive resistance of the masonry, to prevent brittle failure. According to BS 5628:Part 2, the design moment of resistance of a singly reinforced member in tension, M_d, is given by

$$M_d = A_s.f_y.z/\gamma_{ms}$$

which should be less than the design moment of resistance in compression given, for rectangular members, by

$$0.4 f_k.b.d^2/\gamma_{mm}$$

and for flanged members by

$$f_k.b.t_f(d-0.5\ t_f)/\gamma_{mm} \text{ where the lever arm,}$$

$$z = \left(1 - \frac{0.5\ A_s.f_y.\gamma_{mm}}{b.d.f_k.\gamma_{ms}}\right) d$$

The lever arm should not be taken to be greater than $0.95\ d$. A_s is the area of reinforcing steel; b is the width of the section; d the effective depth; f_k the characteristic compressive strength of the masonry; f_y the characteristic tensile strength of the reinforcing steel; γ_{mm} the partial safety factor for strength of masonry equal to 2.0 or 2.3; γ_{ms} the partial safety factor for strength of steel equal to 1.15 and t_f is the thickness of the flange of a flanged member. The effective thickness and width of the flange depends on the type of wall; see **A13.3**.

(d) Walls with pockets or ribs. For walls with pockets or ribs at spacings greater than one metre, the ability of the wall to span horizontally between the pockets or ribs should be checked; see Chapter 6 or Appendix A14 for design of walls in bending. In many cases it is more convenient to rely on arch action; see Example **A13.2**.

(e) Check shear. The design shear strength of the masonry, f_v/γ_{mv}, must be greater than the design stress. The design shear stress, v, is the horizontal design load divided by the area which is given by multiplying the effective depth, d, by the effective width, b; for walls with piers the effective width of the flange may be taken but d must not be taken as more than the actual flange thickness, m; see Fig. A13.1. In most cases it is sufficient to assume that the characteristic shear strength, f_v, is the same as that for unreinforced sections (see **5.3**); γ_{mv} for reinforced masonry is 2.0. In many cases, the shear strength may be enhanced, if required; see BS 5628:Part 2.

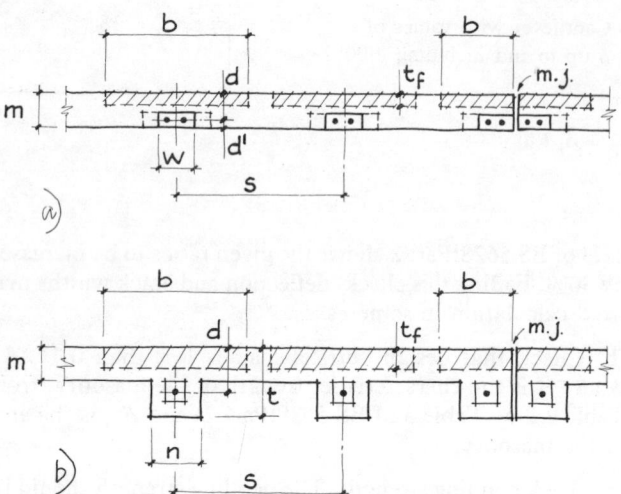

Fig. A13.1 *Plan on (a) simple pocket-type reinforced wall and (b) on pocket-type reinforced wall with piers.*

A13.3 Common types of vertically reinforced walls

(a) Reinforced hollow blockwork walls without piers. These are designed like reinforced rectangular sections. For all strength calculations the width of the reinforced section should be taken equal to the spacing of the reinforcement or three times the thickness of the block if this is less (from Clause 22.4.3 of BS 5628:Part 2); alternatively where all the hollows are infilled the wall can be considered as a simple pocket-type wall. The reinforcement is almost always a normal carbon steel between 6 mm and 32 mm in diameter and surrounded by a concrete infilling mix, often dosed with a plasticiser to make it easier to compact. For reinforced hollow blockwork walls (or pocket-type walls), according to Table 14 of BS 5628:Part 2, with Exposure E2 (moderate/ severe) or Exposure E3 (severe/very severe) the minimum grade of infill concrete required is C30 with a minimum cement dosage of 300 kg/m³.

For C30 concrete, the cover required is 30 mm for E2 and 40 mm for E3 exposure. For E1 exposure, which covers internal work, including the inner leaf of cavity wall construction, C25 concrete with 20 mm cover is adequate.

(b) Walls with piers and simple pocket-type walls. For all strength calculations, effective widths should be used. The effective width of a reinforced flanged member (from Clause 22.4.3 of BS 5628:Part 2) is the same as that of unreinforced masonry (**A3.1**), but with the effective thickness, t_f, used instead of the actual thickness of the flange, m (Fig. A3.2); it may be assumed that the effective widths for unrestrained flanges of reinforced members are the same as those for unreinforced masonry. For simple pocket-type walls, i.e. those without piers, the effective width of the flange can be assumed to be either the pocket spacing, s, or the width of the pocket plus twelve times the effective thickness of the flange, t_f, whichever is less. In all cases, the effective

thickness of the flange is the lesser of m or $0.5d$ where d is the effective depth (Fig. A13.1). Concrete and cover requirements are as those for reinforced hollow blockwork. Simple pocket-type construction has the great advantage that the pocket may be conveniently cleaned out over its height before concreting.

(c) Grouted cavity walls. According to Clause 32.2 of BS 5628:Part 2, carbon steel in a grouted cavity wall requires a minimum cover of 20 mm for exposure E2 (moderate/ severe) with a concrete infill; for exposure E1, covering internal work, a mortar infill is permitted with the same cover.

Because of the requirement for secondary reinforcement, running parallel to the bed joints, mesh reinforcement is commonly used in the cavity; this requires that the mesh be placed after construction of one leaf, with its cavity ties, and before construction of the other leaf. Wherever possible the cavity should be filled with concrete, containing plasticiser, rather than mortar.

Example A13.1 Reinforced blockwork pier on inner skin of cavity wall

Check that the reinforcement provided for the pier shown in Fig. 12.11 *is adequate, if placed at mid-depth.*

Assume that the pier uses a hollow block with a quoted (i.e. average) compressive strength of 8.5 N/mm² and has 57% solid. Hence net strength of block = 15 N/mm² and with a designation (ii) mortar, $f_k = 10.6$ N/mm² (Table 2.2); the infill concrete is a high slump 25 N/mm² mix allowing the blockwork to be considered as solid when built. If reinforcement is provided over the full height of the building so that the wall is continuous over the first-floor support, from Case (vi) of Table A4.1,

design moment = $3.33 \times 2.45/9 = 0.91$ kN-m
Span to effective depth ratio = $2,450/50 = 49$

Compare allowable ratio of 45 for continuous walls from Table 8 of BS 5628:Part 2 (Table A13.1). Bearing in mind the span and the stiffness afforded by the outer leaf of the cavity wall, this is considered to be satisfactory and no further check is made of deflection or cracking. If three 10 mm mild steel reinforcing bars are provided, then $A_s = 236$ mm² and with $\gamma_{mm} = 2.3$,

$$z = \left(1 - \frac{0.5 \times 236 \times 250 \times 2.3}{600 \times 50 \times 10.6 \times 1.15}\right) d = 0.81 d < 0.95 d$$

$$\therefore M_d = \frac{236 \times 250 \times 0.81 \times 50 \times 10^{-6}}{1.15} = 2.1 \text{ kN-m};$$

Compressive strength in bending

$$= \frac{0.4 \times 10.6 \times 600 \times 50^2 \times 10^{-6}}{2.3} = 2.7 \text{ kN-m} \qquad \text{OK}$$

By inspection it is not necessary to consider shear or the interaction of vertical load and bending.

Example A13.2 Simple pocket-type basement retaining wall

A basement is to be provided for a shop consisting of 225 mm thick brickwork walls with pocket-type reinforcement, at 1.125 m centres, resting on a 200 mm thick concrete basement slab with mesh reinforcement (Fig. A13.2). Bricks having a compressive strength of 35 N/mm² with a designation (i) mortar are specified for which $\gamma_{mm} = 2.0$. The axial load in the wall at ground-floor level is 60 kN/m. Check.

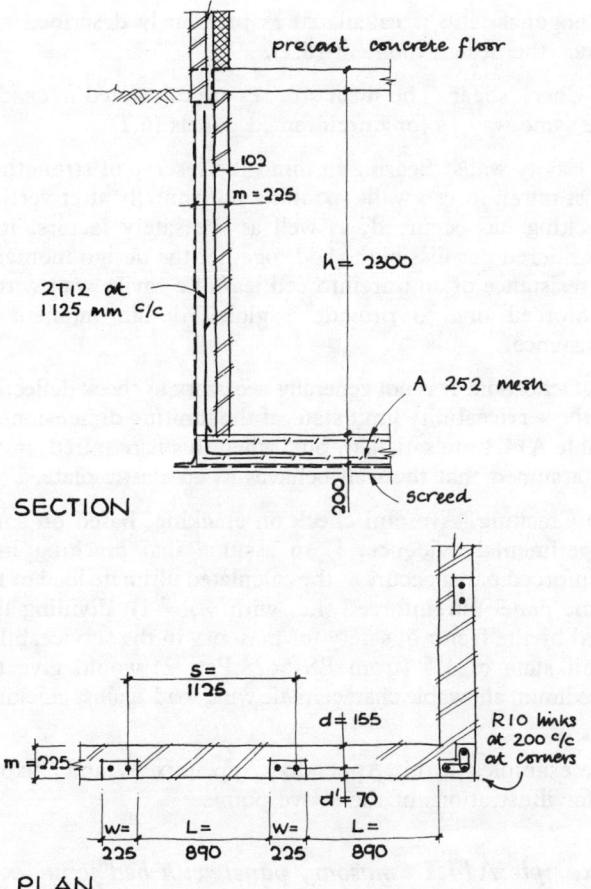

$$\therefore M_d = \frac{226 \times 460 \times 0.82 \times 70 \times 10^{-6}}{1.15 \times 1.125} = 4.6 \text{ kN-m/m}$$
$$\text{OK}$$

The design moment of resistance in compression should also be checked at the bottom and in the span; this is satisfactory however. The design shear stress

$$at\ bottom = \frac{0.67 \times 18}{1.125 \times 1.55} = 0.07 \text{ N/mm}^2$$

$$in\ span = \frac{0.33 \times 18}{1.125 \times 102} = 0.06 \text{ N/mm}^2$$

The design shear strength of unreinforced masonry, ignoring enhancement due to axial load, $f_v/\gamma_{mv} = 0.35/2.0 = 0.17$ N/mm^2. Check the capacity of the unreinforced masonry to span between the reinforced pockets by arch action; the maximum allowable design ud load is given by

$$q = (f_k/\gamma_{mm}).(m^2/L^2)$$
$$= (4.0/2.0).(0.05/0.79) \times 10^3 = 126 \text{ kN/m}^2$$

where f_k is the characteristic strength of the masonry parallel to the bed joints, say equal to 4 N/mm^2, and L is the span between the piers. Compare this with the design horizontal pressure at, say, a depth of $2.30 - 0.89/2 = 1.85$ m which is

$$= 18 \times 0.27 \times 1.85 = 9 \text{ kN/m}^2 \qquad \text{OK}$$

Check the design side thrust at the corner positions which is given by

$$p = (\gamma_f.q.L^2)/(8m)$$
$$= (1.4 \times 9 \times 0.89^2)/(8 \times 0.225) = 5.6 \text{ kN/m}$$

This side thrust can easily be absorbed by the L-shaped pocket, shown in Fig. A13.2, which is tied in at floor levels.

Fig. A13.2 Section and corner plan on basement wall.

Consider the wall in bending with the back face in tension at the bottom but in compression in the span. The effective thickness of the flange, t_f, of the wall is the lesser of

m or $0.5d$ i.e. $= 77$ mm (at bottom) or $= 35$ mm (in span)

and the flange is the lesser of

$w + 12t_f$ or s i.e. $= 1\ 125$ mm (at bottom)
or $= 645$ mm (in span)
L/d ratio $= (2\ 300 + 155)/155 = 15.8 < 35 \qquad$ OK
from Table 2.2, $f_k = 11.4$ N/mm^2
$\therefore 0.1 f_k.A_m = 0.1 \times 11.4 \times 225$
$= 256$ kN/m > 60 kN/m

Design horizontal force on wall, using Case (b) (**5.3**).

$$= 1.4 \times 0.5 \times 18 \times 0.27 \times 2.3^2 = 18.0 \text{ kN/m}$$

and the design moment for triangular-shaped load,

$at\ bottom = 0.134 \times 18.0 \times 2.5 = 6.1$ kN-m/m
$in\ span = 0.060 \times 18.0 \times 2.5 = 2.7$ kN-m/m

Try two 12 mm high yield bars in each pocket then, at bottom of wall,

$$z = \left(1 - \frac{0.5 \times 226 \times 460 \times 2.0}{1125 \times 155 \times 11.4 \times 1.15}\right) d$$

$$= 0.96\ d > 0.95\ d$$

$$\therefore M_d = \frac{226 \times 460 \times 0.95 \times 155 \times 10^{-6}}{1.15 \times 1.125} = 11.8 \text{ kN-m/m}$$

and in the span of the wall,

$$z = \left(1 - \frac{0.5 \times 226 \times 460 \times 2.0}{645 \times 70 \times 11.4 \times 1.15}\right) d$$

$$= 0.82\ d < 0.95\ d$$

A14 Horizontally reinforced walls in bending

A14.1 Introduction

Ordinary masonry walls can be reinforced horizontally, without undue interruption to the laying of the brick or blocks, by purpose-made bed joint reinforcement. Typically, for a 100 mm thick wall, the reinforcement would consist of two parallel, oval-shaped, steel wires spaced 55 mm apart, each with a cross-sectional area of 10 mm^2 and a minimum yield strength of 485 N/mm^2. The bed joint reinforcement can be used to control vertical cracking and to increase the ultimate bending strength of the wall such that it does not fail in a brittle manner. However, the reinforcement does not substantially increase the initial stiffness and may decrease it.

Corrosion is a major concern. Bed joint reinforcement is not as well protected by the mortar as the reinforcement in vertically reinforced walls is by concrete and this means that heavily galvanised or stainless steel reinforcement is required, except if the wall concerned is in a sheltered position. BS 5628:Part 2 requires that the minimum mortar cover to the bed joint reinforcement is 15 mm and only allows the use of ordinary carbon steel in internal walls, which category does not include the inner leaf of external cavity walls. Nevertheless the inner leaves of cavity walls should be reinforced in preference to outer leaves, because of the greater degree of protection afforded there.

A14.2 Structural design

The general design procedures given for the design of horizontally reinforced masonry wall panels in Appendix A of BS 5628:Part 2 are based on those for unreinforced panels given in BS 5628:Part 1: the same partial safety factors on loads, γ_f, and on material strengths, γ_m, are used as in BS 5628:Part 1 including those used to calculate M_d (**A13.2.1**); mortar designations (i), (ii) and (iii) are permissible; and, by convention, the design moments are normally calculated about a vertical axis (failure perpendicular to bed joints) as in the case of unreinforced walls.

Four methods for the design of two-way spanning reinforced masonry walls under horizontal load are given in Appendix A, of which only one, the modified orthogonal ratio method, is considered here. This method is the same one used for unreinforced walls, the bending moment coefficients being obtained from Table A6.1 or calculated from the original yield-line formulae given in Appendix A7. These formulae also allow discrete beams or bands of reinforcement to be treated. A very slight difference, from unreinforced walls, is that for horizontally reinforced walls the orthogonal ratio is calculated as the ratio of the (unreinforced) moment of resistance about a horizontal axis to the (reinforced) moment of resistance about a vertical axis, the latter being given by M_d in **A13.2.1**. Note that, according to Appendix A, the additional bending strength given by this calculation may only be taken into account where the bed joint reinforcement has a minimum total area of 14 mm² and is placed at a maximum spacing of 450 mm; elsewhere the effect of the reinforcement should be discounted.

A14.2.1 Procedure for checking horizontally reinforced wall panels under horizontal load

(a) Check limiting dimensions given in Table A14.1 (after Appendix A of BS 5628:Part 2).

(b) Check the walls without reinforcement, as detailed in Chapter 6. Assuming reinforcement is required, devise a suitable preliminary arrangement.

(c) Design moment of resistance. For walls spanning one-way horizontally or two-ways, the design moment of resistance about a vertical axis is calculated using the formulae given in **A13.2.1** but with γ_m taken from BS 5628:Part1 (**5.1**). It is not normally necessary to calculate the design moment of resistance in compression.

(d) Design moment. For one-way panels the design moment is calculated directly (Table A4.1); for two-way panels the

Table A14.1 Limiting dimensions of laterally-loaded panels with bed joint reinforcement (adapted from Clause A23 of BS 5628:Part 2)

The limiting dimensions of panels should be as follows.

- (a) Panel supported on three edges:
  - (1) two or more sides continuous:
    height × length equal to 1800 t_{ef}^2 or less;
  - (2) all other cases:
    height × length equal to 1600 t_{ef}^2 or less.
- (b) Panel supported on four edges:
  - (1) three or more sides continuous:
    height × length equal to 2700 t_{ef}^2 or less;
  - (2) all other cases:
    height × length equal to 2400 t_{ef}^2 or less.

No dimension should exceed 60 t_{ef} where t_{ef} is the effective thickness as defined in **5.1.1**.

orthogonal ratio is calculated as previously described and hence the design moment (**6.1**).

(e) Check shear. The shear stresses are calculated in exactly the same way as for unreinforced panels (**6.1**).

(f) Cavity walls. Bearing in mind the reserve of strength of most unreinforced walls spanning horizontally after vertical cracking has occurred, as well as the safety factors, it is considered permissible to add together the design moments of resistance of an unreinforced leaf of a cavity wall with a reinforced one to provide a global design moment of resistance.

(g) Deflection. It is not generally necessary to check deflection at the serviceability limit state, if the limiting dimensions in Table A14.1 are satisfied, but, when this is required, it can be assumed that the wall behaves as an elastic plate.

(h) Cracking. A useful check on cracking, based on some experimental evidence, is to assume that cracking in a reinforced panel occurs at the calculated ultimate load of the same panel unreinforced (i.e. with $\gamma_m = 1$); dividing this load by the factor of safety for masonry in the serviceability limit state of 1.5 (from BS 5628:Part 2) would give the maximum allowable characteristic wind load against cracking, W_k.

See examples in this Appendix, Appendix A7 and Chapter 6 for illustrations of the above points.

Example A14.1 masonry panel with bed joint reinforcement above opening

Check that the masonry reinforcement specified in Example 6.6 is adequate.

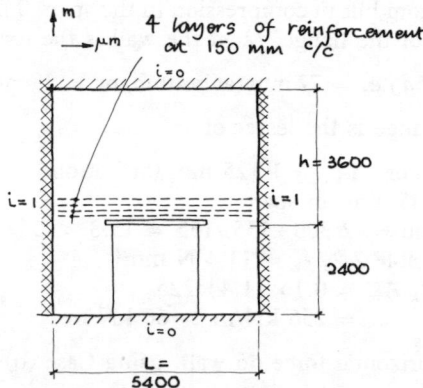

Fig. A14.1 Elevation on reinforced panel.

Four layers of bed joint reinforcement are specified in each leaf at 150 mm centres (Figs 6.12 and **A14.1**). For each leaf with

$f_k = 5.8$ N/mm²; $A_s = 10$ mm²; $d = 78$ mm;

$$z = \left(1 - \frac{0.5 \times 10 \times 485 \times 2.8}{150 \times 78 \times 5.8 \times 1.15}\right) d$$

$$= 0.91\, d < 0.95\, d$$

$$\therefore M_d = \frac{10 \times 485 \times 0.91 \times 78 \times 10^{-6}}{1.15 \times 0.150} = 2.00 \text{ kN-m/m}$$

over reinforced area and

$$M_b = \frac{1.1 \times 1.75}{2.8} = 0.68 \text{ kN-m/m}$$

over unreinforced area;

taking an average value for the moment of resistance about a vertical axis

$$= (2.00 \times 0.6 + 0.68 \times 3.0)/3.6 = 0.90 \text{ kN-m/m}$$

and $M_a = \dfrac{0.55 \times 1.75}{2.8} = 0.34$ kN-m/m

$\therefore \mu = 0.34/0.90 = 0.38$; $h/L = 0.66$;

Assuming Case C of Table A6.1 to be the relevant case, then the bending moment coefficient, $\alpha = 0.043$.

If $\beta = 1\,200/3\,600 = 0.33$ and $w_e = 1.66$ w then,

from **A7.1**, for both leaves

$m = 1.66$ w $\times 0.043 \times 5.4^2 = 2.08$ w kN-m/m
$\therefore$ w $= \gamma_f \cdot W_k = 2 \times 0.90/2.08 = 0.86$ kN/m^2
and max $W_k = 0.86/1.4 = 0.61$ kN/m^2

In fact this panel has a type A failure pattern and reference to **A7.1.2** shows that $w_e = 1.33$ w and $a_e = 1.15a = 1.15h$ for this case so that the answer is conservative in this respect. However, in using Case C of Table A6.1 it has been assumed that on the vertical edges $i = 1$ and this is no longer true for the reinforced panel. The correct value is given by

$i_1 = i_3 = 0.68/0.90 = 0.75$

Using the formula for a type A failure pattern (**A7.2**) with the modifications given above leads to,

$m = 1.33$ w $\times 0.05 \times 5.4^2 = 1.95$ w kN-m/m
and max. $W_k = 1.80/(1.95 \times 1.4) = 0.66$ kN/m^2

The same result can be obtained by treating the reinforced portion as a discrete beam with a moment of resistance,

$M_r = (2.00 - 0.68)\,0.6 = 0.79$ kN-m;
and $k = 0.79/(3.6 \times 0.68) = 0.32$; $i_1 = i_3 = 1$;
$\therefore m = 1.33$ w $\times 0.038 \times 5.4^2 = 1.47$ w kN-m/m
and max. $W_k = 1.36/(1.47 \times 1.4) = 0.66$ kN/m^2

This last calculation would be valid for any width of opening and is still conservative. In the particular case considered, the line load factor, β, must be less than that assumed because of the support available along two edges of the corner panels, either side of the opening. By a similar calculation to those in the examples in **A7**, the maximum characteristic load can be shown to be in excess of 0.80 kN/m^2. OK

Cracking can be assumed to occur at the load corresponding to the ultimate limit state of strength for the unreinforced panel, which, from Example 6.6, is when the characteristic wind load is, at least

$$= 2.8 \times 0.49 = 1.37 \text{ kN/m}^2$$

Applying a partial safety factor of serviceability for masonry, $\gamma_{mm} = 1.5$, then maximum allowable characteristic wind load, for limit state of cracking,

$$W_k = 1.37/1.5 = 0.91 \text{ kN/m}^2 > 0.75 \text{ kN/m}^2$$

OK

In this example both leaves of the wall are reinforced. An alternative arrangement would be with all the reinforcement uniformly spaced in the inner leaf, the outer leaf being designed as an unreinforced panel. In both arrangements, however, the reinforced and unreinforced areas are assumed to develop their full design moments of resistance simultaneously. Only a modest increase in flexural strength has been achieved with this reinforcement; for a more substantial increase, the reinforcement should be placed throughout the height of the wall.

Appendix B   Design details

B1   Typical bonding and movement joint details in walls

Brick wall in stretcher bond

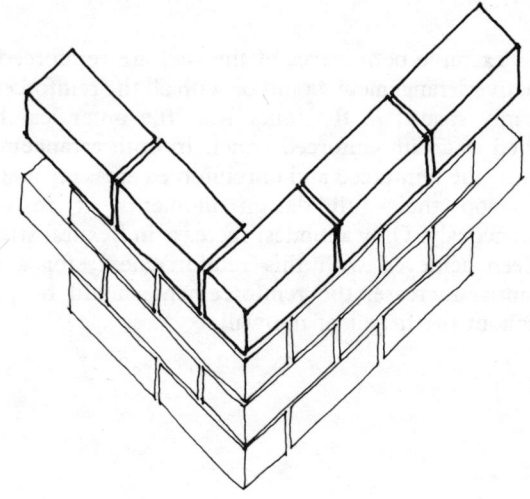

Fig. B1.1   Corner bond.

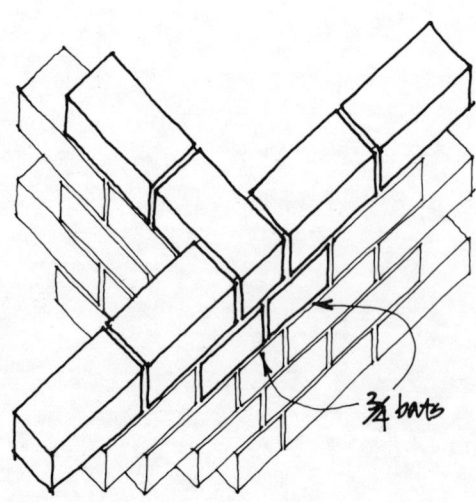

Fig. B1.2   Bond to partition wall.

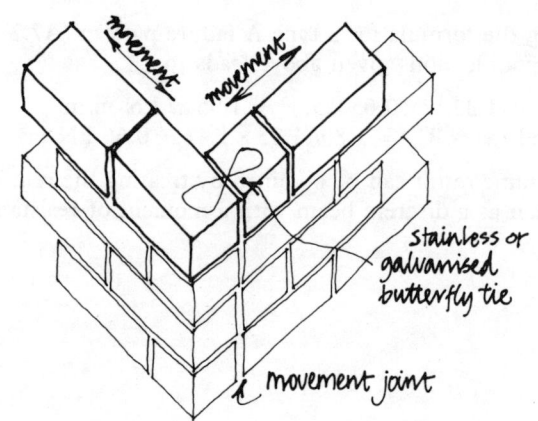

Fig. B1.3   Corner movement joint.

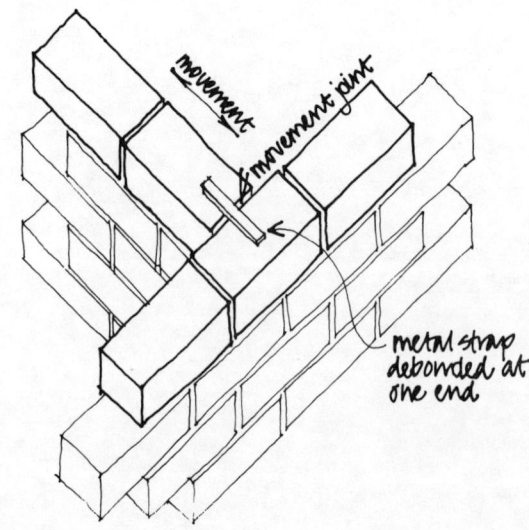

Fig. B1.4   Tie to partition wall.

Brick wall in English bond

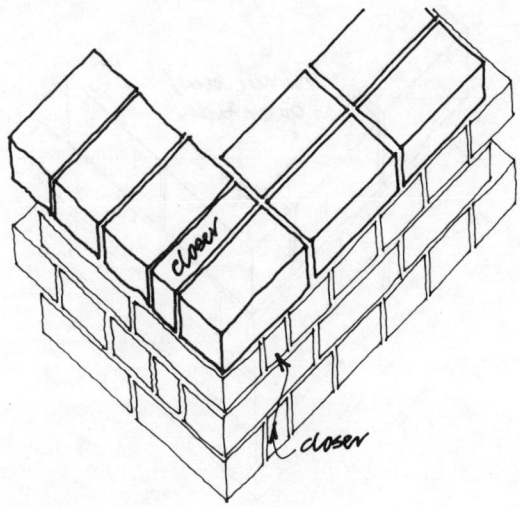

Fig. B1.5   Corner bond.

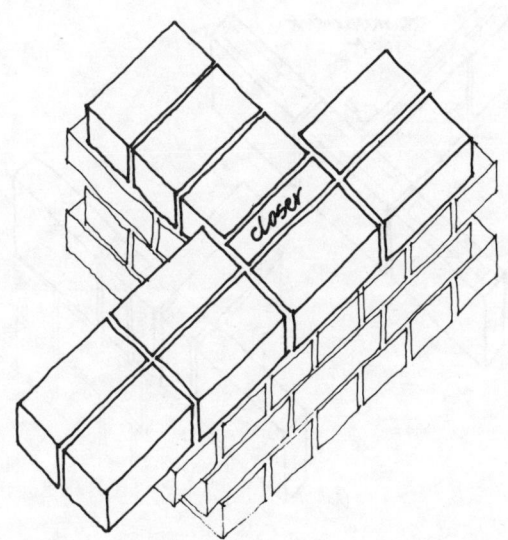

Fig. B1.6   Bond to internal wall.

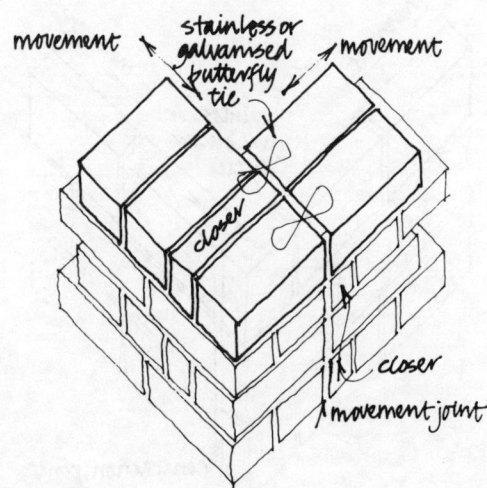

Fig. B1.7   Corner movement joint.

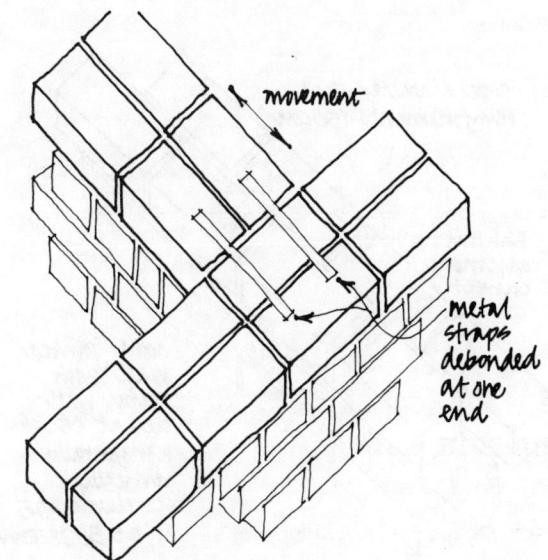

Fig. B1.8   Tie to internal wall.

Brick wall with piers or columns

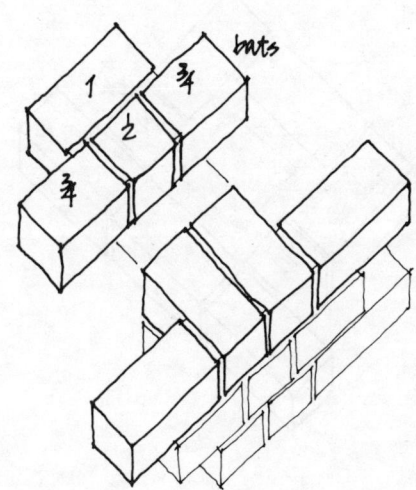

Fig. B1.9   Pier bonded to wall.

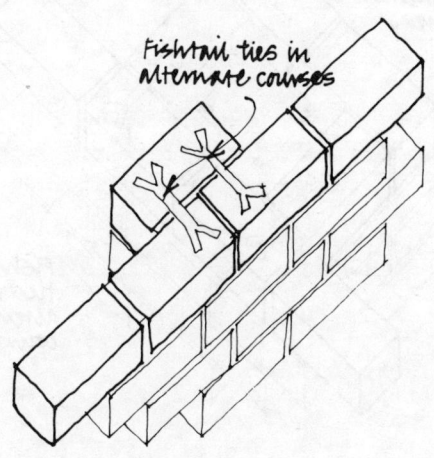

Fig. B1.10   Pier collar-joined to wall.

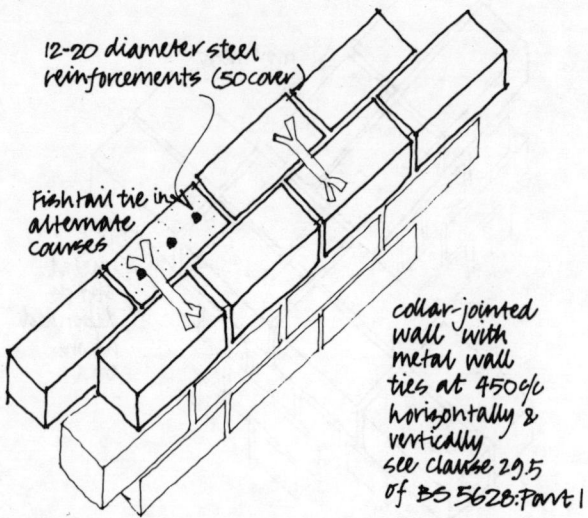

Fig. B1.11   Collar-joined wall with reinforced concrete column.

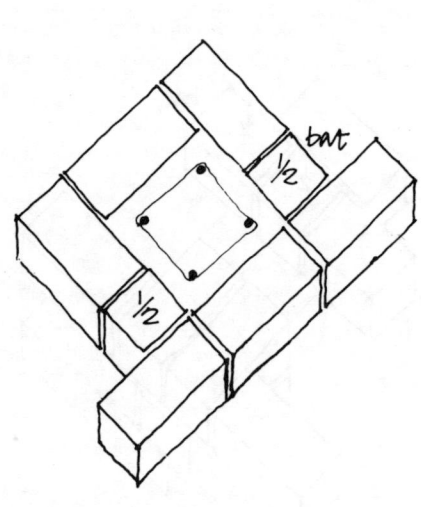

Fig. B1.12   Alternate courses in a stretcher bond wall with a reinforced concrete pier.

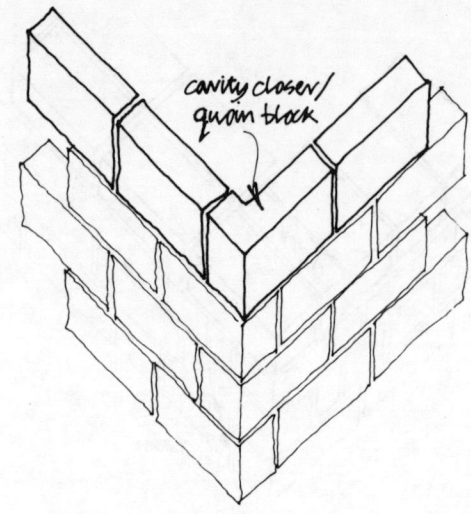

Fig. B1.13   Corner bond.

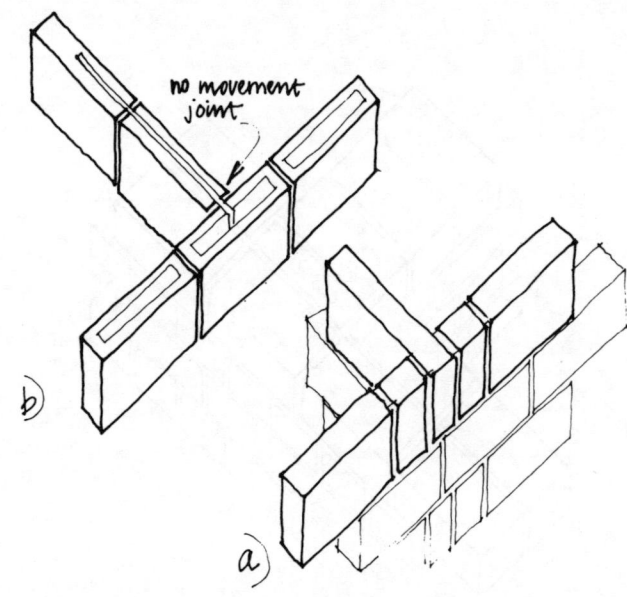

Fig. B1.14   (a) Bond to internal wall, (b) full tie to internal wall.

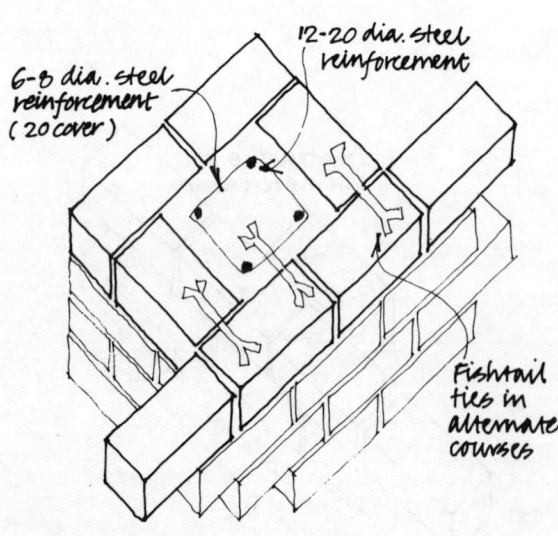

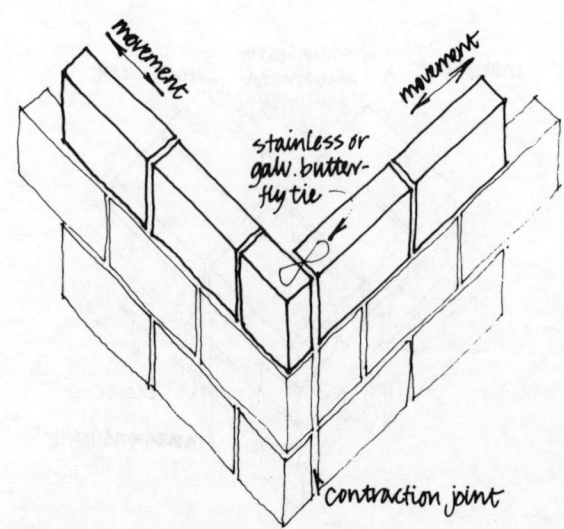

Fig. B1.15   Corner movement joint.

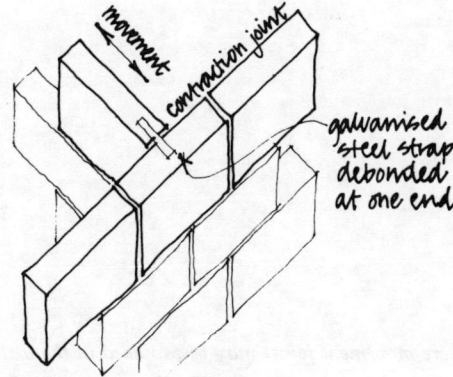

Fig. B1.16 Tie to internal wall.

Concrete block wall with piers

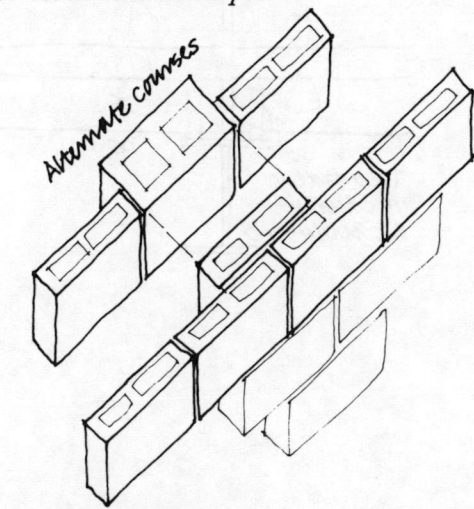

Fig. B1.17 Pier bonded to wall.

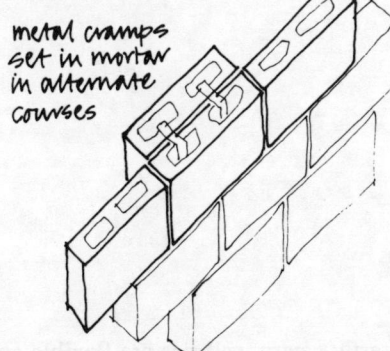

Fig. B1.18 Pier collar-joined to wall.

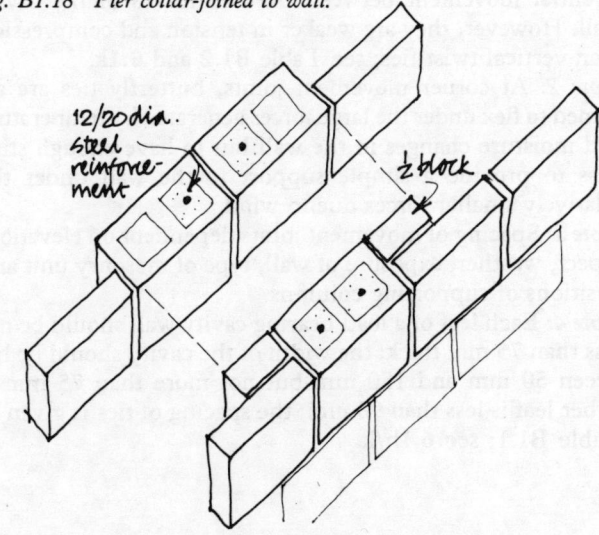

Fig. B1.19 Stretcher bond wall with reinforced hollow block pier.

Movement joints and ties

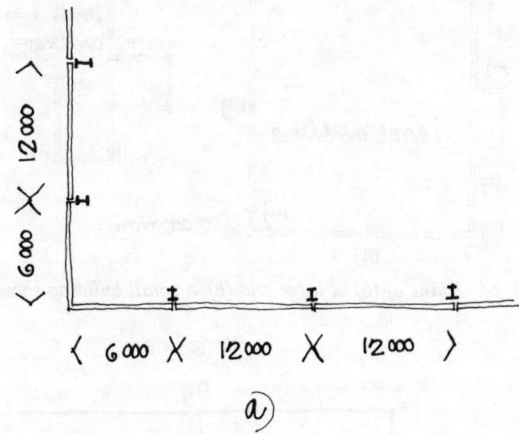

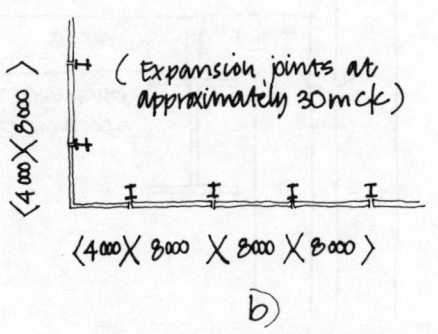

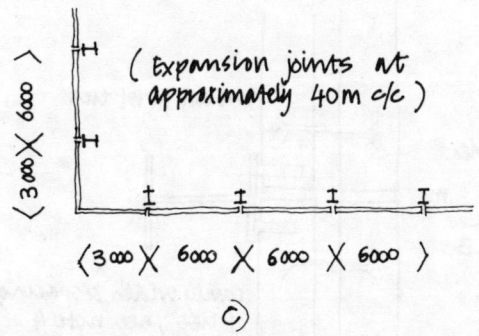

Fig. B1.20 Approximate maximum spacings of movement joints in walls without openings and unrestrained at the top: (a) expansion joints in clay brick wall, (b) contraction joints in calcium silicate brick wall, (c) contraction joints in concrete brick or block wall.

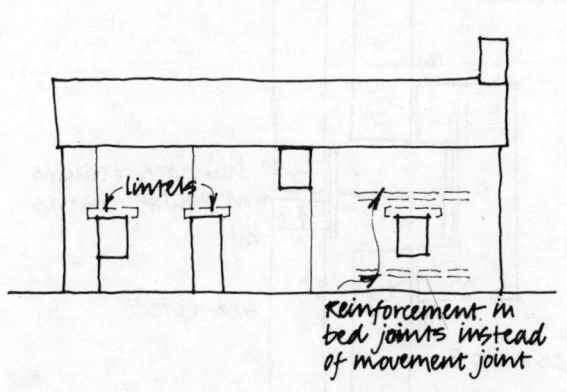

Fig. B1.21 Elevation showing typical places where movement joints may be required.

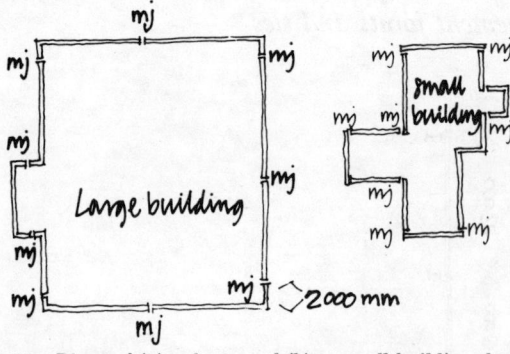

Fig. B1.22 Plans of (a) a large and (b) a small building showing typical places where movement joints may be provided in the walls.

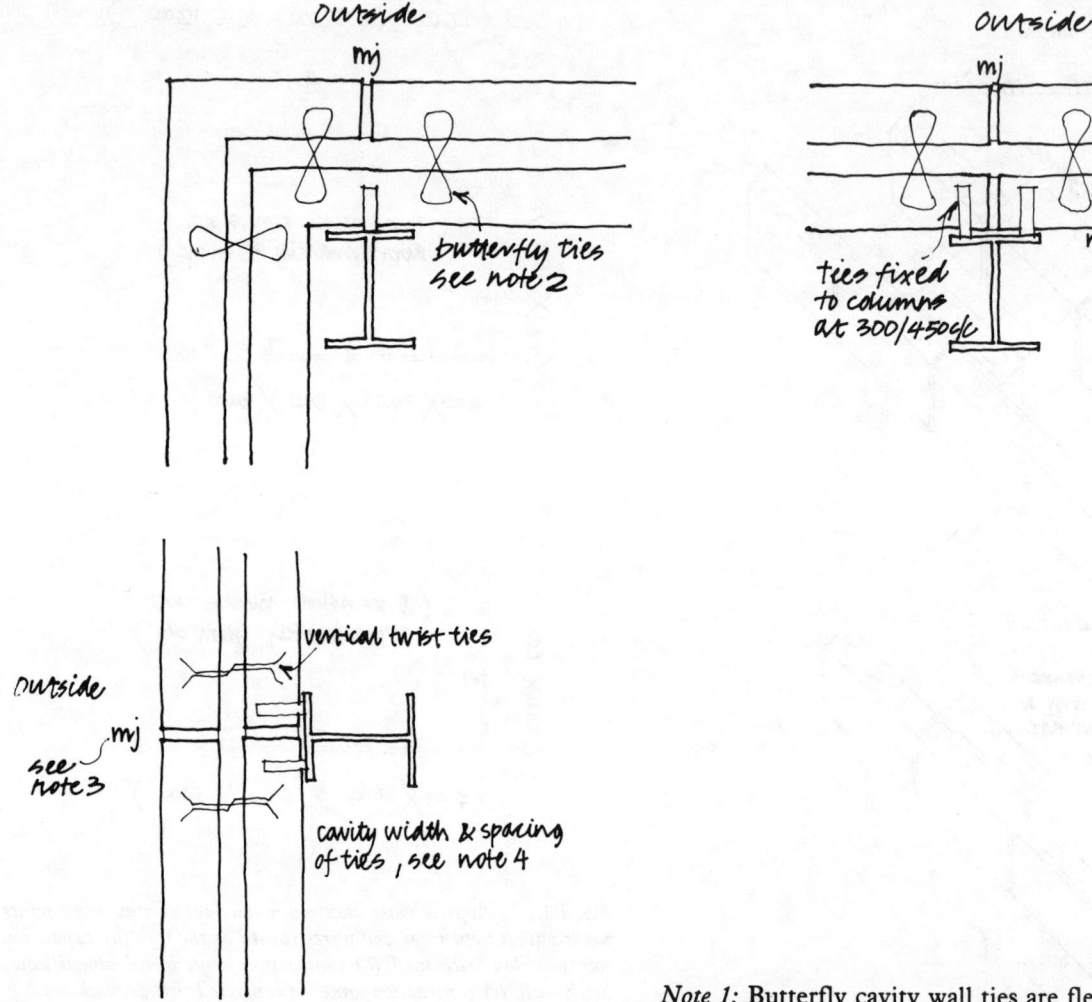

Note 1: Butterfly cavity wall ties are flexible and allow differential movement between the two leaves of the cavity wall. However, they are weaker in tension and compression than vertical twist ties; see Table B1.2 and **6.1**k.

Note 2: At corner movement joints, butterfly ties are assumed to flex under the large forces generated by temperature and moisture changes in the wall but to have enough stiffness to provide a simple support to the wall under the relatively smaller forces due to wind.

Note 3: Spacing of movement joints dependent on elevation, aspect, weather, exposure of wall, type of masonry unit and positions of supporting columns.

Note 4: Each leaf of a load-bearing cavity wall should be not less than 75 mm thick; the width of the cavity should be between 50 mm and 150 mm but not more than 75 mm if either leaf is less than 90 mm; the spacing of ties is given in Table B1.1; see **6.1**h/k.

Fig. B1.23 Typical movement and tie details in masonry clad buildings.

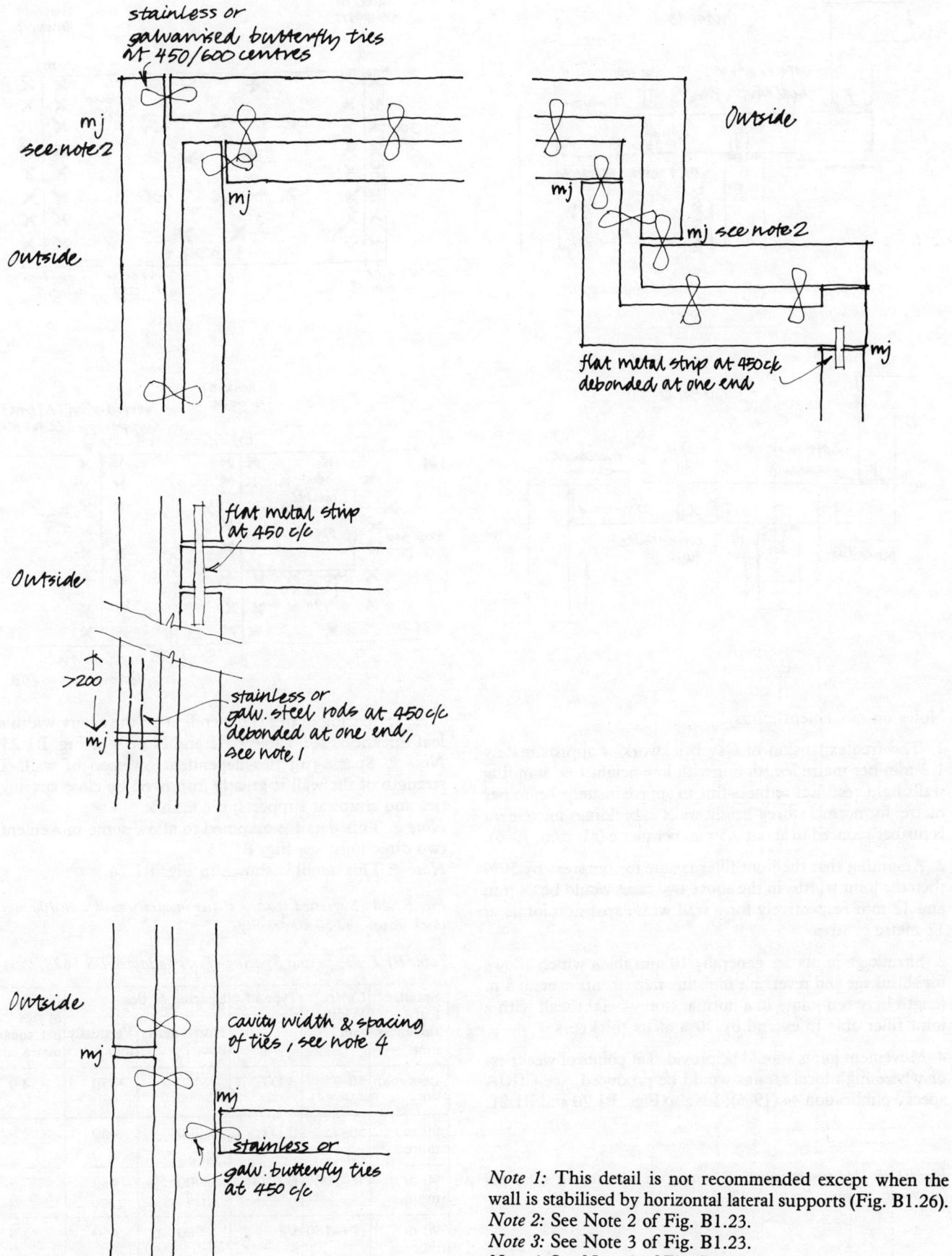

Fig. B1.24 *Typical movement and tie details in small masonry buildings.*

Note 1: This detail is not recommended except when the wall is stabilised by horizontal lateral supports (Fig. B1.26).
Note 2: See Note 2 of Fig. B1.23.
Note 3: See Note 3 of Fig. B1.23.
Note 4: See Note 4 of Fig. B1.23.

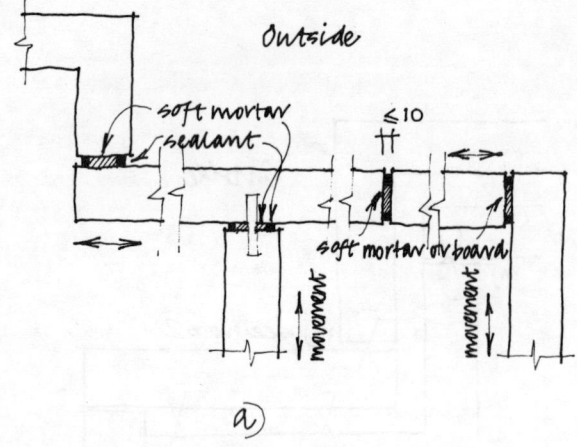

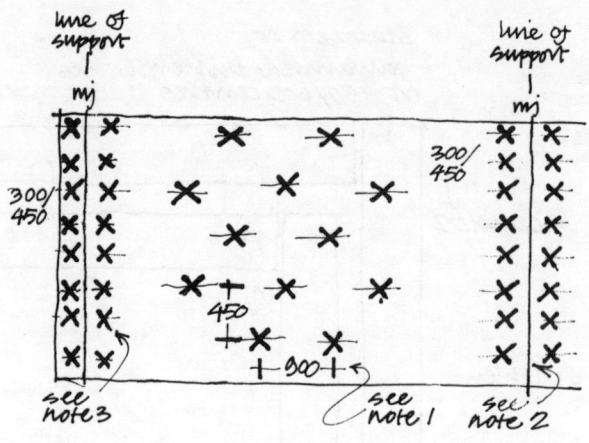

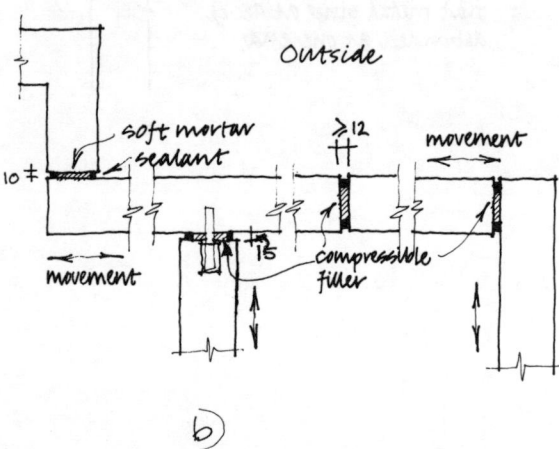

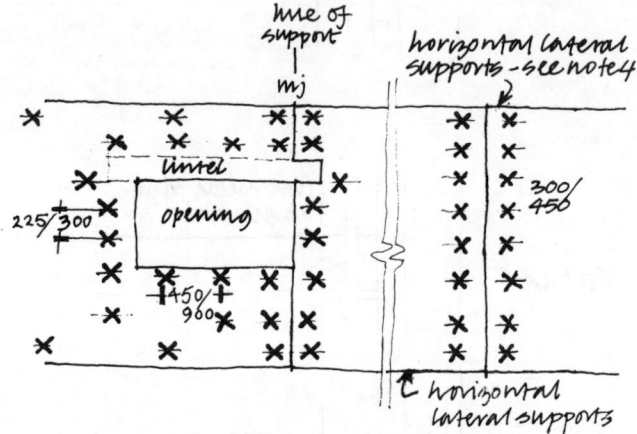

Notes on movement joints:

1. The free expansion of clay brickwork is approximately 1.3 mm per metre length but with low-height free standing walls light restraint reduces this to approximately 1 mm per metre; for normal storey-height walls in buildings movement is further reduced to about 0.5 mm per metre (Morton, 1986).

2. Assuming that the joint filler is able to compress by 50% then the joint widths in the above two cases would be 24 mm and 12 mm respectively for a wall with expansion joints at 12 metre centres.

3. Shrinkage joints are generally 10 mm thick which allows for shrinkage and reversible moisture movements over an 8 m length between joints in a normal storey-height wall with a joint filler able to extend by 30% of its thickness.

4. Movement joints should be provided at points of weakness or where high local strains would be produced; see CIRIA special publication 44 (1986); see also Figs. B1.20 and B1.21.

Fig. B1.25 Movement joints: (a) contraction joints (width ⩽10) and (b) expansion joints (width ⩾12).

Note 1: Spacing of ties is dependent on the cavity width and leaf thickness; see Table B1.1 and Note 4 of Fig. B1.23.
Note 2: Spacing of ties dependent on span of wall. The strength of the wall is greatly improved by close spacing of ties and straps at supports; see **6.1h/k**.
Note 3: This detail is assumed to allow some movement in two directions; see Fig. B1.23.
Note 4: This detail is shown in Fig. B1.24.

Fig. B1.26 Suggested spacing of ties in cavity wall with 100 mm thick leaves and 50 mm cavity.

Table B1.1 Type and Spacing of ties (adapted BS 5628: Part 1)

Smaller leaf thickness mm	Cavity Width mm	Type of tie	Spacing of ties		Number of ties per square metre
			Horizontally mm	Vertically mm	
Less than 90	50–75	B,DT,VT	450	450	4.9
90 or more	50–75	B,DT,VT	900	450	2.5
90 or more	76–100	DT,VT	900	450	2.5
90 or more	101–150	VT	900	450	2.5

Key: B - butterfly tie; DT - double triangle tie; VT - vertical twist tie.

Table B1.2   Characteristic strengths of wall ties used as panel supports in kN (after Table 8 of BS 5628:Part 1)

Type	Characteristic strengths of ties engaged in dovetail slots set in structural concrete	
	Tension	Shear
Dovetail slot types of ties	kN	kN
(a) Galvanised or stainless steel fishtail anchors 3 mm thick, 17 mm min. width in 1.25 mm thick galvanised or stainless steel slots, 150 mm long, set in structural concrete	4.0	5.0
(b) Galvanised or stainless steel fishtail anchors 2 mm thick, 17 mm min. width in 2 mm thick galvanised or stainless steel slots, 150 mm long, set in structural concrete	3.0	4.5
(c) Copper fishtail anchors 3 mm thick, 17 mm min. width, in 1.25 mm copper slots, 150 mm long, set in structural concrete	3.5	4.0

	Characteristic loads in ties embedded in mortar			
	Tension			Shear*
	Mortar designations			Mortar designation
	(i) and (ii)	(iii)	(iv)	(i), (ii) or (iii)
Cavity wall ties†	kN	kN	kN	kN
(a) Wire butterfly type: Zinc-coated mild steel or stainless steel	3.0	2.5	2.0	2.0
(b) Vertical twist type: Zinc-coated mild steel or bronze or stainless steel	5.0	4.0	2.5	3.5
(c) Double-triangle type: Zinc-coated mild steel or bronze or stainless steel	5.0	4.0	2.5	3.0

*Applicable only to cases where shear exists between closely abutting surfaces

†See BS 1243 (1978)

B2   Foundations to masonry walls

Strip footings

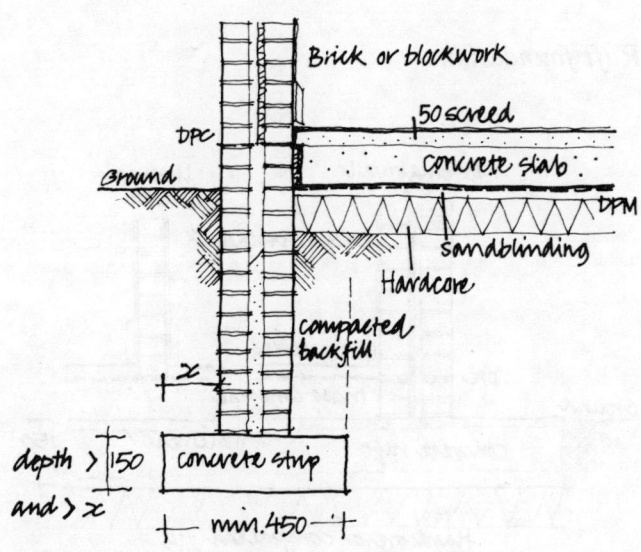

Fig. B2.1   Strip footing.

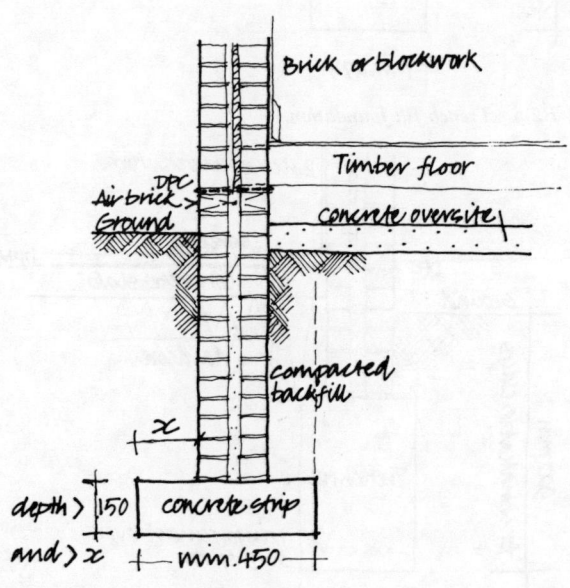

Fig. B2.2   Strip footing with suspended timber floor.

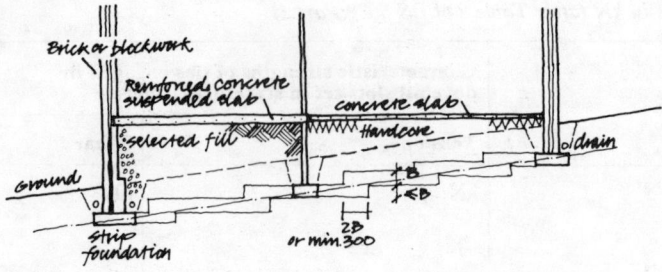

Fig. B2.3 *Strip footing on sloping ground (after Tomlinson et al, 1978).*

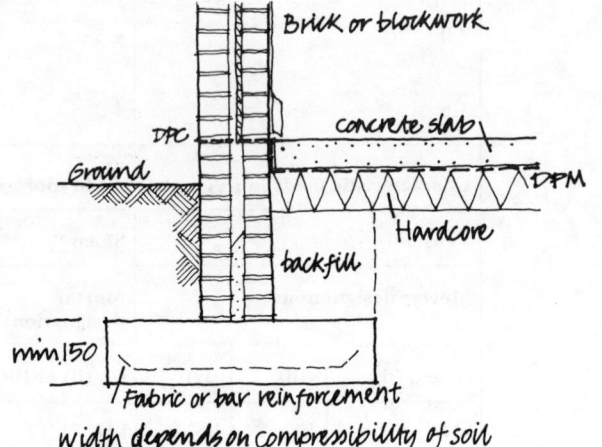

Fig. B2.4 *Wide strip footing.*

Trench fill foundations

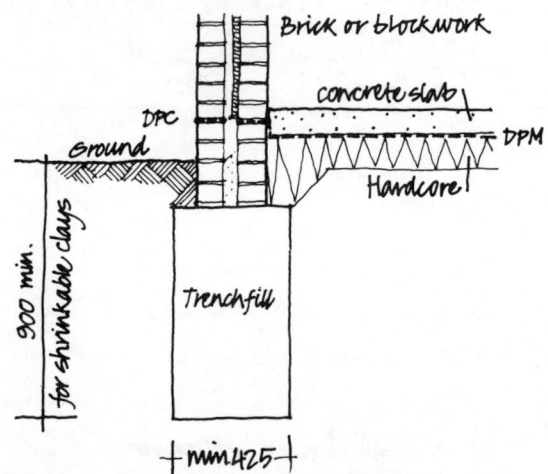

Fig. B2.5 *Trench fill foundation.*

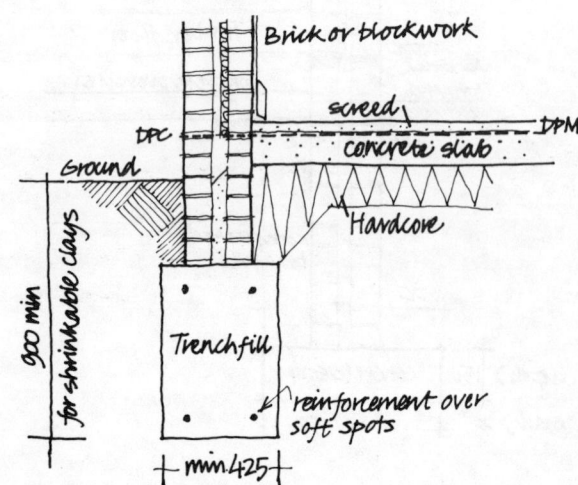

Fig. B2.6 *Trench fill foundation with screed on slab.*

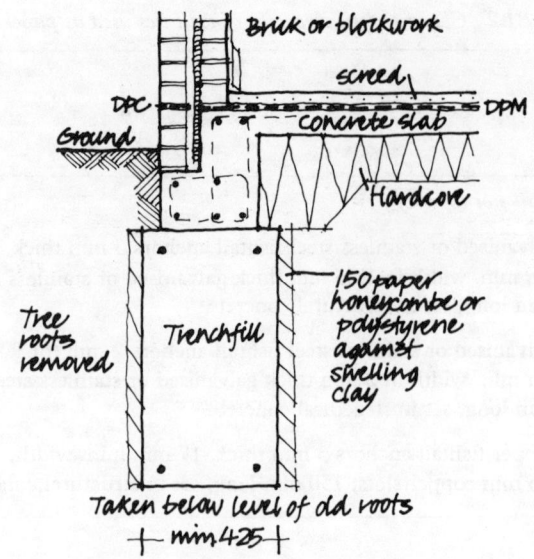

Fig. B2.7 *Trench fill foundation in swelling clay.*

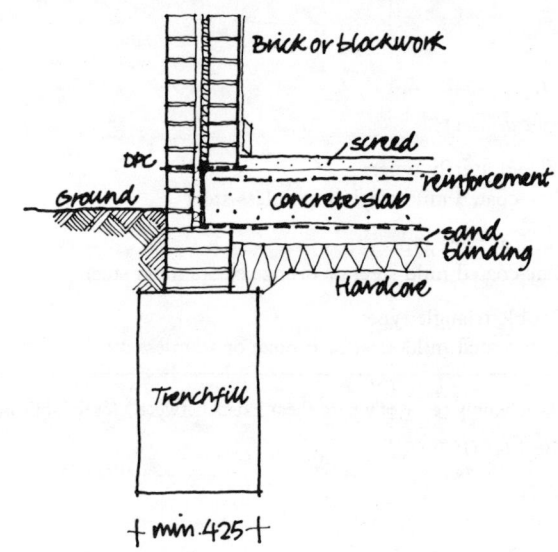

Fig. B2.8 *Trench fill foundation with edge of slab on wall.*

Raft foundations

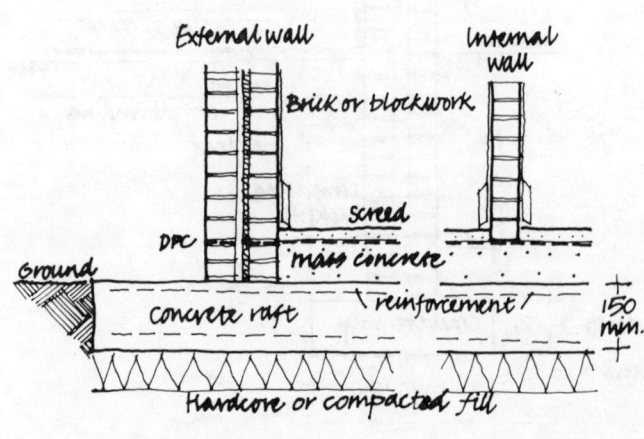

Fig. B2.9 *Flat raft foundation.*

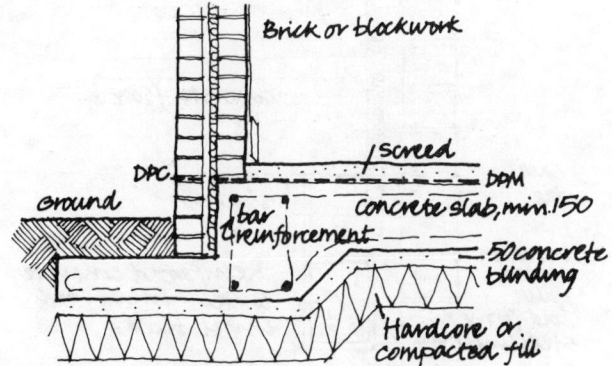

Fig. B2.10   Raft foundation with edge beams.

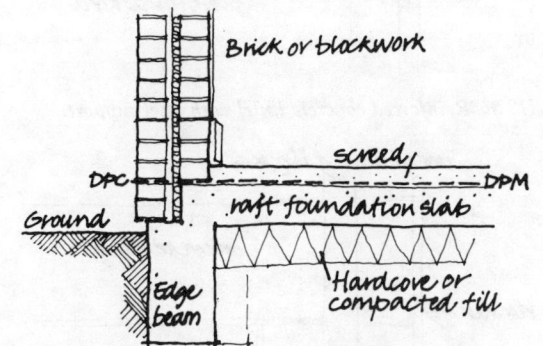

Fig. B2.11   Raft foundation with deep edge beams.

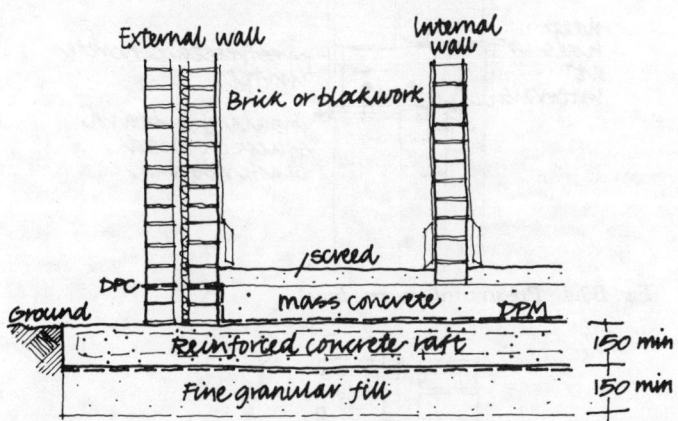

Fig. B2.12   Raft in areas subject to mining subsidence (after Tomlinson et al. 1978).

Piers and piled foundations

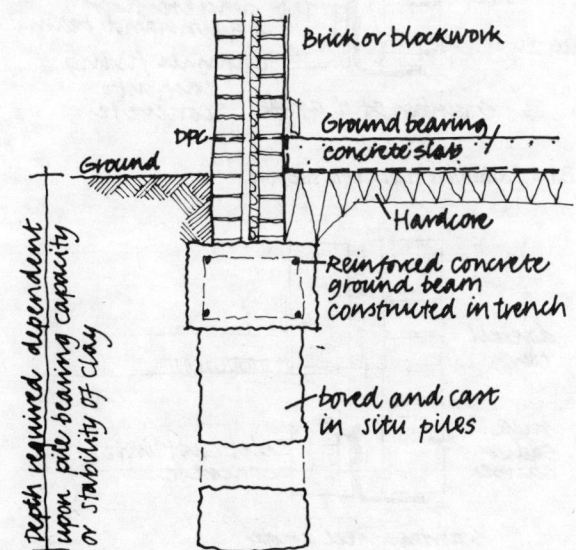

Fig. B2.13   Bored pile foundation and floating slab (after Tomlinson et al. 1978).

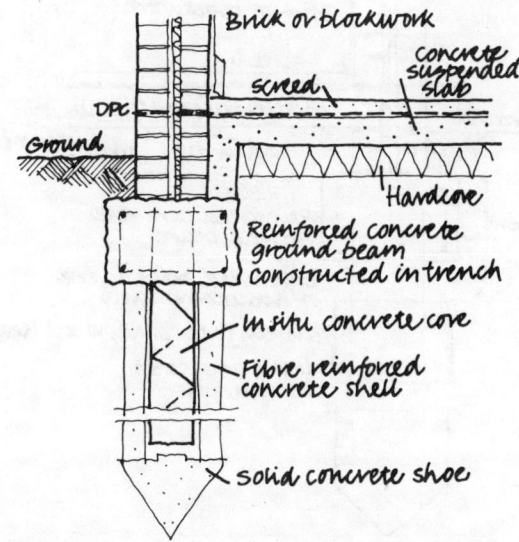

Fig. B2.14   Precast driven segmented pile foundation with suspended slab.

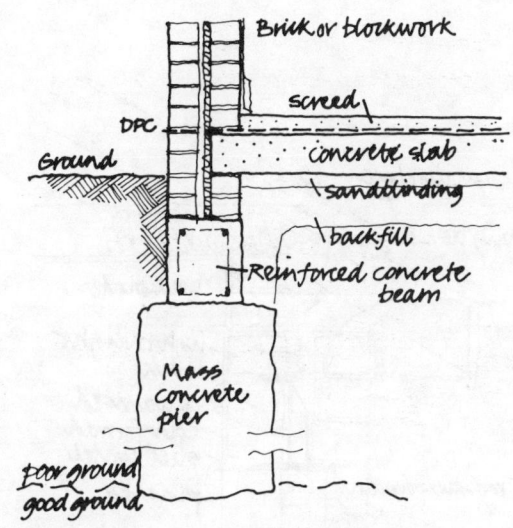

Fig. B2.15   Mass concrete pier foundation.

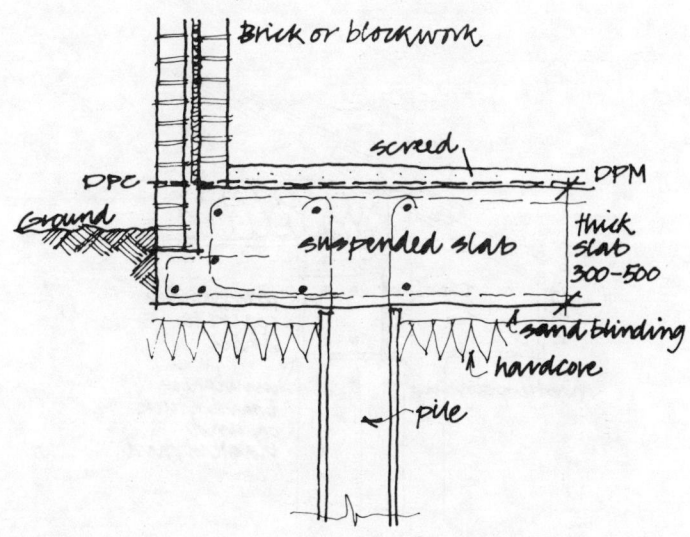

Fig. B2.16   Suspended slab on piles without use of ground beams.

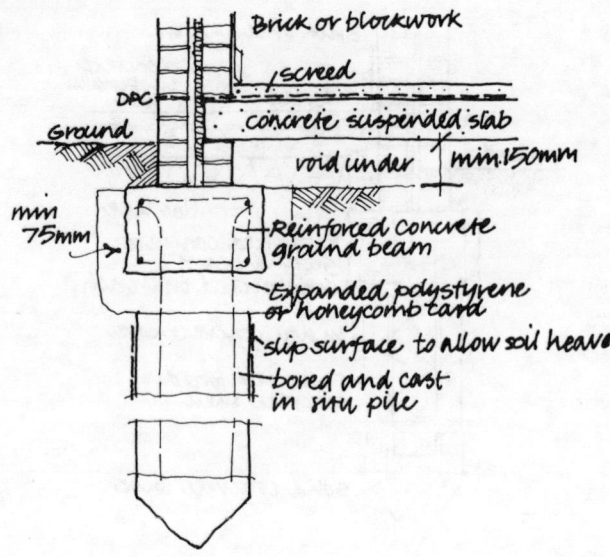

Fig. B2.17   *Bored pile foundation in swelling clay.*

B3   Lintels and other masonry supports

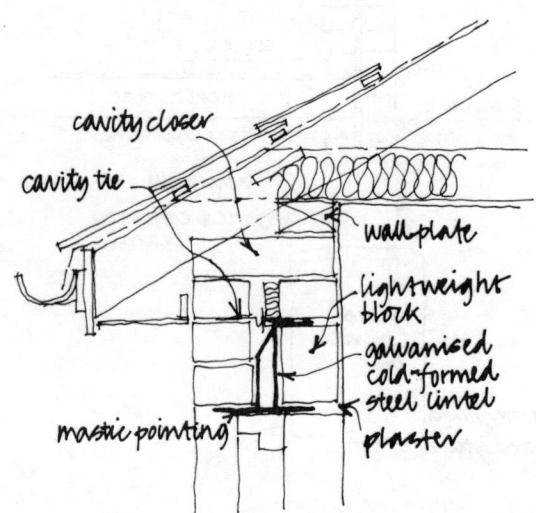

Fig. B3.1   *Steel lintel.*

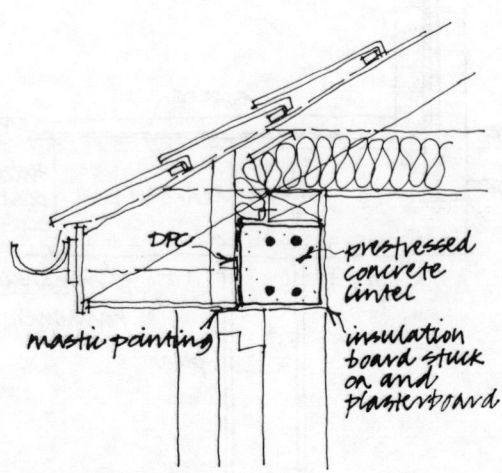

Fig. B3.2   *Pre-stressed concrete lintel.*

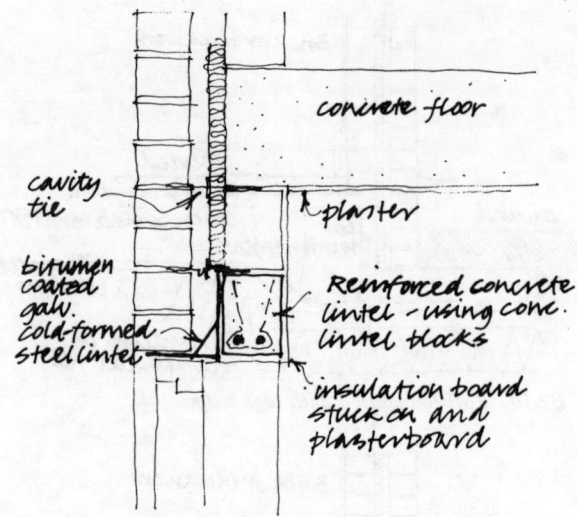

Fig. B3.3   *Reinforced concrete lintel with steel support.*

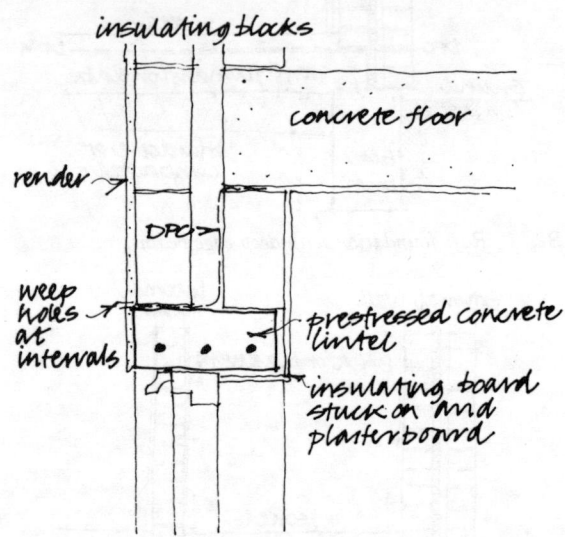

Fig. B3.4   *Pre-stressed concrete lintel.*

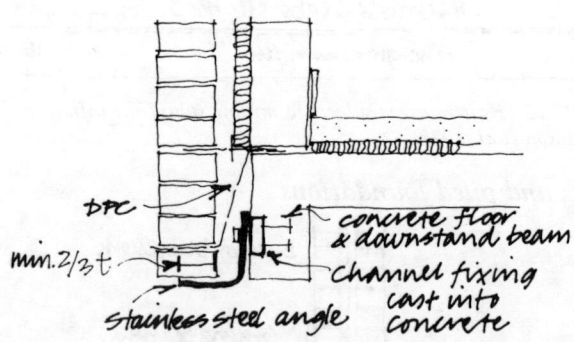

Fig. B3.5   *Stainless steel angle support.*

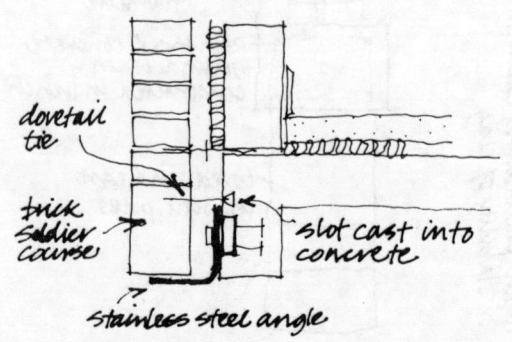

Fig. B3.6   *Stainless steel angle support with dovetail tie.*

B4 Floor and roof to masonry wall details

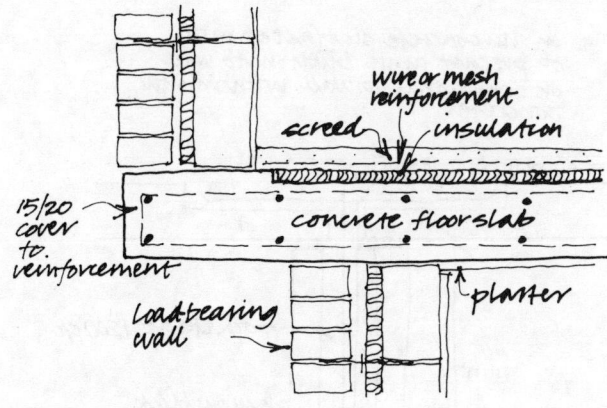

Fig. B3.7 Concrete slab support.

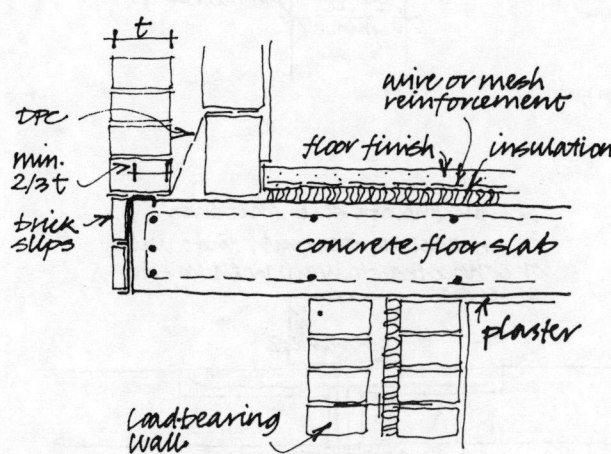

Fig. B3.8 Concrete slab support with brick slips.

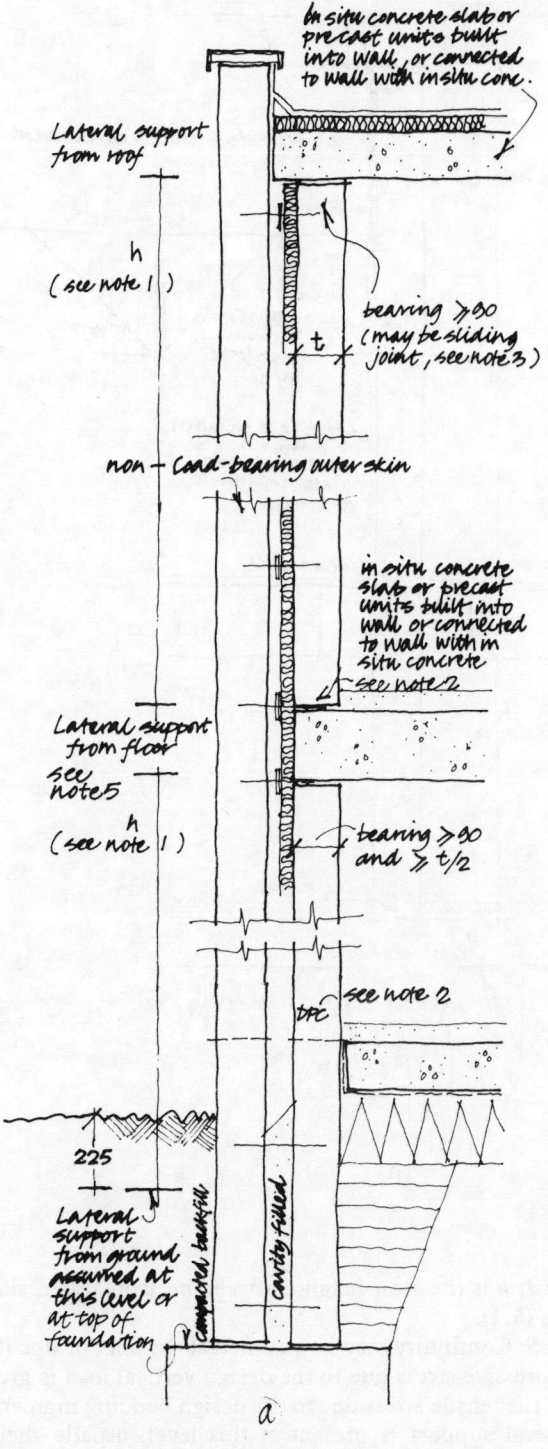

Fig. B4.1 (a)–(d) Section details which are assumed to provide enhanced resistance to lateral movement of walls under vertical load.

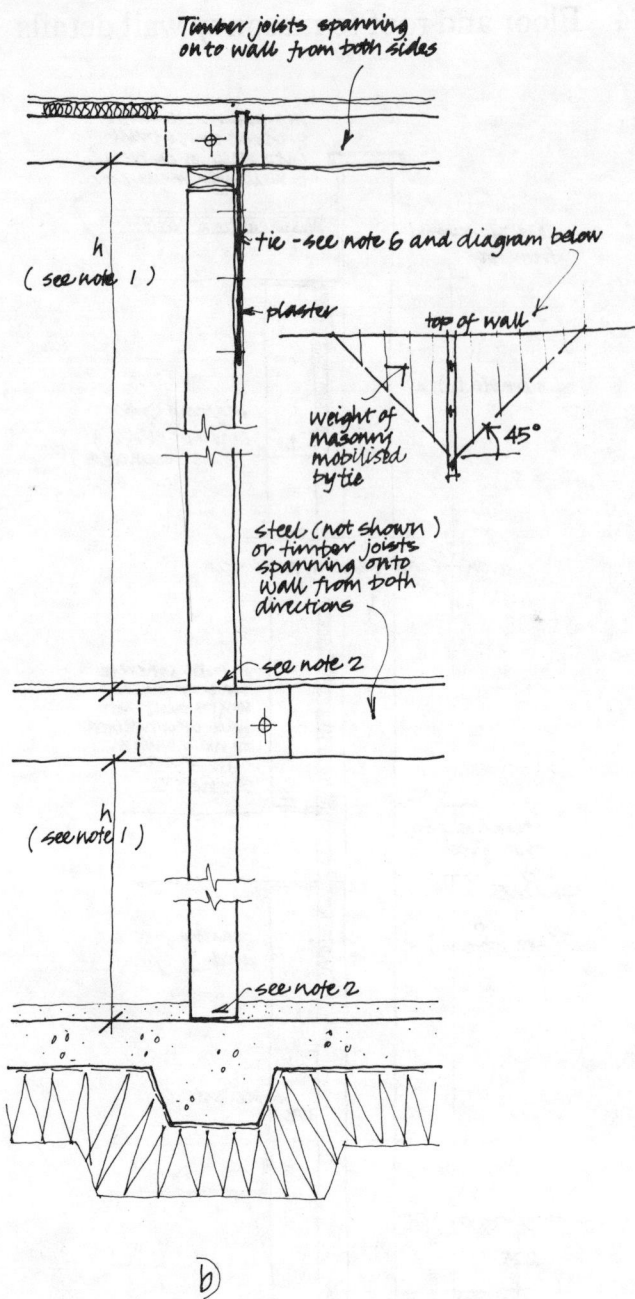

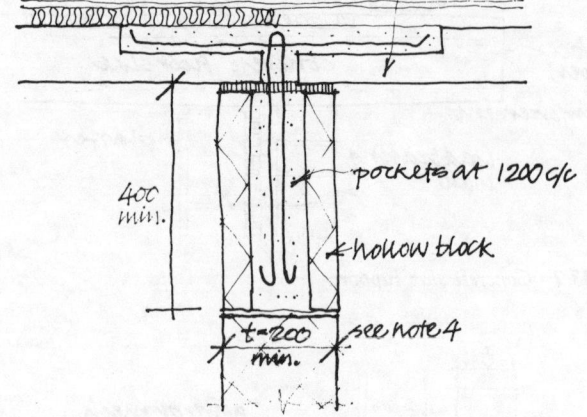

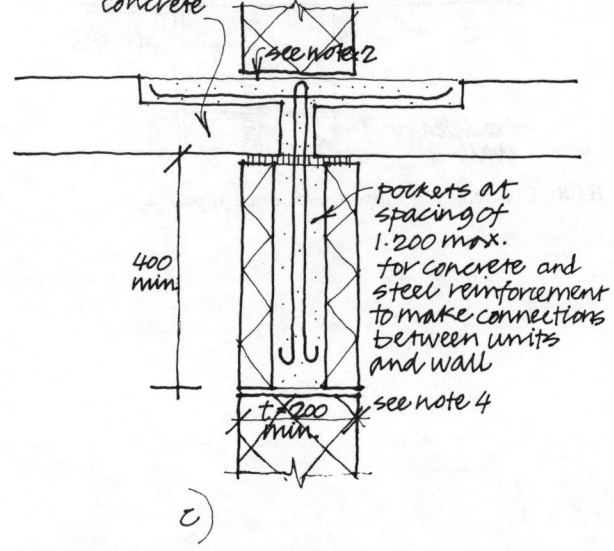

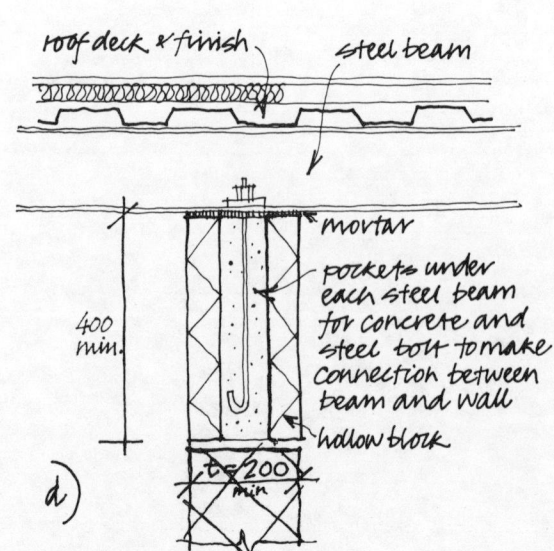

Note 1: h is the clear height between horizontal lateral supports (**5.1**).

Note 2: Continuity may be assumed at the joint or dpc if the compressive stress due to the design vertical load is greater than the tensile stress due to the design bending moment. If a lateral support is present at this level, usually there is enhanced resistance to lateral movement. If there is no continuity at a lateral support, or for convenience, simple resistance to lateral movement may be assumed; see **5.1** and Fig. B4.2.

Note 3: A sliding joint may be required at some roof bearings to allow for temperature movements; it can be made by inserting two layers of slippery material, or a neoprene bearing strip, under the roof slab; see Fig. B4.7.

Note 4: Separate precast units spanning on to the wall must be properly tied to each other and to the wall by steel reinforcement in a structural concrete; for adequate bearing, the wall should have a minimum thickness of 200 mm.

Note 5: Information about enhanced and simple resistance to lateral movement is given in Chapter 5.

Note 6: See Note 4 of Fig. B4.2.

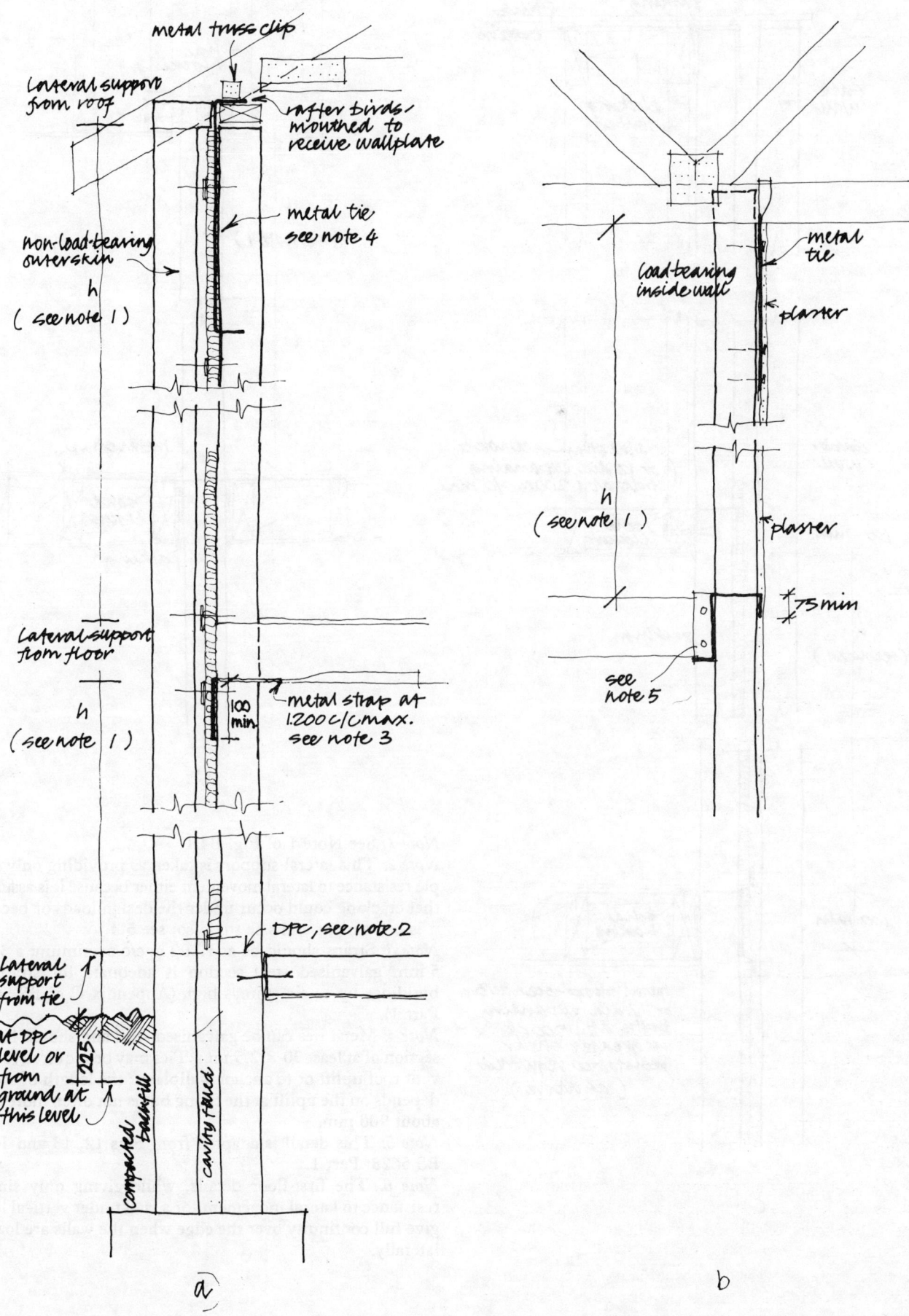

Fig. B4.2 (a)–(d) Section details which are assumed to provide simple resistance to lateral movement of walls under vertical load (timber sloping roofs and timber floors).

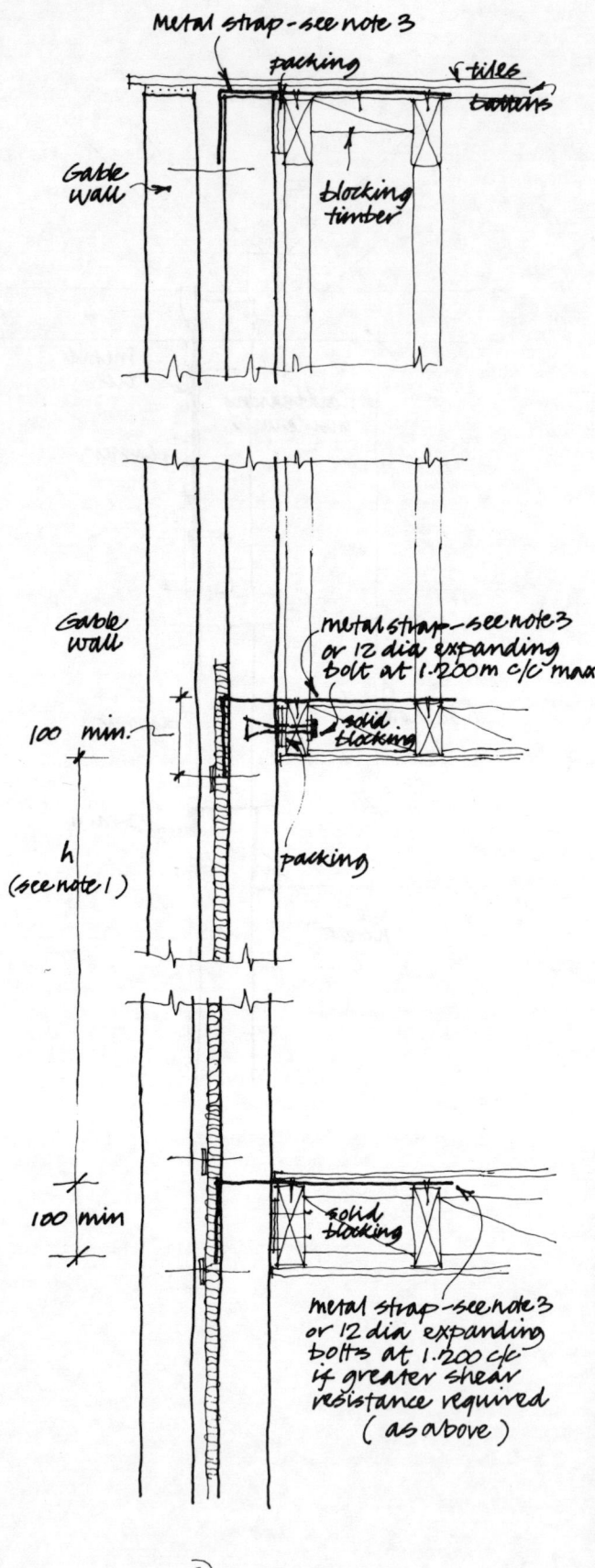

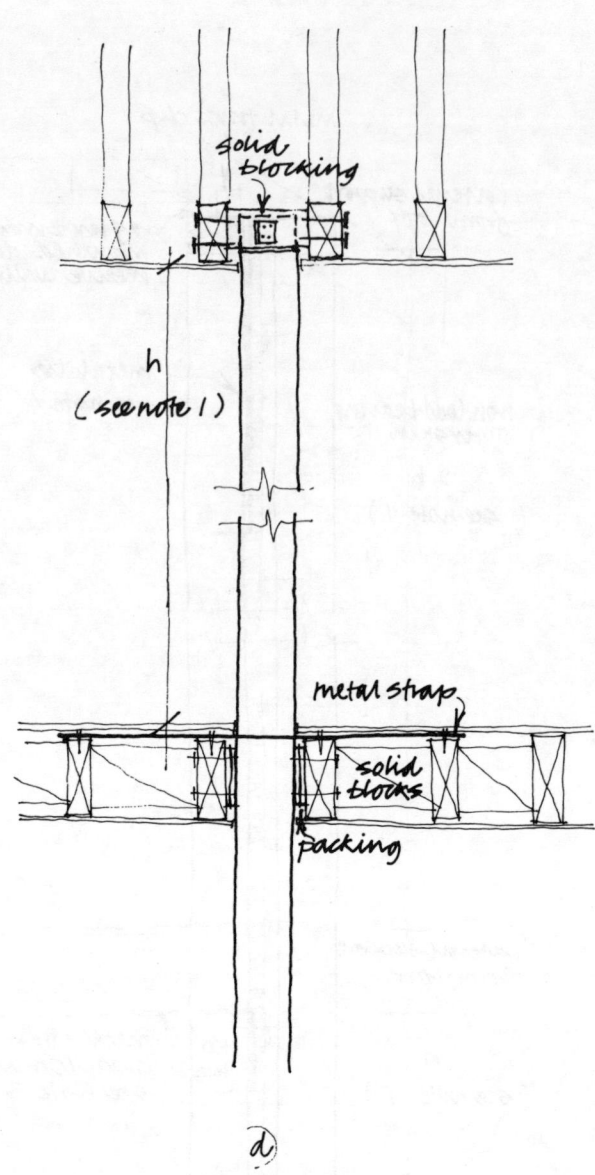

Note 1: See Note 1 of Fig. B4.1.

Note 2: This lateral support is taken as providing only simple resistance to lateral movement either because it is assumed that cracking could occur under the design loads or because it is convenient and safe to do so; see **5.1**.

Note 3: Straps should be at 1.200 m c/c maximum; a 30 × 5 mm galvanised steel section is adequate for straps in buildings up to six storeys high (Appendix C of BS 5628: Part 1).

Note 4: Metal ties can be galvanised steel and should have a section of at least 30 × 2.5 mm. Ties may be required to prevent roof uplift or to anchor wallplates; the length of the tie depends on the uplift at the fixing but is not often more than about 900 mm.

Note 5: This detail is adapted from Figs 12, 13 and 14 of BS 5628: Part 1.

Note 6: The first-floor details, while giving only simple resistance to lateral movement for a wall under vertical load, give full continuity over the edge when the walls are loaded laterally.

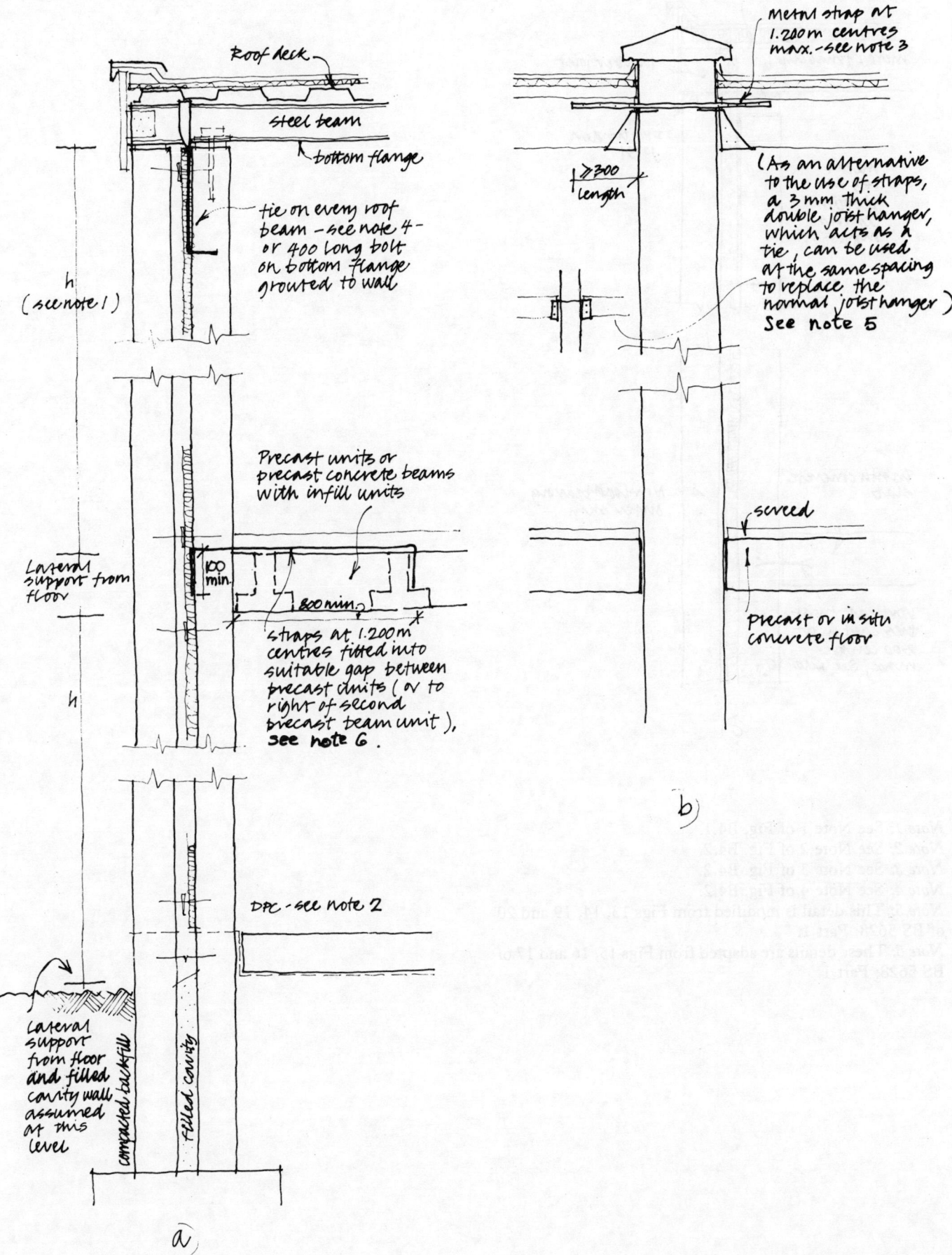

Roof deck.

steel beam

bottom flange

tie on every roof beam - see note 4 - or 400 long bolt on bottom flange grouted to wall

h (see note 1)

Lateral support from floor

Precast units or precast concrete beams with infill units

100 min.

800 min.

straps at 1.200m centres fitted into suitable gap between precast units (or to right of second precast beam unit), **see note 6**.

h

DPC - see note 2

Lateral support from floor **and filled cavity wall** assumed at this level

compacted backfill

filled cavity

ⓐ

Metal strap at 1.200m centres max.- see note 3

≥ 300 length

(As an alternative to the use of straps, a 3mm thick double joist hanger, which acts as a tie, can be used at the same spacing to replace the normal joist hanger) **See note 5**

screed

precast or in situ concrete floor

ⓑ

Fig. B4.3 (a)–(c) Section details which are assumed to provide simple resistance to lateral movement of walls under vertical load (flat roofs and concrete floors).

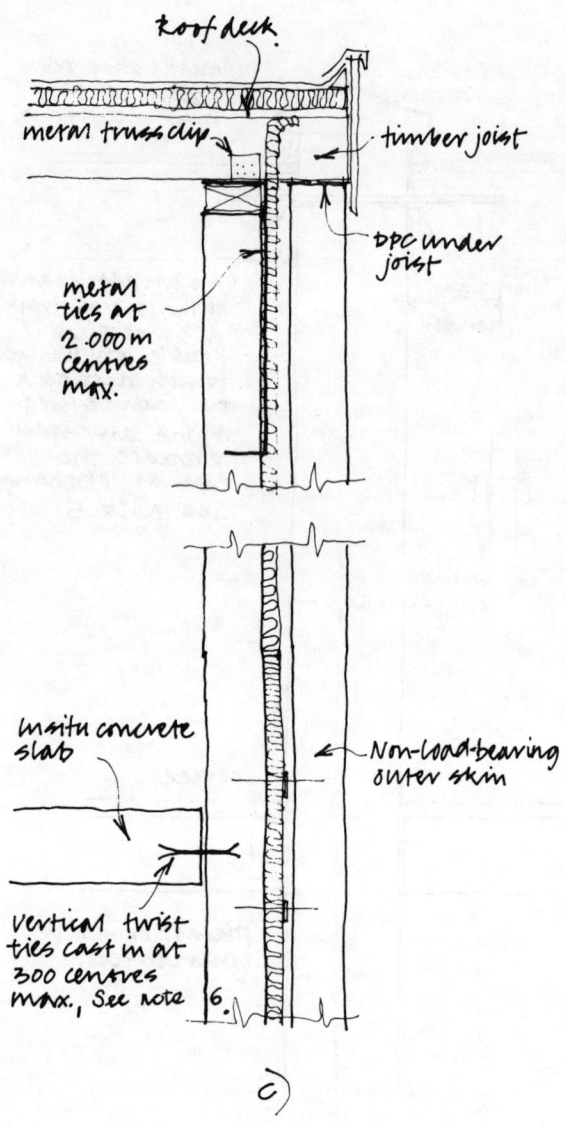

roof deck.

metal truss clip

timber joist

metal ties at 2.000m centres max.

DPC under joist

insitu concrete slab

Non-load-bearing outer skin

Vertical twist ties cast in at 300 centres max., See note 6.

c)

Note 1: See Note 1 of Fig. B4.1.
Note 2: See Note 2 of Fig. B4.2.
Note 3: See Note 3 of Fig. B4.2.
Note 4: See Note 4 of Fig. B4.2.
Note 5: This detail is modified from Figs 13, 14, 19 and 20 of BS 5628: Part 1.
Note 6: These details are adapted from Figs 15, 16 and 17 of BS 5628: Part 1.

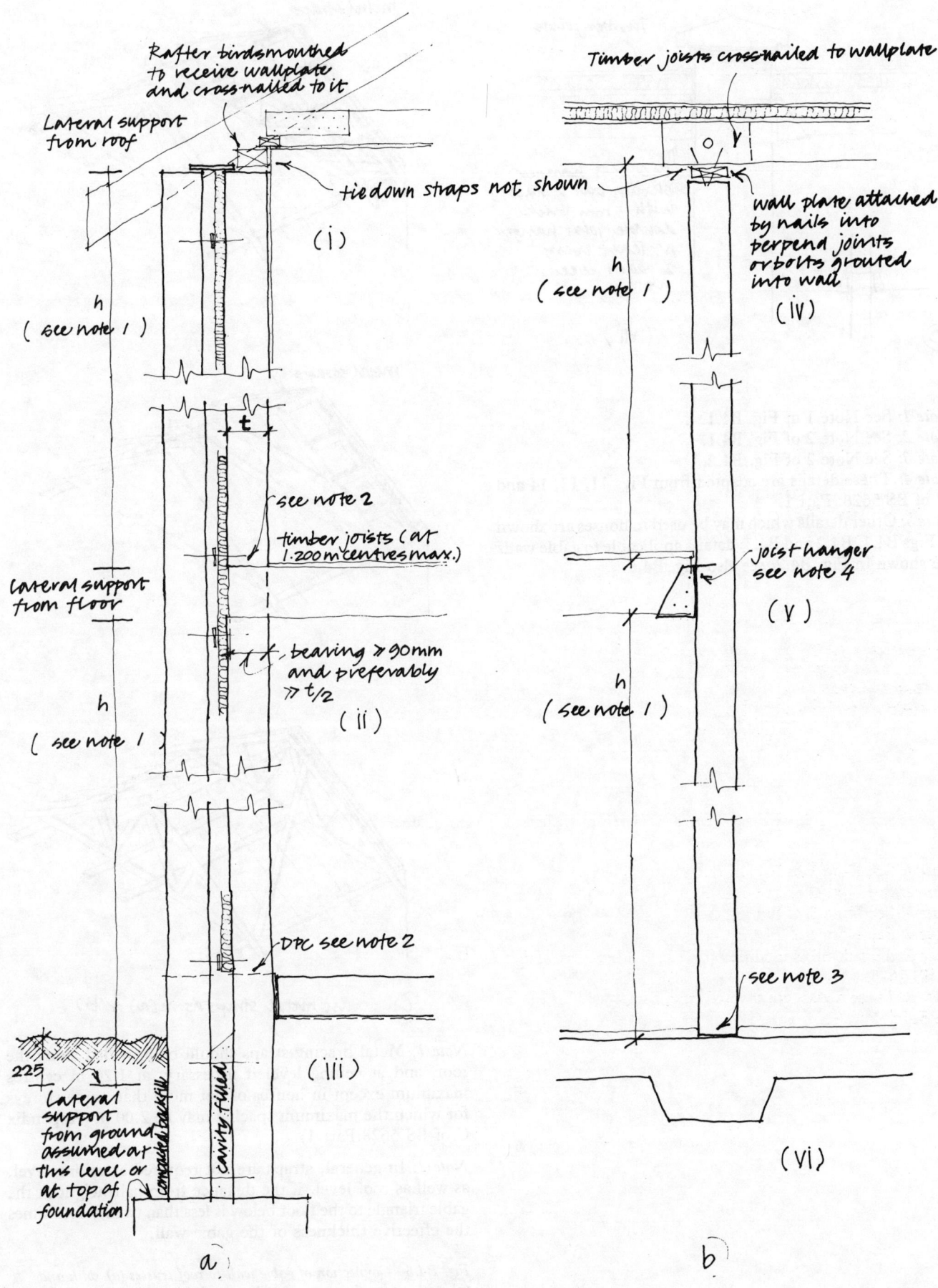

Fig. B4.4 (a)–(c) Section details for houses not more than three storeys high providing either simple resistance (details (i),(v),(vi),(vii)) or enhanced resistance (details (ii),(iii),(iv)) to lateral movement of walls under vertical load.

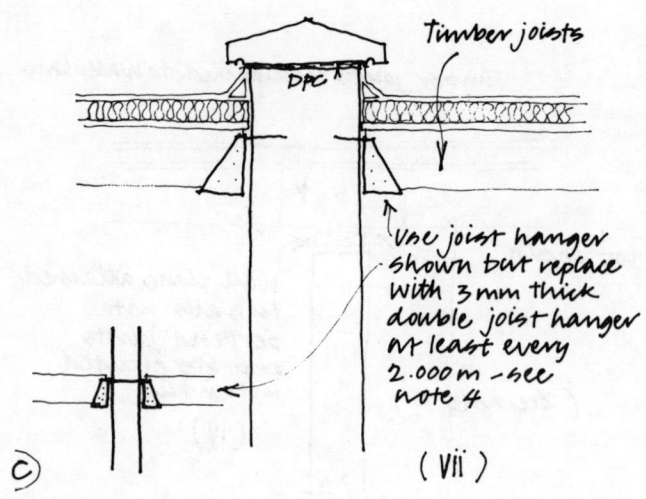

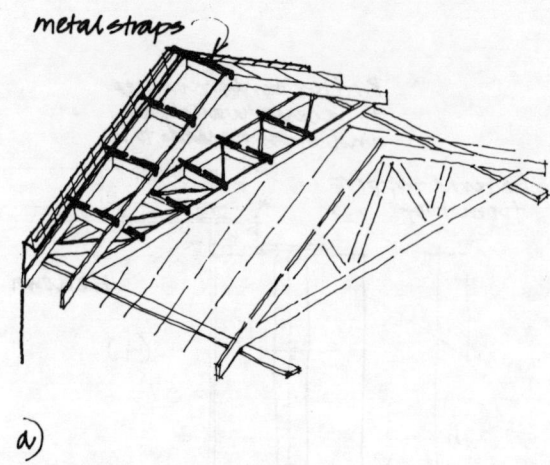

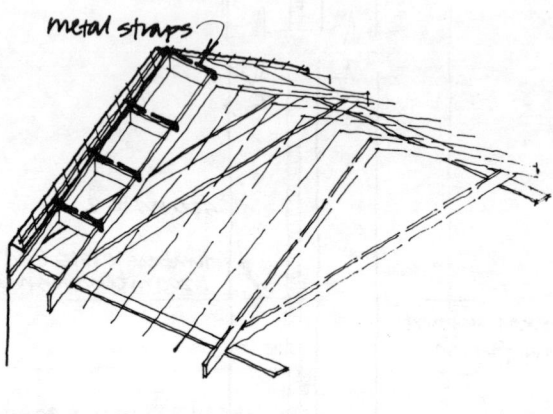

Note 1: See Note 1 of Fig. B4.1.
Note 2: See Note 2 of Fig. B4.1.
Note 3: See Note 2 of Fig. B4.2.
Note 4: These details are adapted from Figs 11, 13, 14 and 20 of BS 5628: Part 1.
Note 5: Other details which may be used in houses are shown in Figs B4.1, B4.2 and B4.3; details applicable to gable walls are shown in Fig. B4.2; see also Fig. B4.5.

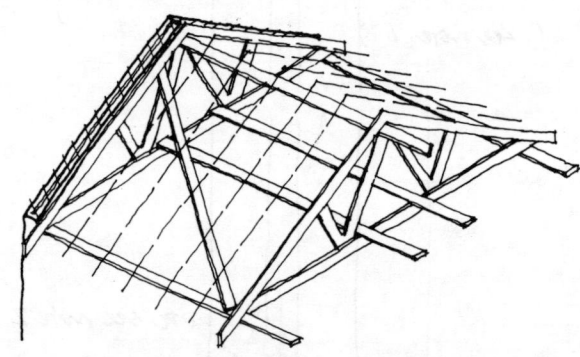

Note 1: Metal bracing straps should be provided along the roof, and at ceiling level if necessary, at 1.20 m centres maximum except in houses of not more than three storeys for which the maximum spacing may be 2.00 m (Appendix C of BS 5628:Part 1).

Note 2: In general, straps are not required at ceiling level, as well as roof level, if the distance from half-height of the gable triangle to the floor below is less than twenty-four times the effective thickness of the gable wall.

Fig. B4.5 Connection of gable wall to roof trusses (a) with and (b) without bracing truss at ceiling level; (c) trusses next to gable wall showing bracing on one side and minimum number of longitudinal binders.

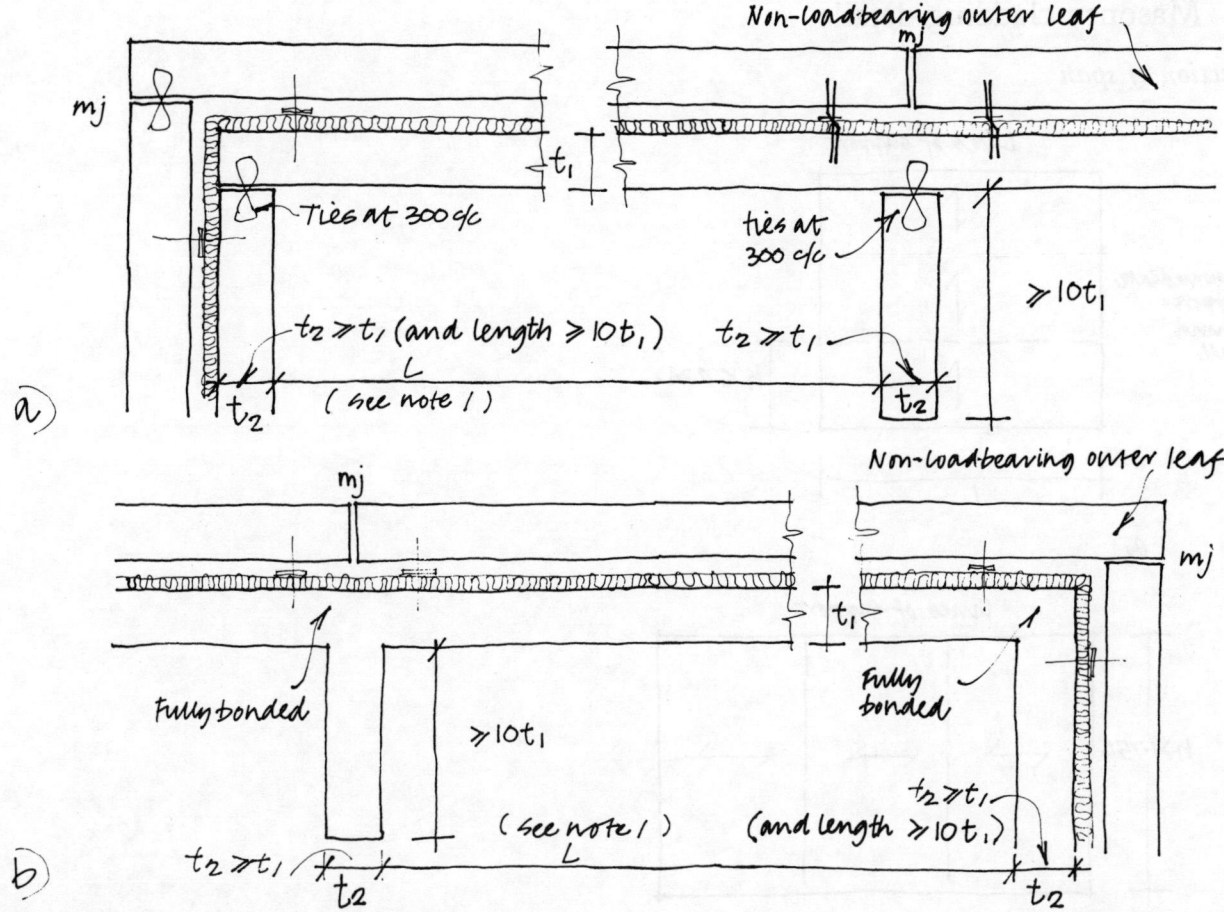

Fig. B4.6 Plan details which are assumed to provide either (a) simple resistance or (b) enhanced resistance to lateral movement of walls under vertical load.

Note 1: L is the clear length between vertical lateral supports **(5.1).**

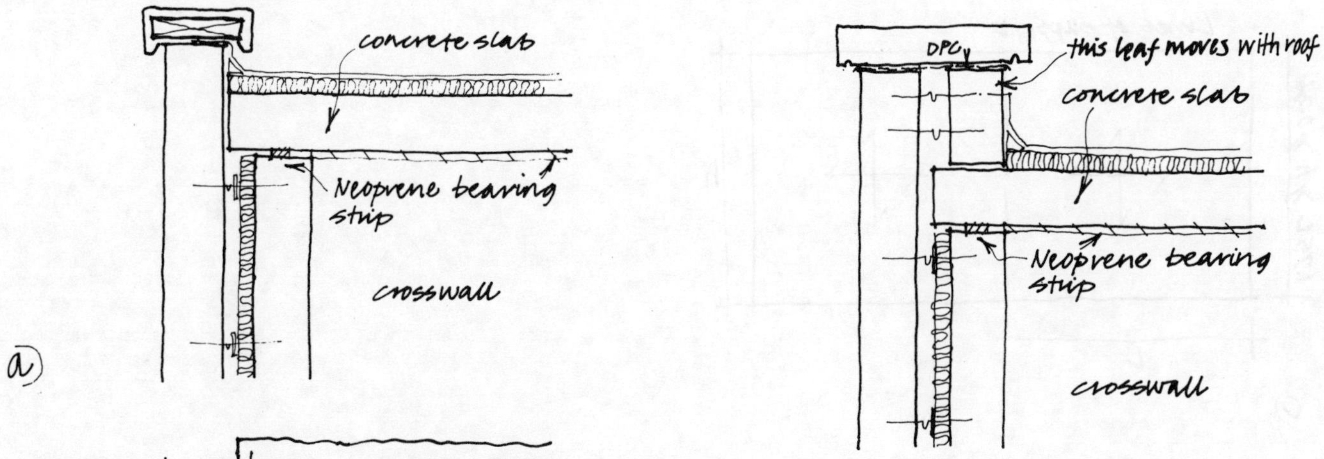

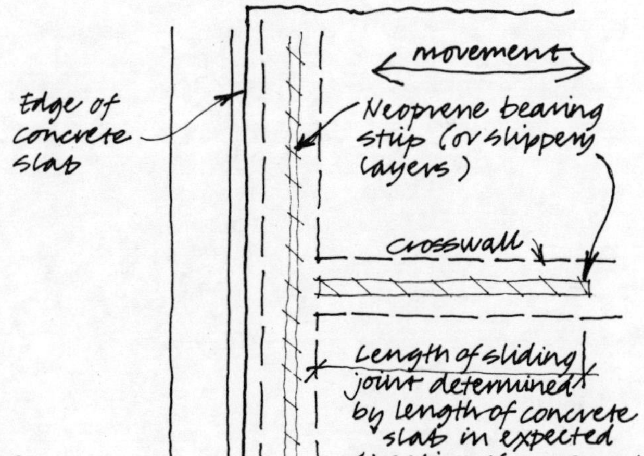

Fig. B4.7 Movement joint details to prevent cracking at junction of wall and concrete roof slab: (a) sections and (b) plan.

B5 Masonry cladding details

Direction of span

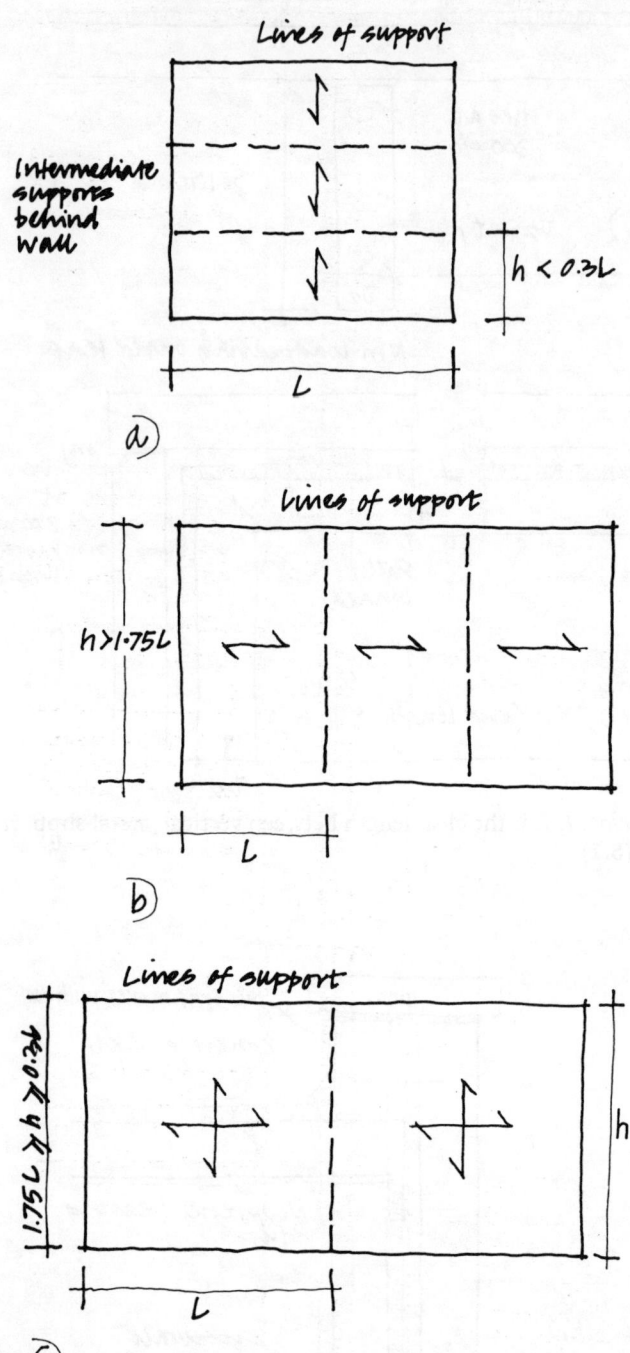

a)

b)

c)

Note 1: Lines of lateral support are assumed, for calculation, to provide either sufficient moment and shear restraint for there to be continuity at the edge or sufficient shear restraint for there to be a simple support at the edge or some intermediate condition between these two extremes (**6.1**); edges of wall panels with lines of support not meeting these conditions should be taken as free edges.

Note 2: For design of masonry wall panels, see Chapter 6.

Fig. B5.1 Direction of span of masonry panels: (a) vertically spanning, (b) horizontally spanning, and (c) two-way spanning, depending on panel proportions.

Load-bearing walls

Note 1: *h* is the clear height between lateral supports.

Note 2: Continuity may be assumed across the dpc or joint if the compressive stress at this level due to the design vertical load is such that the allowable design flexural strength of the wall is not exceeded (**6.1g**); in places where this is not true the lateral support to the edge of the wall panel, which must then be present at that level, should be taken as providing a simple support to the panel; see Fig. B5.3.

Note 3: These conditions provide simple or enhanced resistance to lateral movement for a wall under vertical load (Clause 28.2.2.2 of BS 5628: Part 1 (**5.3**)); however, they can also be taken to give full continuity at the edge for a laterally loaded wall panel (Note 2 and Fig. B4.4).

Note 4: The outer leaf of an external cavity wall should be supported every 3 storeys or every 9 m, whichever is less (**2.4**); details such as some of those in Fig. B5.4 would be suitable. Alternatively the outer leaf may be uninterrupted over its full height, for any height of building, if the calculated differential movement between inner and outer leaves at the top is limited to 30 mm; no calculation is necessary if the building is both less than 4 storeys high and less than 12 m in height (Clause 29.2 of BS 5628: Part 1).

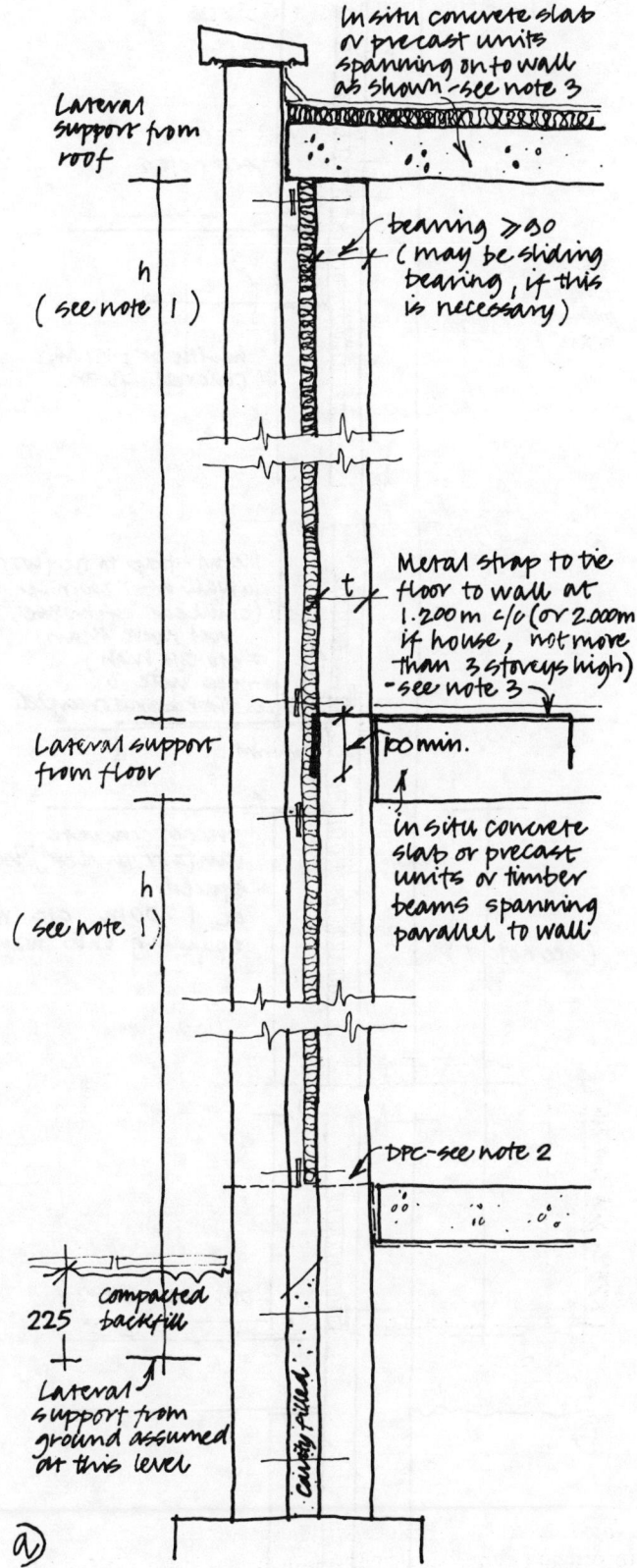

Fig. B5.2 (a)–(b) Section details which are assumed to provide full continuity at the edges of laterally loaded wall panels (load-bearing walls).

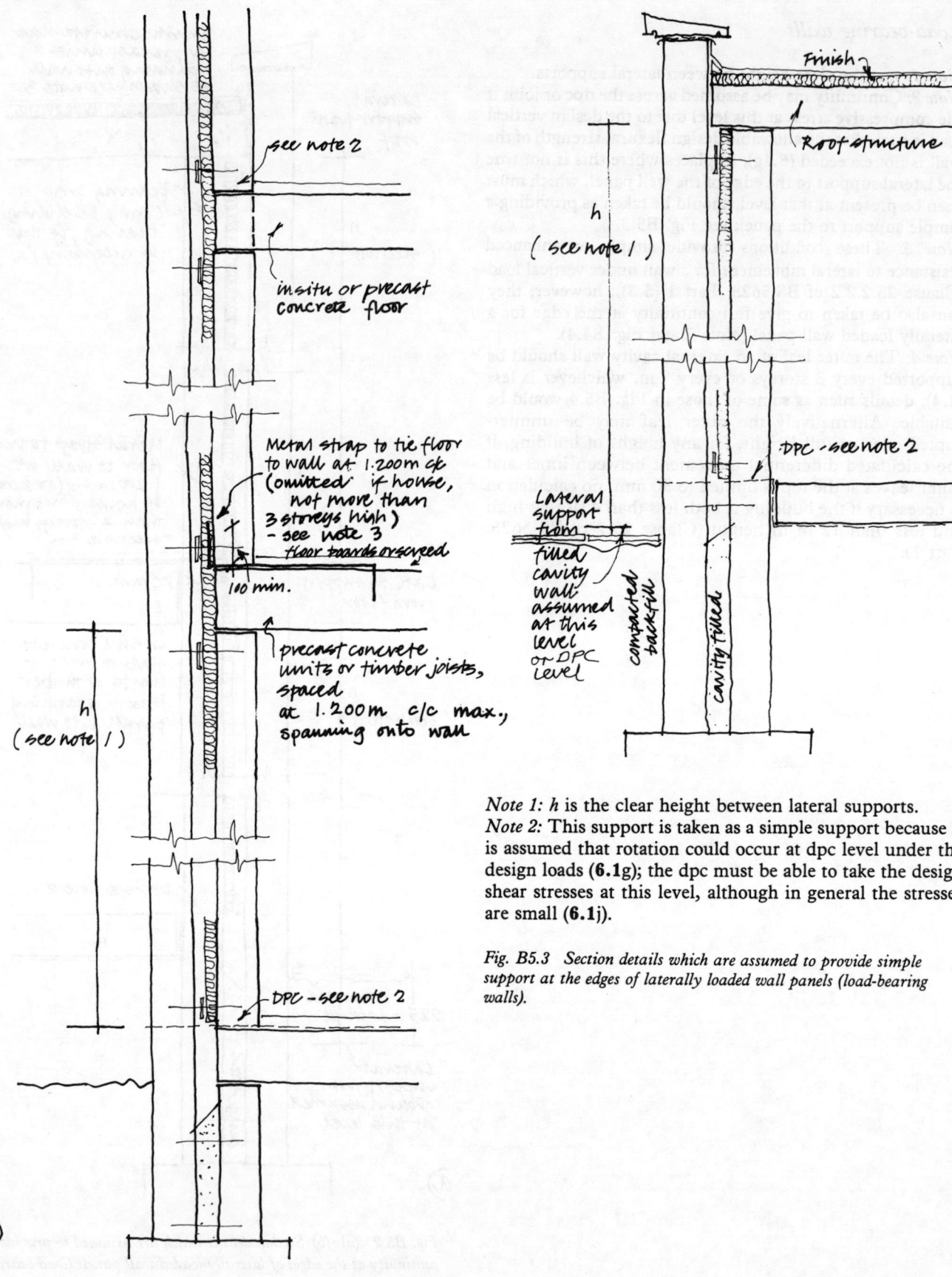

see note 2

insitu or precast concrete floor

Metal strap to tie floor to wall at 1.200m c/c (omitted if house, not more than 3 storeys high) – see note 3

floor boards or screed

100 min.

precast concrete units or timber joists, spaced at 1.200m c/c max., spanning onto wall

h
(see note 1)

DPC – see note 2

Finish

Roof structure

h
(see note 1)

DPC – see note 2

Lateral support from filled cavity wall assumed at this level or DPC level

compacted backfill

cavity filled

b

Note 1: h is the clear height between lateral supports.
Note 2: This support is taken as a simple support because it is assumed that rotation could occur at dpc level under the design loads (**6.1**g); the dpc must be able to take the design shear stresses at this level, although in general the stresses are small (**6.1**j).

Fig. B5.3 Section details which are assumed to provide simple support at the edges of laterally loaded wall panels (load-bearing walls).

Non-load-bearing walls

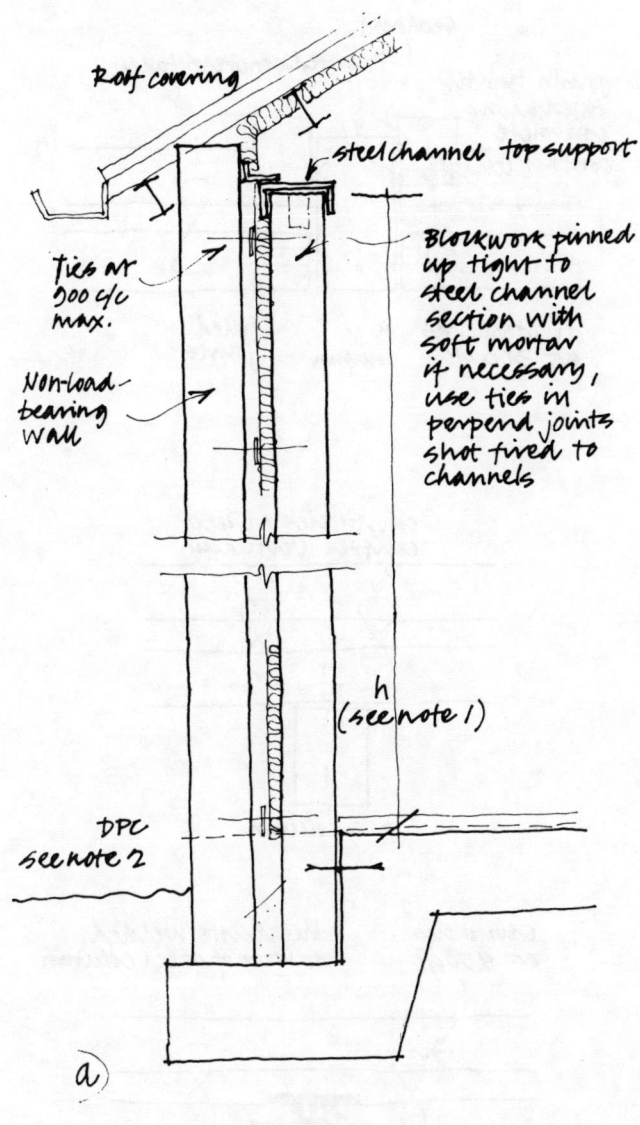

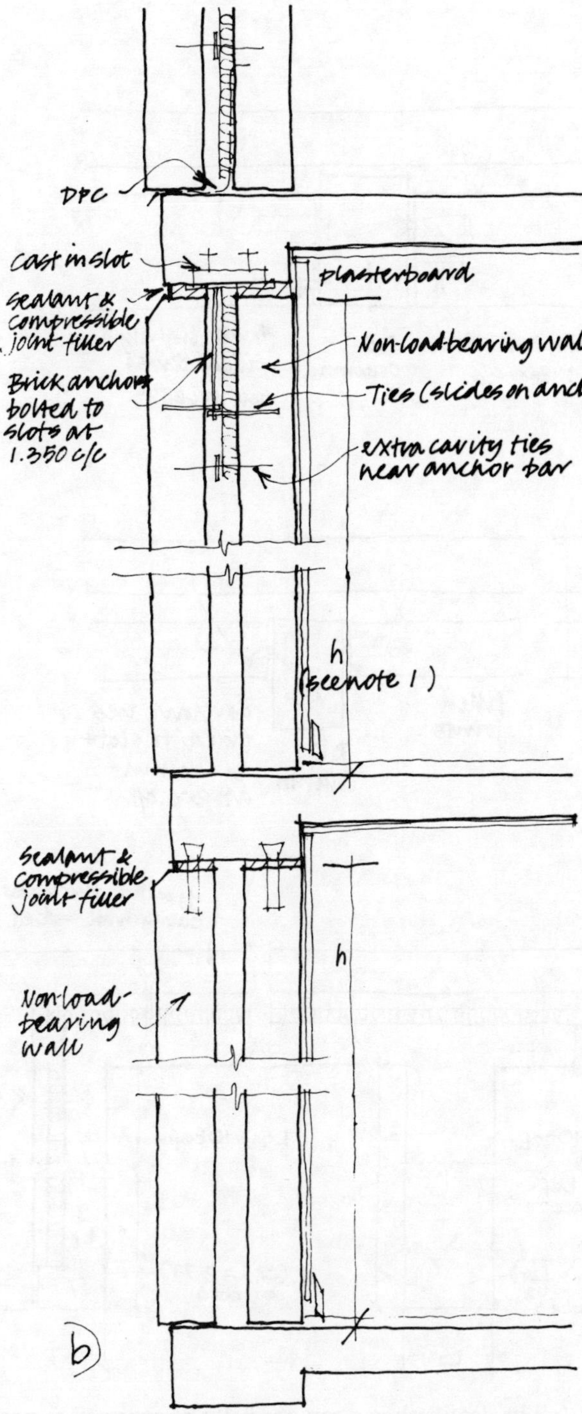

(a) labels:
- Roof covering
- steel channel top support
- Ties at 300 c/c max.
- Non-load-bearing wall
- Blockwork pinned up tight to steel channel section with soft mortar if necessary, use ties in perpend joints shot fired to channels
- h (see note 1)
- DPC see note 2

(b) labels:
- DPC
- Cast in slot
- Sealant & compressible joint filler
- Brick anchors bolted to slots at 1.350 c/c
- Plasterboard
- Non-load-bearing wall
- Ties (slides on anchor)
- extra cavity ties near anchor bar
- h (see note 1)
- Sealant & compressible joint filler
- Non-load-bearing wall
- h

(c) labels:
- DPC
- Sealant & compressible joint filler
- metal angle
- Brick anchors bolted to slab at 1.350 c/c
- 2t/3
- expanding bolt/ toothed
- soft joint
- plaster
- ties with slotted holes

Note 1: See Note 1 of Fig. B5.3.
Note 2: See Note 2 of Fig. B5.3.

Fig. B5.4 (a)–(c) Section details which are assumed to provide simple support at the edges of laterally loaded wall panels (non-load-bearing walls).

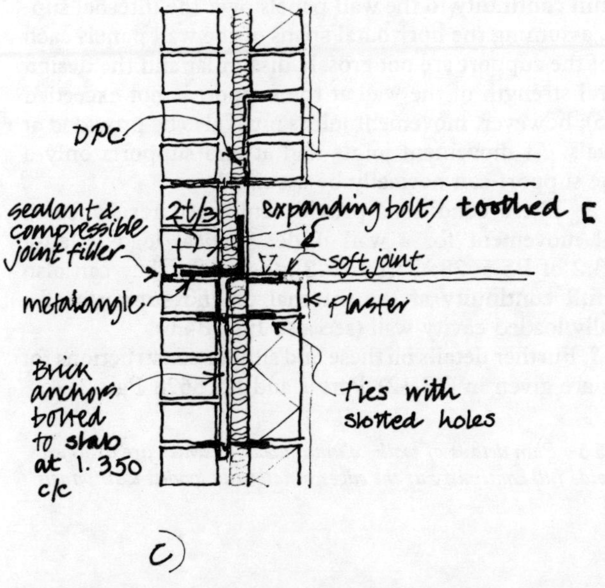

Wall to column joints

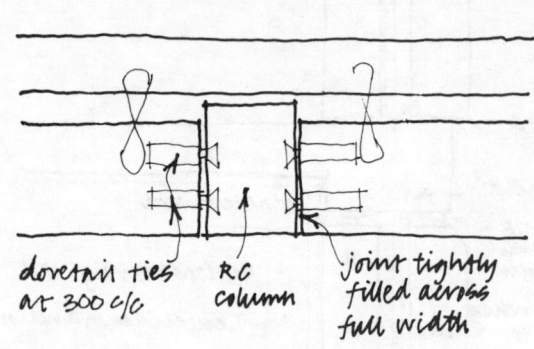

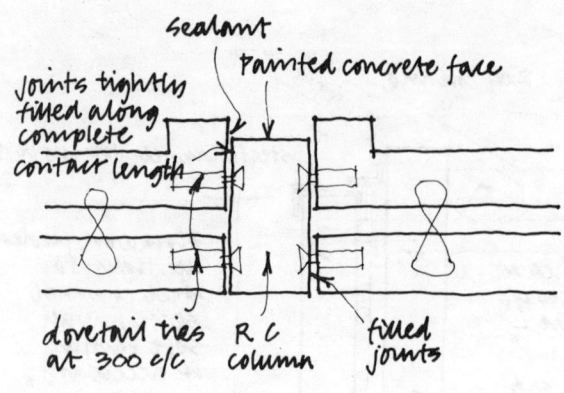

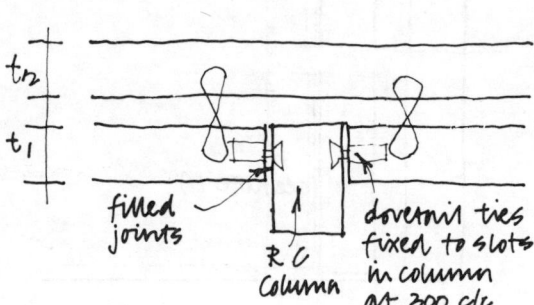

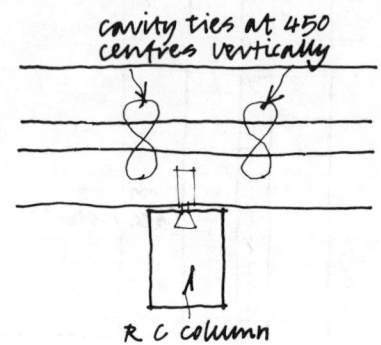

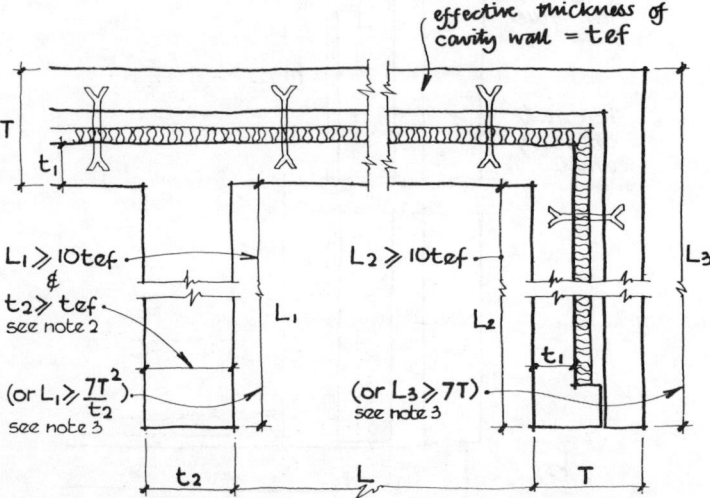

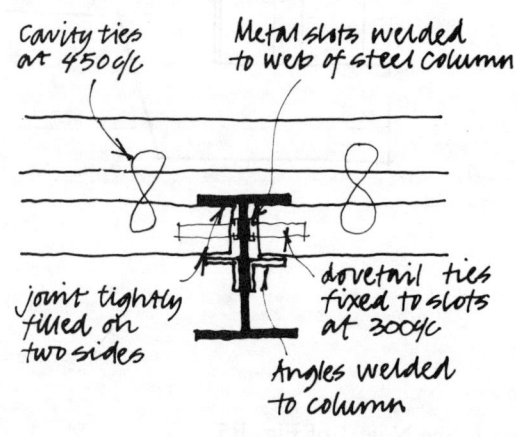

Note 1: The details shown can normally be expected to provide full continuity to the wall panels over the internal supports, assuming the horizontal spans of the wall panels each side of the support are not grossly dissimilar and the design flexural strength of the wall at the support is not exceeded (Ch. 6); however, movement joints must also be provided at intervals. At movement joints and at end supports only a simple support can normally be assumed.

Note 2: These conditions provide enhanced resistance to lateral movement for a wall under vertical load (Clause 28.2.3.2 of BS 5628: Part 1 (**5.3**)); however, they can also give full continuity at the internal or end supports of a laterally loaded cavity wall (see also Fig. B4.6).

Note 3: Further details on these and similar constructions for fixity are given in CP 121 Part 1 and BS 5628 Part 3.

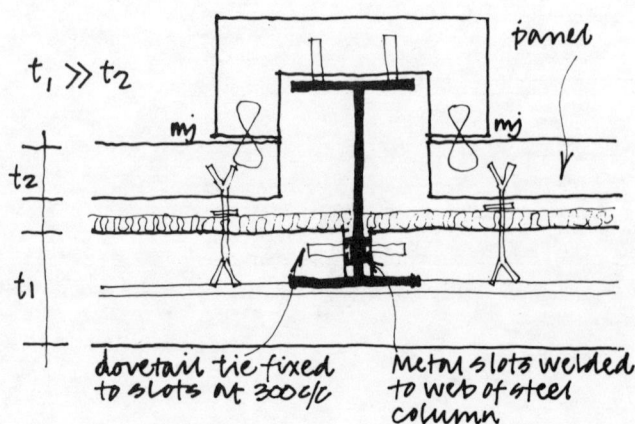

Fig. B5.5 Plan details of wall–column junctions which are assumed to provide full continuity at the edges of laterally loaded wall panels.

Masonry facing to concrete wall

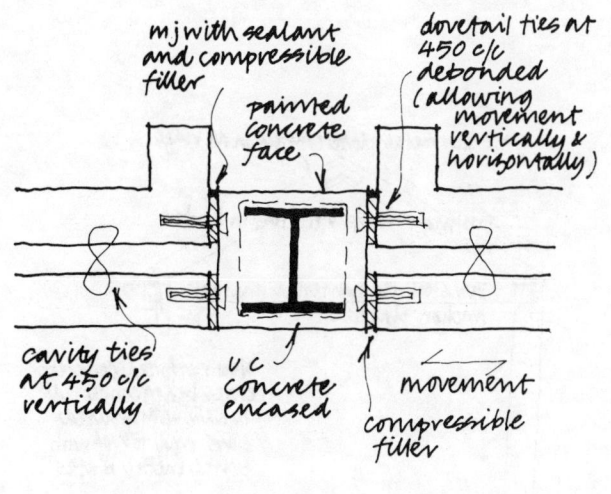

mj with sealant and compressible filler

painted concrete face

dovetail ties at 450 c/c debonded (allowing movement vertically & horizontally)

cavity ties at 450 c/c vertically

u c concrete encased

movement

compressible filler

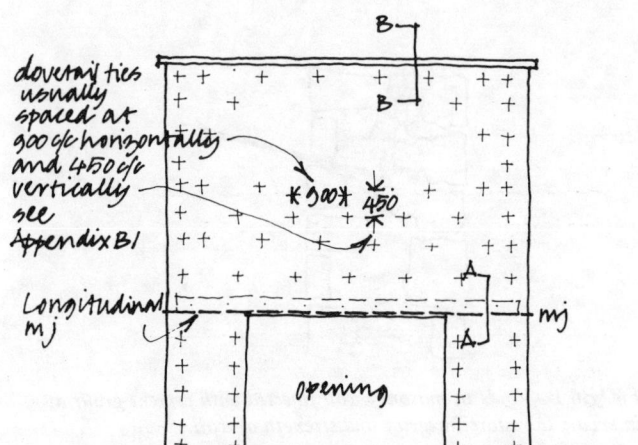

dovetail ties usually spaced at 900 c/c horizontally and 450 c/c vertically see Appendix B1

Longitudinal mj

mj

opening

a)

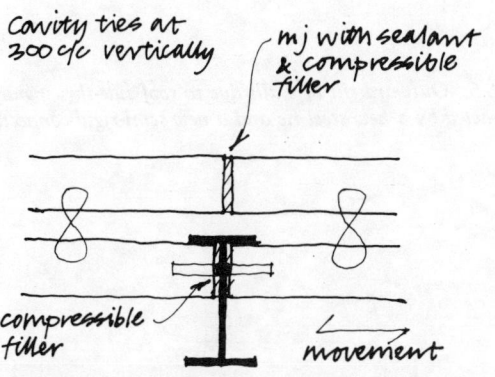

Cavity ties at 300 c/c vertically

mj with sealant & compressible filler

compressible filler

movement

Ties shot fixed to columns at 300c/c and debonded allowing movement in plane of wall

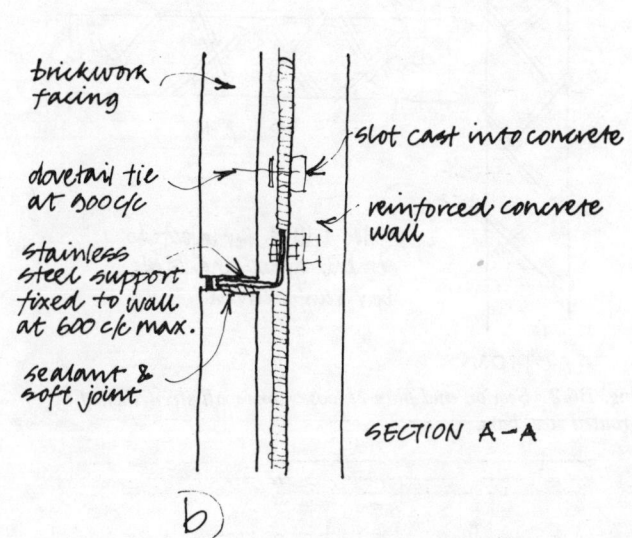

brickwork facing

slot cast into concrete

dovetail tie at 300 c/c

reinforced concrete wall

stainless steel support fixed to wall at 600 c/c max.

sealant & soft joint

SECTION A–A

b)

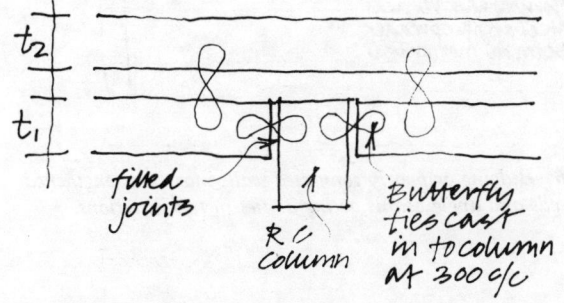

t_2

t_1

fixed joints

R C Column

Butterfly ties cast in to column at 300 c/c

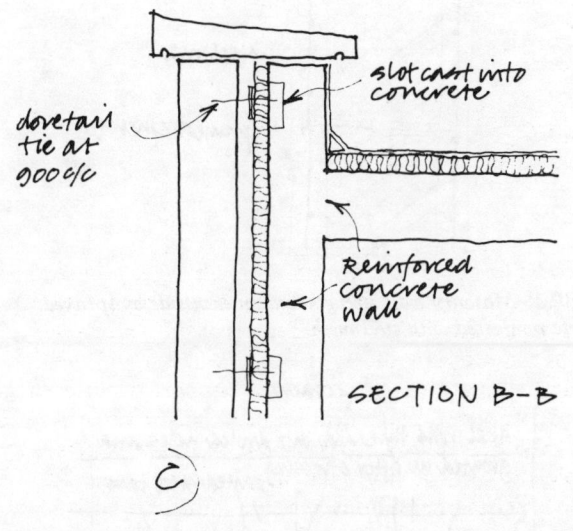

slot cast into concrete

dovetail tie at 900 c/c

reinforced concrete wall

SECTION B–B

c)

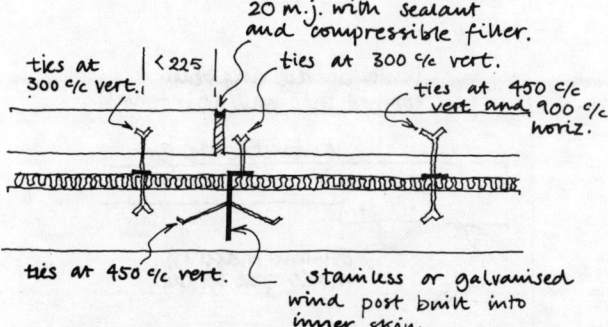

For small spans or internal walls

20 m.j. with sealant and compressible filler.

ties at 300 c/c vert.

< 225

ties at 300 c/c vert.

ties at 450 c/c vert and 900 c/c horiz.

ties at 450 c/c vert.

stainless or galvanised wind post built into inner skin.

Fig. B5.6 Plan details of wall–column junctions which are assumed to provide simple support at the edges of laterally loaded wall panels.

Fig. B5.7 Concrete wall with masonry facing: (a) elevation, (b) and (c) sections.

B6 Alteration and repair details

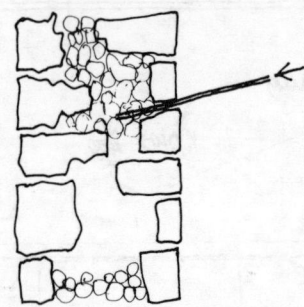

Fig. B6.1 *Voids in masonry wall injected with cement grout at intervals to restore integrity and strength of wall.*

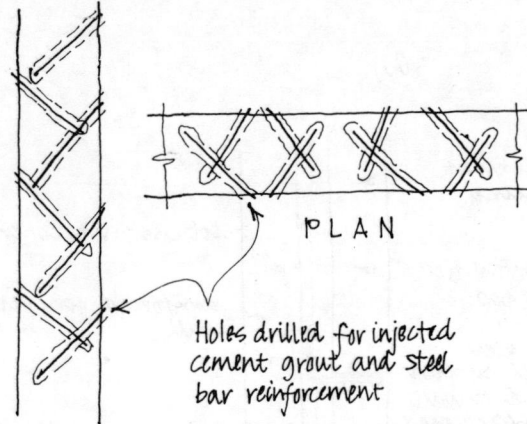

PLAN

SECTION

Holes drilled for injected cement grout and steel bar reinforcement

Fig. B6.2 *Section and plan of loose stone wall strengthened by grouted steel bars.*

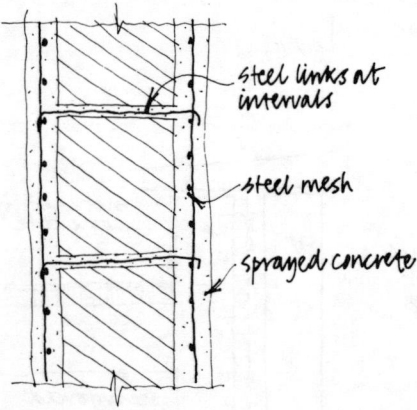

steel links at intervals

steel mesh

sprayed concrete

Fig. B6.3 *Masonry wall strengthened on each side by sprayed concrete reinforced with steel mesh.*

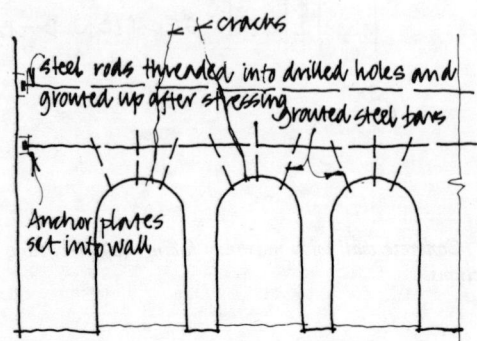

cracks

steel rods threaded into drilled holes and grouted up after stressing

grouted steel bars

Anchor plates set into wall

Fig. B6.4 *Elevation on openings showing cracks in wall which may be arrested or closed by light pre-stressing of steel rods.*

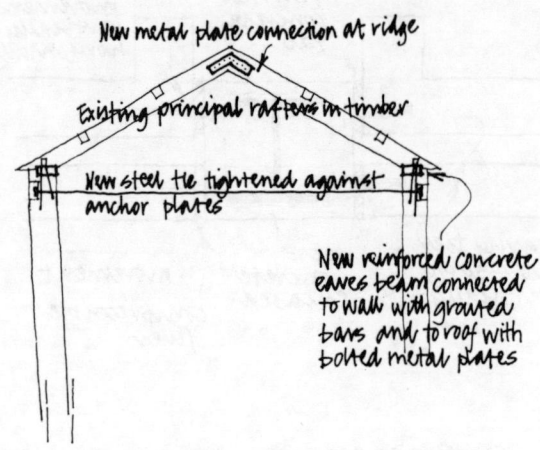

New metal plate connection at ridge

Existing principal rafters in timber

New steel tie tightened against anchor plates

New reinforced concrete eaves beam connected to wall with grouted bars and to roof with bolted metal plates

Fig. B6.5 *Outward tilt of walls due to roof side-thrust may be counteracted by a new steel tie and a new semi-rigid connection at the apex.*

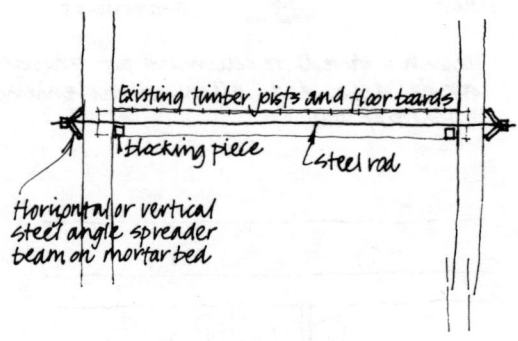

Existing timber joists and floor boards

blocking piece

steel rod

Horizontal or vertical steel angle spreader beam on mortar bed

Fig. B6.6 *Bulging or poorly connected walls may be strengthened by steel rods and angle beams acting as ties in two directions.*

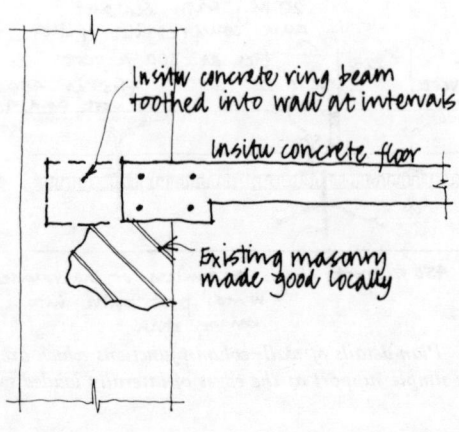

Insitu concrete ring beam toothed into wall at intervals

Insitu concrete floor

Existing masonry made good locally

Fig. B6.7 *New reinforced concrete floor supported by existing masonry wall.*

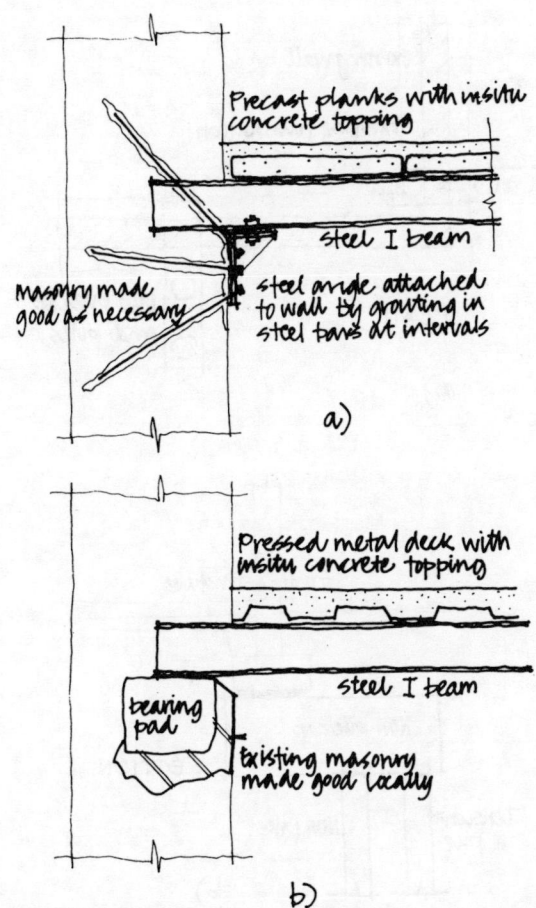

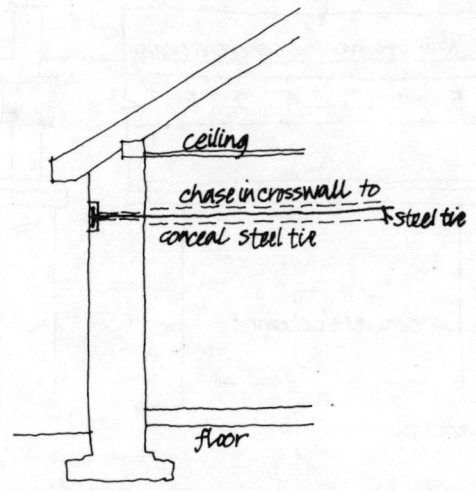

Fig. B6.10 Wall leaning outwards stabilised by ties across the building.

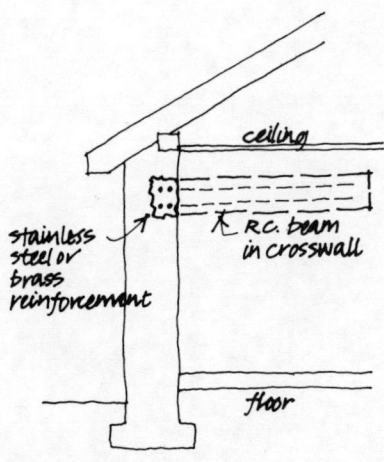

Fig. B6.8 New floor supported by steel beam and existing masonry wall: (a) using steel angle bearing plate and (b) using concrete bearing pad.

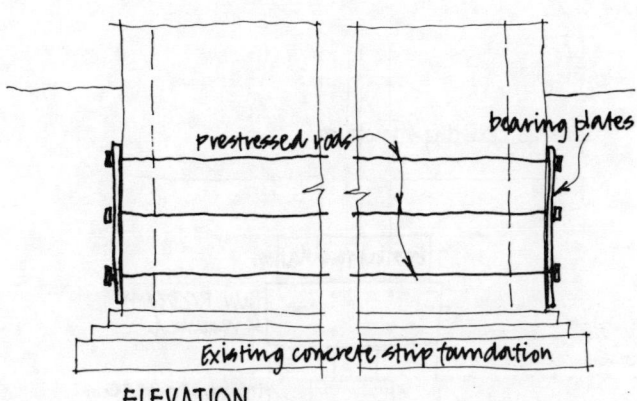

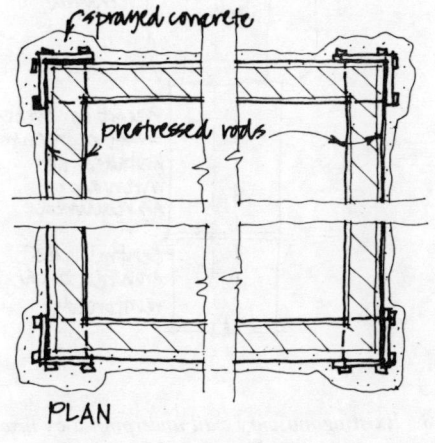

Fig. B6.9 Wall leaning outwards stabilised by r.c. eaves beam returned into crosswalls.

Fig. B6.11 Masonry walls pre-stressed by external steel rods to close cracks and provide stiff superstructure in order to limit differential movement.

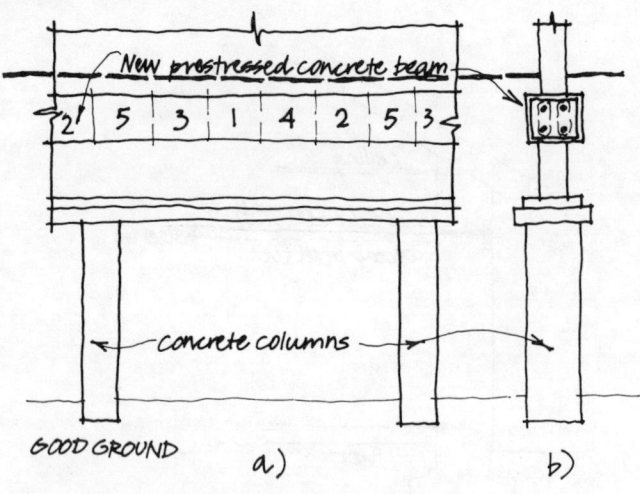

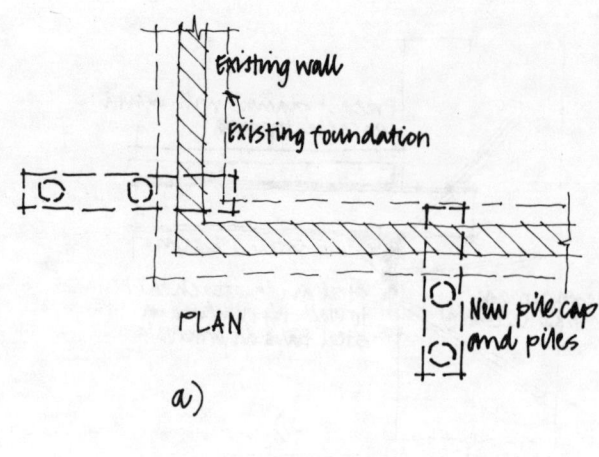

Fig. B6.12 *Existing masonry walls stiffened by new pre-stressed concrete beam; concrete casings are used as temporary supports to the wall, then filled with in situ concrete and pre-stressed; existing foundations may be underpinned by concrete columns, as necessary.*

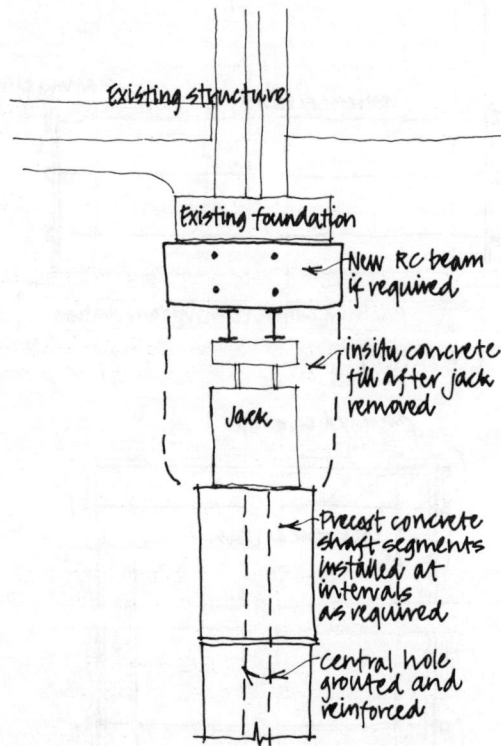

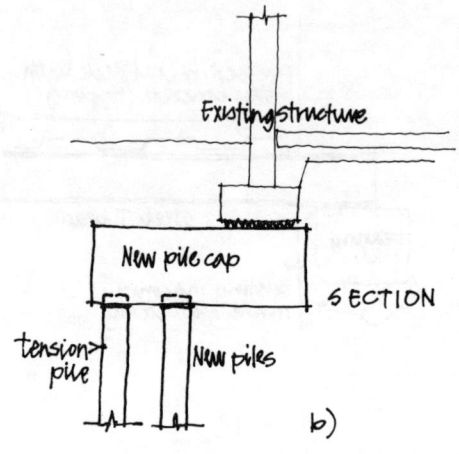

Fig. B6.14 *Pile cap cantilevered from piles to support unstable wall; wall may be jacked against new pile cap if necessary: (a) plan and (b) section.*

Fig. B6.13 *Existing masonry wall underpinned by new piles; piles made of precast segments ('Miga' piles) jacked against existing building with further segments added until pile reaches firm set.*

References

BSI 1978 *BS 187: Specification for calcium silicate (sandlime and flintlime) bricks.* British Standards Institution

BSI 1985 *BS 3921: Specification for clay bricks.* British Standards Institution

BSI 1992 *BS 5628: part 1: Code of practice for use of masonry Part 1 the structural use of unreinforced masonry.* British Standards Institution

BSI 1985 *BS 5628: Part 2: Code of practice for use of masonry Part 2 the structural use of reinforced and prestressed masonry.* British Standards Institution

BSI 1985 *BS 5628: Part 3: Code of Practice for use of masonry Part 3 materials and components, design and workmanship.* British Standards Institution

BSI 1991 *BS 5837: 1991 Guide for trees in relation to construction.* British Standards Institution

BSI 1981 *BS 5930: 1981 Code of practice for site investigations.* British Standards Institution

BSI 1985 *BS 5977: Part 1: 1981 Lintels Part 1 method for assessment of load.* British Standards Institution

BSI 1984 *BS 6073: Part 1: 1981 Specification for precast concrete masonry units.* British Standards Institution

BSI 1989 *BS 6399: Loading for buildings Part 1 Code of practice for dead and imposed loads.* British Standards Institution

BSI 1988 *BS 6399: Loading for buildings Part 3 Code of practice for imposed roof loads.* British Standards Institution

BSI 1988 *CP 3: Chapter V: Part 2: 1972 loading Part 2 wind loads.* British Standards Institution

BSI 1986 *BS 8004: 1986 Code of practice for foundations.* British Standards Institution

Building Research Establishment 1972 *Digest 64 Soils and foundations: 2.* HMSO

Building Research Establishment 1980 *Digest 67 Soils and foundations: 3.* HMSO

Building Research Establishment 1980 *Digest 242 Low-rise buildings on shrinkable clay: part 3.* HMSO

Building Research Establishment 1990 *Digest 251 Assessment of damage in low-rise buildings.* HMSO

Burland J B, Wroth C P 1975 Settlement of buildings and associated damage. In proc. conf. *Settlement of structures,* Cambridge. Pentech Press

CIRIA 1986 *Movement and cracking in long masonry walls:* special publication 44. CIRIA.

Davies S R, Ahmed A E 1980 A graphical solution of composite wall beams. *International Journal of Masonry Construction* 1: 29–33

Haseltine B A, Moore J F A 1981 *Handbook to BS 5628: structural use of masonry.* Brick Development Association

Haseltine B A, Tutt J N 1991 *Handbook to BS 5628:Part 2: Section 1 reinforced masonry.* Brick Development Association

Hendry A W 1986 *Calculation of eccentricities in loadbearing walls: EFN 3.* Brick Development Association

Hendry A W 1990 *Structural masonry.* Macmillan Press

Heyman J 1982 *The masonry arch.* Ellis Horwood

Institution of Structural Engineers 1989 *Soil-structure interaction: the real behaviour of structures.* Institution of Structural Engineers

Johansen K W 1972 *Yield-line formulae for slabs* trans P M Katborg. Cement and Concrete Association

Jones L L, Wood R H 1967 *Yield-line analysis of slabs.* Thames and Hudson, Chatto and Windus

Moore J F A 1978 Stability of low-rise masonry construction. In sym. on *Stability of low-rise buildings of hybrid construction,* London. Institution of Structural Engineers

Morton J 1985 *Accidental damage, robustness and stability:* DG15. Brick Development Association

Morton J 1986 *Designing for movement in brickwork:* DN10. Brick Development Association

Page A W, Hendry A W 1988 Design rules for concentrated loads on masonry. *The Structural Engineer* 66 (17) 273–282. Institution of Structural Engineers

Roberts J J, Tovey A K, Cranston W B, Beeby A W 1983 *Concrete masonry designer's handbook.* Eyre and Spottiswoode

Roberts J J, Edgell G J, Rathbone A J 1986 *Handbook to BS 5628:Part 2.* Palladian Publications

Schneider R R, Dickey W L 1987 *Reinforced masonry design.* Prentice-Hall

Sutherland R J M 1978 Principles for ensuring stability. In sym. on *Stability of low-rise buildings of hybrid construction,* London. Institution of Structural Engineers

Tomlinson M J, Driscoll R, Burland J B 1978 *Foundations for low-rise buildings.* BRE current paper 61/78. Building Research Establishment

Wood R H 1952 *Studies in composite construction: Part 1: The composite action of brick panel walls supported on reinforced concrete beams.* National Building Studies research paper 13. HMSO

Index